# Lecture Notes in Management and Industrial Engineering

This book series provides a means for the dissemination of current theoretical and applied research in the areas of Industrial Engineering and Engineering Management. The latest methodological and computational advances that can be widely applied by both researchers and practitioners to solve new and classical problems in industries and organizations contribute to a growing source of publications written for and by our readership.

The aim of this book series is to facilitate the dissemination of current research in the following topics:

- Strategy and Entrepreneurship
- Operations Research, Modelling and Simulation
- Logistics, Production and Information Systems
- Quality Management
- Product Management
- Sustainability and Ecoefficiency
- Industrial Marketing and Consumer Behavior
- Knowledge and Project Management
- Risk Management
- Service Systems
- Healthcare Management
- Human Factors and Ergonomics
- Emergencies and Disaster Management
- Education

Viktor Koval · Yigit Kazancoglu ·
Elena-Simina Lakatos
Editors

# Circular Business Management in Sustainability

## Proceedings of the 2nd International Conference on Sustainable, Circular Management and Environmental Engineering (ISCMEE 2022), October 19–20, 2022, İzmir, Turkey

Springer

*Editors*
Viktor Koval
Izmail State University of Humanities
Izmail, Ukraine

Yigit Kazancoglu
Department of Logistics Management
Yaşar University
İzmir, Turkey

Elena-Simina Lakatos
Institute for Research in Circular Economy
and Environment "Ernest Lupan"
Technical University of Cluj-Napoca
Cluj-Napoca, Romania

ISSN 2198-0772 ISSN 2198-0780 (electronic)
Lecture Notes in Management and Industrial Engineering
ISBN 978-3-031-23462-0 ISBN 978-3-031-23463-7 (eBook)
https://doi.org/10.1007/978-3-031-23463-7



This Springer imprint is published by the registered company Springer Nature Switzerland AG
The registered company address is: Gewerbestrasse 11, 6330 Cham, Switzerland

# Preface

This book compiles extended versions of a selection of the papers presented at the 2nd International Conference on Sustainable, Circular Management and Environmental Engineering (ISCMEE 2022) held in Izmir, Turkey, on October 19–20, 2022. The conference was organized by Yaşar University, Institute for Research in Circular Economy and Environment "Ernest Lupan" (IRCEM) in partnership with the Department of the Human Dimensions of Global Change of Global Change Research Institute of the Czech Academy of Sciences, Czech Republic.

Sustainable development can be defined as development that meets the needs of the present without compromising the ability of future generations to meet their own needs. For this reason, the Sustainable, Circular Management and Environmental Engineering is a potential frame for natural resource development leading to social cohesion, economic competitiveness and stability, use of resources and economic development, safeguarding biodiversity and the ecosystem. The aim of the ISCMEE 2022 conference is to explore current trends in sustainable development, circular management and environmental engineering, stimulate dialogue and develop new perspectives in the field of environmental protection. Conceptual, empirical and methodological studies as well as country case studies on various topics of sustainable development were presented at the conference. It is expected that the reports and presentations will cover both theoretical, methodological and practical aspects of sustainable development and the circular economy. ISCMEE 2022 focuses on the themes of Circular Economy and Environmental Sustainability and in this context, the aim of ISCMEE 2022 is to provide a platform for researchers, entrepreneurs, higher education teachers and authorities to present their research results and development activities in the following conference topics:

Sustainable Development and Bioengineering
Circular Economy and Business Engineering
Energy Policy and Environmental Engineering

The conference was attended by leading researchers, engineers and scientists from Bulgaria, Canada, Romania, Moldova, Turkey, United Kingdom, India, United

# An Exploratory Study on Implementing Circular Economy in Rural Family Businesses

Elena-Simina Lakatos[1,2](✉), Roxana-Maria Druta[1,2], Andrea Szilagyi[1,2], Andreea-Loredana Birgovan[1,2], Lucian-Ionel Cioca[3], and Sorin Daniel Vatca[4]

[1] Institute for Research in Circular Economy and Environment "Ernest Lupan", 71-73 Dorobantilor Avenue, 400609 Cluj-Napoca, Romania
Simina.lakatos@ircem.ro

[2] Faculty of Industrial Engineering, Robotics and Product Management, Technical University of Cluj-Napoca, 400641 Cluj-Napoca, Romania

[3] Faculty of Engineering, Lucian Blaga University of Sibiu, 550024 Sibiu, Romania

[4] Faculty of Agriculture, University of Agricultural Sciences and Veterinary Medicine Cluj-Napoca, 400372 Cluj-Napoca, Romania

**Abstract.** Despite recent efforts to promote family businesses as a means of socio-economic development, the allocation and expenditure of resources in this regard remains inequitable for rural areas. The purpose of this study is to investigate what the perceived needs of family businesses in rural regions, as exemplified by Romania, are, with a focus on circular economy. A total of 155 responses were collected via an web-based questionnaire. Our findings show that product and service marketing training and support (47.7%), as well as financial management training (41,3%) are two of the most important training requirements for engaging in entrepreneurial action. Furthermore, respondents favor training in practices that enable the circular economy, such as repair and reuse. More than half of the sample (58.6%) reported that they would implement actions to reduce waste minimization if they were to start a family business in a rural area. From a practical standpoint, local stakeholders might take this finding into consideration to create training programs that are especially adapted for the rural setting and its unique economic and social conditions. Finally, limitations and further research directions are discussed.

**Keywords:** Family businesses · Rural areas · Circular economy · Rural development · Sustainability

## 1 Introduction

Given the increasing volatility of the availability of human resources and jobs globally, family businesses are differentiated as an important generator of technological and innovative development. Therefore, many countries prioritize family businesses as a strategic direction for economic and social development [1]. In heavily urbanized regions, the prioritization of family businesses is evident through the support and attention given to start-ups, and in general to entrepreneurs beginning their career who need assistance for


V. Koval et al. (Eds.): ISCMEE 2022, Lect. Notes in Managem., Industrial Engineering, pp. 1–14, 2023.
https://doi.org/10.1007/978-3-031-23463-7_1

the development of new family businesses. However, in the case of rural family businesses, the efforts of local authorities do not stand out as much as in urban regions, where the model of innovative, high-tech family businesses predominates. This comes in spite of the fact that rural regions cover 57% of the European Union's territory and 24% of the population [2]. The lack of a well-developed entrepreneurial sector can be attributed to several factors such as lack of resources, networking opportunities, or expertise [3, 4]. The literature published so far constantly highlights that rural areas face far more obstacles to the development of an entrepreneurial sector. The problems generally steam from the elderly resident population of these areas- a factor closely linked with mass emigrations to urban centers and as well as from the lack of proximity to markets and services [5, 6].

Thus, the objective of this research is to analyze the current situation regarding the state of family businesses in rural Romania. Specifically, the research questions (RQ) are as follows:

- RQ1: What are the main aspects of the rural socio-economic environment that should be improved from an entrepreneurial perspective?
- RQ2: What are the entrepreneurial training needs for starting a family business in rural area?
- RQ3: What are the needs for entrepreneurial development in terms of family business financing?
- RQ4: What are the intentions for the development of sustainable family business aimed at sustainability and environmental protection?

## 2 Theoretical Framework

This section will further deepen the discussion about the concept of family businesses and circular economy in rural areas based on the existing body of literature. The main question addressed is if rural family business is a concept of its own or is it just that family businesses occur in urban as well as in rural areas? Contemporary perspectives on family businesses shifted from firm-based approaches build on rational action aimed at profit maximization toward an approach that conceptualizes family businesses as part of a dynamic and territorially defined network, that recognize the importance of the relationship between the entrepreneur and the local society [7–9]. Although debates are still undergoing there is a profound regional and local substrate to understand family businesses, such as the local environment concerning business and physical infrastructure, culture and social support can significantly impact the entrepreneur possibility of success in that rural locality [10, 11].

### 2.1 Challenges of Rural Areas

Although the differences between rural and urban areas are becoming less evident, there are still numerous limitations that the rural world has to overcome [12]. Whilst the trend of merging public administrations and pool public services into towns gains growing popularity, rural areas are in peril of becoming marginalized areas hidden from public and private stakeholders [13]. Furthermore, this marginalization translates into the struggle of novel business models to thrive because of insufficient involvement in new policies or reservations about embracing change [14–16]. Therefore, there are specific challenges in rural areas that are far less frequent or nonexistent in entrepreneurship in urban areas [8].

Firstly, rural areas have to deal with a mono-industrial profile of economy. This means that rural areas are prone to be highly engaged in only one industry, implying a sector-specific risk. In a framework like this, the diversity of local economies is reduced and the innovation is hindered by old industrial traditions. Thus, in order to assure their functioning in areas with limited demand, entrepreneurs must seek markets and business-to-business services outside rural regions [8, 17].

Moreover, achieving economies of scale in rural areas seems like a difficult task as population size is shrinking and emigration increasing. This trend, together with limited local demand and the geographic isolation causing limited access to a diversified human resources pool contributes to a difficult environment for profit or further investments [18]. Without such economies, products and services must be sold at higher prices in order to make profit, often beyond the possibilities of the average consumer [19]. Alternatively, small businesses that are unable to offer competitive prices become vulnerable in front of large regional and national discounters situated in their proximity [20, 21].

## 2.2 Rural Family Businesses and Sustainability

All the challenges discussed in the previous section reach the conclusion that rural development is needed in order to improve social and economic conditions. However, another important discussion is about how can this development be achieved in a sustainable manner. The main goals of sustainable rural development are natural resources management and preservation [22], increased rate of employment and level of well-being [23], diversification of economic activities, including agriculture production [24] and preservation of social and cultural traditions [25, 26]. Therefore, sustainable rural development represents a multidimensional process of change consisting of three main pillars, all equally important: economic growth, natural resources preservation and improved social environment. Circular economy can be achieved through various pathways in urban settlements, but the situation concerning rural areas is different [27]. Agriculture is still the key pathway that rural areas rely on, despite its inability to solely generate sufficient revenue and to contribute to the labour market [28, 29].

The concept of circular economy is also relevant in this discussion, as it is becoming increasingly used in both rural and urban development strategies as a tool to achieve sustainable communities [30, 31]. When put into practice, the circular economy can minimize environmental and societal impacts of anthropogenic activities in rural areas based on the conservation of stocks of resources (including slowing raw material in-flows and minimizing waste generation) [32–34].

In this context, new rural development paradigms emerged, in which agriculture is placed among other rural and non-rural activities that are important to the construction of resilient rural communities [35]. This means that interconnected practices such as organic farming, agri-tourism, the commercialization of local specific products, rural-tourism, landscape management, etc. lead to novel forms of social cohesion, this revealing the multi-facetted nature of rural development [36–38]. Figure 1 presents the means through which family businesses can foster circular economy in rural areas.

**Fig. 1.** Circular economy and Family businesses [39]

Circular economy in the rural context definitely represents a challenge for decision makers as they deal with a high level of complexity generated by different yet interconnected economic activities, social environments and cultural traditions.

## 2.3 Rural Family Businesses in Romania

Covering more than 85% of Romania's total surface with a population of 8.900.241 of inhabitants in 2021 the rural family businesses in Romania is marked predominantly by a strong activity in agriculture [40]. Among the opportunities of rural areas in Romania are the recent trend of increasing agricultural exports, a cultural heritage and a rich biodiversity closely linked to the use of extensive traditional land management practices. Yet, besides the agriculture activities, rural tourism is still an underexploited alternative for the diversification of activities in order to obtain additional income for the rural population. Furthermore, switching the focus on organic agriculture would cause less damage to biodiversity compared to traditional practices [41], while generating much greater compatibility with nature preservations regulations, for example Natura 2000 regulations [42]. Nonetheless, besides all the efforts targeted at circular economy, it is worth mentioning that the development of basic local infrastructure and services still remain a major challenge for many rural areas facing poverty. Basic local infrastructure and services represent the starting point in the local business environment as access to utilities, goods and/or services increases the attractiveness of the area [43].

# 3 Materials and Methods

## 3.1 Sample and Procedure

A total of 155 questionnaires were collected via an online platform. The participants included 102 females (65.8%) and 53 males (34.2%). All the respondents declared their interest in starting a business in a rural area. The average age interval of the sample was 18–29 years (51%). All participants gave their consent to participate in the study and allowed their data to be used for research purposes. Data collection was completely anonymous and voluntary. The questionnaire information was kept confidential.

## 3.2 Studied Variables

Perceived rural development needs refer to those developments needed in rural area in order to create a proper environment for businesses to thrive, for example: de-creasing the emigration rate among adults active in the labour market, creating markets for products obtained in rural areas, investments/donations for the modernization of schools, kindergartens, sanitary points in rural areas, etc. The respondents had to rate on 5-point Likert scale, where 1 = Strongly Disagree, 5 = Strongly Agree, the degree to which they consider the item to be important in creating a suitable environment for family businesses in rural areas. Perceived training needs refers to the learning/training needs considered essential for starting an entrepreneurial process in rural areas. Among the needs that respondents had to assess were: elaboration of the business plan, product and services marketing, financial management, etc. Likewise, all the items were on a 5-point Likert Scale and respondents had to assess their importance in the process of starting a business in a rural area. Estimated financial needs refer to the financial resources needed for starting a business in a rural area. Respondents selected from the amounts of 150.000–200.000 €, 100.000–150.000 €, 80.000–100.000 €, 60.000–85.000 €, 40.000–60.000 €, >40.000 € which sum they considered fit for starting a business in a rural community from their perspective. Perceived importance of sustainability practices in business refer to disponibility of engaging in practices related to business management that contribute to the circular economy of the environment such as: use of technologies to minimize carbon di-oxide emissions, use of renewable energy in enterprises, introduction of efficient re-source management systems (water, soil, air), biodiversity protection. Respondents were asked to rate on a 5 point Likert Scale their importance.

# 4 Results

## 4.1 Perceived Rural Development Needs

Table 1 shows the rural development needs measured and the frequency of agreement and disagreement of the sample.

The most important needs rated were" financial, psychological and social support for single-parent families or children left alone by parents working abroad" –40.6% strongly agreed," investments/donations for the modernization of schools, kindergartens, sanitary points in rural areas" –36.1% strongly agreed and" creating jobs

**Table 1.** Frequency analysis of perceived rural development needs.

| Items | M | SD | Valid Percents % | | | | |
|---|---|---|---|---|---|---|---|
| | | | Strongly Disagree | Disagree | Neither agree, nor disagree | Agree | Strongly Agree |
| 1. Education in family businesses and business management in rural areas | 3.1 | 0.959 | 2.6 | 5.2 | 19.4 | 43.9 | 29.0 |
| 2. Markets for products obtained in rural areas | 3.98 | 0.911 | 2.6 | 2.6 | 18.7 | 45.8 | 30.3 |
| 3. Stimulating interest in the processing and marketing of agricultural products | 3.68 | 0.972 | 2.6 | 9.0 | 25.2 | 43.9 | 19.4 |
| 4. Creating jobs for young graduates | 4.07 | 0.905 | 1.9 | 4.5 | 12.3 | 47.1 | 34.2 |
| 5. Decreasing the emigration rate among adults active in the labor market | 4.18 | 0.724 | – | 3.2 | 9.0 | 54.2 | 33.5 |
| 6. Investments for the modernization of schools, kindergartens, sanitary points in rural areas | 4.08 | 0.918 | 1.3 | 6.5 | 11.0 | 45.2 | 36.1 |

(*continued*)

for young graduates" –34.2% strongly agreed. These results confirmed not only the upward trend of emigration characterizing rural areas nowadays [44], but also the lack of basic infrastructure and services reflected in the fact that the majority of respondents considered that investments or donation in educational and sanitary units are needed in order to create a suitable environment for family businesses in rural areas [45].

**Table 1.** (*continued*)

| Items | M | SD | Valid Percents % | | | | |
|---|---|---|---|---|---|---|---|
| | | | Strongly Disagree | Disagree | Neither agree, nor disagree | Agree | Strongly Agree |
| 7. Financial, psychological and social support for single-parent families or children left alone by parents working abroad | 4.21 | 0.831 | 1.3 | 3.2 | 8.4 | 46.5 | 40.6 |

This section describes the results concerning the perceived training/learning needs of the respondents in relation with the process of engaging in an entrepreneurial process. Training and support in product and service marketing was one of the most important rated item by the sample –47.7% strongly agreed in this sense, as it can be seen in Table 2. Training in financial management was also a high rated item as 41.3% of the respondents strongly agreed that support is needed in this direction. Overall, the results show a declared need in training in various stages of the entrepreneurial process.

This question addressed the problem of financing. The respondents were asked to choose what sum would best fit their expenses for starting a business in a rural area. More than half of the sample represented by men (50.9%) reported that they need between 100.000–150.000 €, compared to 23.5% estimating that they need less than 40.000 € to start a business in a rural area (Table 3).

## 4.2 Perceived Importance of Sustainability Family Business Practices

This question explored what are the intentions of the respondents for the development of sustainable businesses aimed at environmental protection. The respondents were asked to rate on a 5-point Likert Scale what business practices that adhere to the circular economy they would consider important to implement if they were to start a business in a rural area (Table 4). Most respondents strongly agreed (48.3%) that the use of renewable energies is a practice they would consider, followed by introduction of efficient resource management –46.7%.

Another aspect investigated related to sustainability was the intention of implementing circular economy practices in rural businesses. More than half of the sample (58.7%) reported that they would implement actions to reduce waste minimization if they were to start a family business in a rural area. Also, product repair was a popular variant among respondents –42.6% declared they would engage their business in such activities. Product reuse was chosen only by 33.5% of the sample, being the least favorite option (Fig. 2).

**Table 2.** Perceived training needs refers to the learning/ training needs reported among the respondents.

| Items | M | SD | Valid Percents % | | | | |
|---|---|---|---|---|---|---|---|
| | | | Strongly Disagree | Disagree | Neither agree, nor disagree | Agree | Strongly Agree |
| 1. Elaboration of the business plan | 4.08 | 1.06 | 1.3 | 9.7 | 16.1 | 25.2 | 47.7 |
| 2. Human resources management | 3.92 | 1.05 | 1.3 | 9.0 | 23.9 | 27.1 | 38.7 |
| 3. Financial management | 4.05 | 1.0 | 1.9 | 8.4 | 13.5 | 34.8 | 41.3 |
| 4. Product and Services marketing | 4.20 | 0.923 | 1.3 | 3.9 | 14.8 | 32.9 | 47.1 |
| 5. Social activity management | 4.14 | 0.983 | 0.6 | 7.1 | 16.8 | 28.4 | 47.1 |
| 6. Digitizing the business | 4.17 | 0.981 | 0.6 | 6.5 | 17.4 | 25.8 | 49.7 |

**Table 3.** Estimated financial needs.

| Items | Valid Percents % | | | | |
|---|---|---|---|---|---|
| | Women | Men | Employee | Entrepreneur | Student |
| 150.000–200.000€ | 8.8 | 9.4 | 8.3 | 8.3 | 9.8 |
| 100.000–150.000€ | 22.5 | 50.9 | 35.0 | 54.2 | 17.6 |
| 80.000–100.000€ | 19.6 | 17.0 | 21.7 | 20.8 | 15.7 |
| 60.000–85.000€ | 11.8 | 3.8 | 1.7 | 4.2 | 15.7 |
| 40.000–60.000€ | 4.9 | 11.3 | 5.0 | 8.3 | 3.9 |
| >40.000€ | 23.5 | 7.5 | 18.3 | 4.2 | 27.5 |

# 5 Discussion

## 5.1 Practical and Theoretical Implications

Family businesses are positioned as an instrument for tackling the ascending peripherality and low economies of the rural region that is becoming an ever more serious factor given its potential rule as a catalyst for mitigating climate change and achieving sustainability [46]. As in the case of many Eastern European Countries, in Romania the

**Table 4.** Perceived importance of sustainability business practices.

| Items | M | SD | Valid Percents % | | | | |
|---|---|---|---|---|---|---|---|
| | | | Strongly Disagree | Disagree | Neither agree, nor disagree | Agree | Strongly Agree |
| 1. Use of technologies to minimize carbon dioxide emissions | 4.14 | 0.978 | 1.7 | 3.3 | 31.7 | 20.0 | 43.3 |
| 2. Use of renewable energy in enterprises | 4.13 | 1.099 | 1.7 | 6.7 | 26.7 | 16.7 | 48.3 |
| 3. Introduction of efficient resource management systems (water, soil, air) | 4.01 | 1.156 | 5.0 | 8.3 | 30.0 | 10.0 | 46.7 |
| 4. Biodiversity protection | 3.98 | 1.162 | 5.0 | 5.0 | 30.0 | 15.0 | 45.0 |

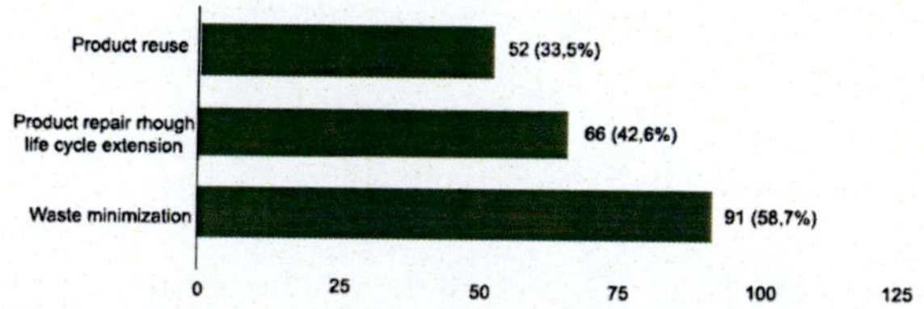


**Fig. 2.** Intentions towards implementing circular economy practices.

rural areas represent the less-favored and least-developed regions that still rely a great deal on agriculture to thrive, contributing frequently to negative growth rates, soaring and mounting rural poverty [47]. Starting from the purpose of exploring perceptions of rural family businesses among a sample with a stated intention to start a business in a rural community, this paper explored what are the current perceptions towards rural family businesses in Romania. Confirming the literature published before [16, 48, 49], one of the main needs of rural areas that has to be addressed is emigration of the labor force, reflected on the sample perception of giving support to the families left behind.

Therefore, being able to retain and attract skilled people capable of contributing to economic growth [50] remains an ideal that makes sense in theory but it is difficult to put into practice due to the intrinsically complex problem. Another important aspect worth mentioning is that the majority of respondents considered training an learning strongly important for starting an activity in rural area. Aspects such as elaboration of a business plan, financial management, product marketing are essential learning needs that have to be fulfilled, as other papers published before [51–53] confirm that training is a significant predictor of successful entrepreneurial approaches. In this sense, local stakeholders can take this result into consideration in order to prepare courses and trainings specifically tailored for the rural environment and its economic and social peculiarities.

### 5.2 Limitations and Future Research Directions

The present study has some limitations that must be acknowledged when interpreting the results obtained. First of all, the cross-sectional design of the study does not allow us to draw any inferences or assumptions about causality among the variables investigated. As far as the sample is concerned, the number of questionnaires collected does not grant representiveness at national level. Also, the respondents were selected based on the declared intention to become entrepreneurs in rural areas and not on the basis of an actual entrepreneurial activity currently carried out in a rural locality. This can be problematic as entrepreneurial intents does not always translate into action [54].

Regarding future research directions, there are still questions and challenges that require an answer, such as: Are rural areas in Romania prepared to host an environment for family businesses? Do the attempts to develop infrastructure and cultural amenities stimulating rural family businesses sufficiently? How do local rural authorities position themselves in the broad context of circular economy? Is there a need for policy change that tackle rural development? Such questions can constitute the basis of future studies needed to deepen our understanding of how can family businesses function optimally in rural regions.

## 6 Conclusions

From a theoretical point of view and the results of this present study, we conclude that overall, sustainable rural development means increasing the wealth-generating capacity of rural communities while maintaining and preserving the community' built and natural environment to support this capacity. Concerning the first research question that addressed the main aspects of the rural socio-economic environment that require improvement, we find out that support for single-parent families or children left alone by parents working abroad, investments for the modernization of educational and sanitary infrastructure and job creation for young graduates are the current development needs characterizing rural regions. The second research question explored the entrepreneurial training needs of respondents. Our findings suggest that support in product and service marketing and financial management are two main topics that deserves attention from local stakeholders. Last but not least, regarding the intentions of sustainable practices, we found out that waste minimization, product repair and the use of renewable energy are practices that arouse the interest of respondents with a declared intent of family businesses.

**Acknowledgment.** This paper was financially supported by the Project "Network of excellence in applied research and innovation for doctoral and postdoctoral programs/InoHubDoc", project co-funded by the European Social Fund financing agreement no. POCU/993/6/13/153437.

## References

1. Soleymani, A., Farani, A.Y., Karimi, S., Azadi, H., Nadiri, H., Scheffran, J.: Identifying sustainable rural entrepre-neurship indicators in the Iranian context. J. Clean. Prod. **290**, 125186 (2021). https://doi.org/10.1016/j.jclepro.2020.125186
2. EC: Commission Staff working document a view of employment, growth and innovation in rural areas. European Commission, Brusssels (2012)
3. Muñoz, P., Kimmitt, J.: Rural family businesses in place: an integrated framework. Family Bus. Reg. Dev. **31**(9–10), 842–873 (2019). https://doi.org/10.1080/08985626.2019.1609593
4. Dhewanto, W., Ratnaningtyas, S., Permatasari, A., Anggadwita, G., Prasetio, E.A.: Rural family businesses: towards collaborative participative models for economic sustainability. Family Bus. Sustain. Issues **8**(1), 705 (2020). https://doi.org/10.9770/jesi.2020.8.1(48)
5. Brown, D.L., Schafft K.A.: Rural people & communities in the 21st century: resilience & transformation. Polity Press, Cambridge (2011)
6. Pato, M.L., Teixeira, A.A.: Twenty years of rural family businesses: a bibliometric survey. Sociol. Rural. **56**(1), 3–28 (2016)
7. Davidsson, P.: The family businesses research challenge. Edward Elgar, Cheltenham (2007)
8. Fortunato, M.W.P.: Supporting rural family businesses: a review of conceptual developments from research to practice. Community Dev. **45**(4), 387–408 (2014)
9. Audretsch, D.B.: Entrepreneurship policy and the strategic management of places. In: The Emergence of Entrepreneurship Policy, pp. 20–38 (2003). https://doi.org/10.1017/cbo9780511610134.002
10. Flora, C.B., Flora, J.: Entrepreneurial social infrastructure: a necessary ingredient. Ann. Am. Acad. Pol. Soc. Sci. **529**, 48–58 (1993)
11. Markley, D., Macke, D.: Entrepreneurial support organizations. Center for Rural Family Businesses, Lincoln, NE (2003). (Monograph)
12. Abreu, I., Mesias, F.J.: The assessment of rural development: identification of an applicable set of indicators through a Delphi approach. J. Rural. Stud. **80**, 578–585 (2020). https://doi.org/10.1016/j.jrurstud.2020.10.045
13. Richter, R.: Rural social enterprises as embedded intermediaries: the innovative power of connecting rural commu-nities with supra-regional networks. J. Rural. Stud. **70**, 179–187 (2019). https://doi.org/10.1016/j.jrurstud.2017.12.005
14. Fischer-Tahir, A., Naumann, M. (eds.).: Peripheralization: the making of spatial dependencies and social injustice. Springer VS, Wies-baden (2013)
15. Moraru, R.I., Babut, G.B., Cioca, L.I.: Study of methane flow in caved goafs ajacent to longwall faces in Valea Jiului coal basin. In: The 13 th International Multidisciplinary Scientific Geoconference, SGEM 2013, Geoconference on Science and Technologies in Geology, Exploration and Mining, SGEM 2013, vol I , pp.731-+

16. Shucksmith, M., Brown, D.L.: Rural studies: the challenges ahead. In Routledge international handbook of rural studies, pp. 693–705. Routledge (2016). https://doi.org/10.4324_9781315753041
17. Lichtenstein, G.A., Lyons, T.S.: Incubating new enterprises: a guide to successful practice. The Aspen Institute, Washington, DC (1996)
18. Dabson, B.: Supporting rural family businesses. Rural Development Paper Series. Federal Reserve Bank of Kansas City, Kansas City, MO (2001)
19. Dabson, B.: Family businesses as rural economic development policy. In: Walzer, N. (ed.) Family Businesses and Local Economic Development, pp. 21–38. Lexington Books, Lanham, MD (2007)
20. Szilagyi, A., Cioca, L.I., Bacali, L., Lakatos, E.S., Birgovan, A.L.: Consumers in the circular economy: a path analysis of the underlying factors of purchasing behaviour. Int. J. Environ. Res. Public Health **19**(18), 11333 (2022). https://doi.org/10.3390/ijerph191811333
21. Dabson, B.: Family businesses as rural economic development policy. In: Walzer, N. (ed.), Family Businesses and Local Economic Development, pp. 21–38. Lexington Books, Lanham, MD (2007). https://doi.org/10.1080/15575330.2015.1133683
22. Stathopoulou, S., Psaltopoulos, D., Skuras, D.: Rural family businesses in Europe: a research framework and agenda. Int. J. Entrep. Behav. Res. (2004). https://doi.org/10.1108/13552550410564725
23. Emerson, H.J., Gillmor, D.A.: The rural environment protection scheme of the Republic of Ireland. Land Use Policy **16**(4), 235–245 (1999). https://doi.org/10.1016/s0264-8377(99)00018-6
24. Shortall, S., Shucksmith, M.: Rural development in practice: issues arising in Scotland and Northern Ireland. Community Dev. J. **36**(2), 122–133 (2001). https://doi.org/10.1093/cdj/36.2.122
25. Baldock, D., Dwyer, J., Lowe, P., Petersen, J.E., Ward, N.: The nature of rural development: towards a sustainable integrated rural policy in Europe. Synthesis Report (2001)
26. Kneafsey, M.: Rural cultural economy: tourism and social relations. Ann. Tour. Res. **28**(3), 762–783 (2001)
27. Boggia, A., Rocchi, L., Paolotti, L., Musotti, F., Greco, S.: Assessing rural circular economy potentialities using a dominance-based rough set approach. J. Environ. Manag. **144**, 160–167 (2014). https://doi.org/10.1016/j.jenvman.2014.05.021
28. Pugliese, P.: Organic farming and sustainable rural development: a multifaceted and promising convergence. Socio-logia ruralis **41**(1), 112–130 (2001)
29. Cudečka-Puriņa, N., Atstāja, D., Koval, V., Purviņš, M., Nesenenko, P., Tkach, O.: Achievement of sustainable development goals through the implementation of circular economy and developing regional cooperation. Energies **15**, 4072 (2022). https://doi.org/10.3390/en15114072
30. Muizniece, I., Zihare, L., Pubule, J., & Blumberga, D.: Circular economy and bioeconomy interaction development as future for rural regions. Case study of Aizkraukle Region in Latvia. Environ. Climate Technol. **23**(3) (2019)
31. Lakatos, E.S., et al.: Conceptualizing core aspects on circular economy in cities. Sustainability **13**(14), 7549 (2021). https://doi.org/10.3390/su13147549
32. Pacurariu, R.L., Vatca, S.D., Lakatos, E.S., Bacali, L., Vlad, M.: A critical review of EU key indicators for the transition to the circular economy. Int. J. Environ. Res. Public Health **18**(16), 8840 (2021). https://doi.org/10.3390/ijerph18168840
33. Perey, R., Benn, S., Agarwal, R., Edwards, M.: The place of waste: changing business value for the circular economy. Bus. Strateg. Environ. **27**(5), 631–642 (2018). https://doi.org/10.1002/bse.2068

34. Rada, E.C., Ragazzi, M., Torretta, V., Castagna, G., Adami, L., Cioca, L.I.: Circular economy and waste to energy. In: Conference on Technologies and Materials for Renewable Energy, Environment and Sustainability (TMREES), Technologies and Materials for Renewable Energy, Environment and Sustainability (2018)
35. Ellis, F., Biggs, S.: Evolving themes in rural development 1950s–2000s. Dev. Policy Rev. **19**(4), 437–448 (2001)
36. Van der Ploeg, J.D., Renting, H., Brunori, G., Knickel, K., Mannion, J., Marsden, T., De Roest, K., Sevilla-Guzmán, E., Ventura, F.: Rural development: from practices and policies towards theory. Sociologia ruralis **40**(4), 391–408 (2000)
37. Yankovyi, O., Koval, V., Lazorenko, L., Poberezhets, O., Novikova, M., Gonchar, V.: Modeling sustainable economic development using production functions. Estud. Econ. Apl. **39**, (2021). https://doi.org/10.25115/eea.v39i5.5090
38. Atstaja, D., Koval, V., Grasis, J., Kalina, I., Kryshtal, H., Mikhno, I.: Sharing model in circular economy towards rational use in sustainable production. Energies **15**, 939 (2022). https://doi.org/10.3390/en15030939
39. Van der Ploeg, J. D., Renting, H., Brunori, G., Knickel, K., Mannion, J., Marsden, T., De Roest, K., Sevilla-Guzmán, E., Ventura, F.: Rural development: from practices and policies towards theory. Sociologia ruralis **40**(4), 391–408 (2000). https://doi.org/10.1111/1467-9523.00165
40. National Institute of Statistics: Tempo Online Database (2021). http://statistici.insse.ro:8077/tempo-online/#/pages/tables/insse-table
41. Bengtsson, J., Ahnstrom, J., Weibull, A.C.: The effects of organic agriculture on biodiversity and abundance: a metaanalysis. J. Appl. Ecol. **42**(2), 261–269 (2005). https://doi.org/10.1111/j.1365-2664.2005.01005.x
42. Mikulcak, F., Newig, J., Milcu, A.I., Hartel, T., Fischer, J.: Integrating rural development and biodiversity conservation in Central Romania. Environ. Conserv. **40**(2), 129–137 (2013). https://doi.org/10.1017/S0376892912000392
43. Academy of Economic Studies. Studiu Privind Potențialul-Socio-Economic de Dezvoltare al Zonelor Rurale [Study Regarding the Socio-Economic Potential of Rural Areas]; Study Conducted for the Ministry of Agriculture and Rural Development from Romania; Academy of Economic Studies: Bucharest, Romania (2014). http://www.madr.ro/docs/dezvoltare-rurala/programare-2014-2020/studiu-potential-socio-economic-de-dezvoltare-zone-rurale-ver-10.04.2015.pdf. Accessed 4 July 2022
44. Gutoiu, G.: Spatial and temporal patterns of polarization and population decline in Romania's demography: 1992–2018. Revista Universitară de Sociologie **15**(2), 41–58 (2019)
45. Pavel, A., Moldovan, B., Neamtu, B., Hintea, C.: Are investments in basic infrastructure the magic wand to boost the local economy of rural communities from Romania? Sustainability **10**(10), 3384 (2018). https://doi.org/10.3390/su10103384
46. Kitchen, L., Marsden, T.: Creating sustainable rural development through stimulating the eco-economy: beyond the eco-economic paradox? Sociol. Rural. **49**(3), 273–294 (2009). https://doi.org/10.1111/j.1467-9523.2009.00489.x
47. Piperno, F.: The impact of female emigration on families and the welfare state in countries of origin: the case of Romania. Int. Migr. **50**(5), 189–204 (2012). https://doi.org/10.1111/j.1468-2435.2010.00668.x
48. Brădăţan, C.: The interplay between family and emigration from Romania. Migration Lett. **11**(3), 368–376 (2014). https://doi.org/10.33182/ml.v11i3.230
49. Simionescu, M.: Macroeconomic determinants of permanent emigration from Romania. A Bayesian approach. J. Int. Stud. **9**(2) (2016). https://doi.org/10.14254/2071-8330.2016/9-2/13
50. Glaub, M., Frese, M.: A critical review of the effects of family businesses training in developing countries. Enterp. Dev. Microfinance J. **22**(4), 335 (2011)

51. Gielnik, M.M., Frese, M., Kahara-Kawuki, A., Wasswa Katono, I., Kyejjusa, S., Ngoma, M., Munene, J., Namatovu-Dawa, R., Nansubuga, F., Orobia, L., Oyugi, J., Dlugosch, T.J.: Action and action-regulation in family businesses: evaluating a student training for promoting family businesses. Acad. Manag. Learn. Educ. **14**(1), 69–94 (2015). https://doi.org/10.5465/amle.2012.0107
52. Gielnik, M.M., Uy, M.A., Funken, R., Bischoff, K.M.: Boosting and sustaining passion: a long-term perspective on the effects of family businesses training. J. Bus. Ventur. **32**(3), 334–353 (2017). https://doi.org/10.1016/j.jbusvent.2017.02.003
53. Schlaegel, C., Koenig, M.: Determinants of entrepreneurial intent: a meta–analytic test and integration of competing models. Family Bus. Theory Pract. **38**(2), 291–332 (2014). https://doi.org/10.1111/etap.12087
54. Cha, M.S., Bae, Z.T.: The entrepreneurial journey: from entrepreneurial intent to opportunity realization. J. High Technol. Manag. Res. **21**(1), 31–42 (2010). https://doi.org/10.1016/j.hitech.2010.02.005

# Is Investment Portfolio Construction Sustainable in the Circular Economy Paradigm—The Case of ESG Investment?

Nikola Iliev[1], Marin Marinov[2], Valentin Milinov[2], and Mariana Petrova[2,3](✉)

[1] Paysafe Group, 90 Tsarigradsko Shose Blvd., 1784 Sofia, Bulgaria
[2] 2 Em. Chakarov Str., Tsenov Academy of Economics, 5250 Svishtov, Bulgaria
mm_p@abv.bg
[3] St. Cyril and St, Methodius University of Veliko Tarnovo, 5003 Veliko Turnovo, Bulgaria

**Abstract.** A circular economy is a "closed-loop" economic process, in which raw resources, products, and services keep their value after being processed, incorporated into production, and used by economic agents. Reiterating CE towards investing and portfolio management, one can wonder what could be their relationship. The study looks at investment caveats when facing CE's aim of a "closed cycle"-based economy. The authors bridge the gap between the POV of investing and its interpretation by circular economists as hazardous, exploitable, and speculative, through the concept of ESG investment. The research team gives examples of existing ESG investments, crafted to CE values, like for example Circular Economy Funds, analyzed and assessed on both investment characteristics and embodiment of CE principles. Custom portfolios are constructed with attention to measuring performance against the market and Blackrock. Also, the team studies specific KPIs of BlackRock and custom portfolios in terms of CE embodiment. The findings are as follows: 1. BlackRock is easily beatable by using simple portfolio construction strategies like Naïve Diversification or P/E Weight Scoring. 2. Weight Scoring fails at delivering additional returns over naïve diversification. 3. Lagrangian optimal weight portfolio underperforms against BlackRock 4. The top 33% of stocks in BlackRock are enough to outperform the whole fund.

**Keywords:** ESG score · BlackRock ETF · Circular economy values and principles · Quantitative assessment · Lagrangian multipliers · MSCI sustainability characteristics

## 1 Introduction

The paper's topic formulates a question regarding the importance of Investment Portfolio Construction within the Circular Economy (shortened to CE from now on). To answer this, the authors analyze the key values and principles of Circular Economy, but also through the fundamentals of the Investment process and Portfolio construction in particular. The scientific study is focused on intertwining said CE values and principles with


V. Koval et al. (Eds.): ISCMEE 2022, Lect. Notes in Managem., Industrial Engineering, pp. 15–42, 2023.
https://doi.org/10.1007/978-3-031-23463-7_2

investment theory and practice for shading light on the topic of investing in portfolios to benefit from and towards CE.

Thus, the research gap targets existing CE investment opportunities, by trying to evaluate them from the points-of-view of both CE core values and principles, but also through the lens of the investment process. These existing investment opportunities are analyzed and assessed both in terms of how well they perform as investments, but also as ways to contribute towards CE. Additionally, the authors construct their custom investment portfolios, that seek to follow the key values and principles of CE and evaluate them against the existing investment opportunities.

The CE framework is built around a list of values. A key value for a sustainable society built on CE is the ability to maintain and balance environmental quality and economic prosperity. This expands from a macro scale towards the meso- and micro-level where it shapes the value of the individual's physical and social well-being by offering them equity for a quality of life that at least meets the standard for human rights. From an environmental quality point of view, one can define the value of encompassing the use of natural resources within reasonable limits and in a way that enhances natural capital within and across future generations. From the economic prosperity point-of-view, one can describe the value of equal and fair access and mindful utilization of natural resources concerning their limits, instead of added profit.

The values listed are built around a list of pillar principles, one principle of which is the reciprocal flow of natural resources between the environment and the society embedded within it for their mutual sustainable co-existence. The idea is to not just extract value from but also to add value to the natural capital at a rate that is greater to allow for the regenerative and absorptive capacity of the environment. This could be expanded towards the next principle of promoting natural resources used sufficiently, and efficiently, and in differentiation between progress and unsustainable use.

This translates into the principle of design for circularity, where not just products, but production cycles, supply chains, distribution, and use are built around closing resource flow loops, minimization of raw material use, and waste production with recycling, reuse, and reintegration in mind. In terms of use, this defines the principle of consumption transformation, where production and profit-driven consumerism is replaced by responsible demand-driven use based on sharing, service-based, and experience-based consumption.

For this to be possible, another principle is to be followed—coordinated participatory multi-level change. CE is only possible when every social actor participates—government, industries, the civic sector, consumers, and academia. Also, this expands to the principle of mobilization of diversity, development of a plurality of ideas and solutions, the culture of knowledge exchange, resilience against uncertainty, and precaution against non-sustainable solutions.

To be possible, this would require adherence to the principle of policy and system reassessment in terms of translating from a short-term way of thought e.g., GDP as a measure of prosperity, towards a long-term measure of prosperity encompassing environmental, social, and governmental terms. This principle serves as a bridge between the CE paradigm and the investment theory and practice, through the definition of so-called ESG investments, to be explained further in the text.

In 2015 the European Commission adopted the Circular Economy Action Plan (CEAP) as a comprehensive body of legislative but also non-legislative actions, that aims to transition the economy of the European Union from one built around a linear model to-wards one built around a circular model.

In December 2019 the Von der Leyen commission unveiled the "European Green Deal" [17], as an ambitious plan for the transformation of the Economy of the European Union into a fair, sustainable and prosperous one, by making Europe the first climate-neutral continent. By this time all 54 actions, mapped out within CEAP were already implemented allowing the EU to be recognized as a global leader in terms of CE policymaking.

At the beginning of 2021, the European Parliament (EP) adopted the "New Circular Economy Action Plan" [18] to expand the list of measures for achieving a carbon-neutral, environmentally sustainable, toxic-free, and fully CE by the year 2050 further emphasizing on the definition of CE [19] as an alternative model of production and consumption, encompassing the share, lease, reuse, repair, refurbish and recycle of any existing material or product for as long as possible, by which one seeks to extend its life cycle.

CE departs from the traditional and linear economic model of gathering, making, consuming, and disposing of a product, that not only assumes an abundance of materials and energy, but also plans for product obsolescence, by purposely creating it with a shorter life span, thus encouraging repurchase. CE prioritizes both material and energy scarcity on one hand and the impact of their extraction and production on the environment on the other. It encourages waste prevention, eco-design, and re-use which saves money for the producer, dealer, and end user.

## 2 Literature Review

While innovative and opportunistic, the CE concept still lacks purposeful representation within some fields of the academic, economic, and industrial spheres of knowledge and influence. A perfect example is a field of investing and portfolio construction where the topic of circularity is often discussed indirectly, through the use of other existing principles and concepts. Such is the example of the ESG concept, translating into Environmental, Social, and Governmental principles.

Even if both concepts are not fully interchangeable in the broad sense, for the research on the topic of investing and portfolio construction, the authors assume their interchangeability. The reasoning behind this is simple—existing literature, concerning the topic of Circular Economy, doesn't seem to bother with its specifications within the investment field. A perfect example is the work of Kirchherr et al. [29] which conceptualizes the definition of CE through a semiological analysis of citations from over 110 different studies and publications. In all of the examples given, there are zero mentions of the investing field or the investment process as a whole.

Still, some authors try to build a connection between CE and investing. Morea et al. [35] try to bridge the environmental issues, expressed within the ESG concept and CE, by arguing that, as quote: "CE is a component of the "E" factor—in that it can generate positive and measurable environmental impacts—then it can be associated with the ESG score" The authors argue that this assumption allows them to methodologically overcome the lack of a specific score related to CE. The authors agree with this assumption and follow it throughout the research paper.

In Braun [7] a question is posed "ESG and CE: do they know each other" for which the author gives a very purposeful answer, as a quote: "ESG concerns … investors, who intend to assess potential investments, not only on financial aspects but non-financial factors as well … and to diversify their portfolios to attract green investments… CE involves many actors … start-ups, small- and medium-sized enterprises … large industries to decrease consumption levels of natural resources, lowering carbon footprints and coming up with efficient tools to improve waste management." At this point the author perceives a clear match—one group needs access to green capital (CE projects), while the other has the willingness to allocate its capital to serve ecological and social purposes in the framework of business solutions. The authors follow the same logic, and thus firmly believe that the ESG concept is a good reflection of the CE concept within the investment process.

It is important to distinguish between types of ESG investments. Green bonds, for example, share similarities with traditional bonds, with the difference that the above are issued for the specific purpose of financing projects, classified as friendly for the environment (Nguyen et al. [35]; Reboredo and Ugolini [40]; Saeed et al. [43]. Part of the scientific community expands the ESG investment universe by including instruments emitted by fin-tech companies and cryptocurrencies.

Dutta et al. [15] and Le et al. [31] explore investing in gold, crypto, fin-tech, and green bonds during the COVID-19 pandemic. The first study defines bitcoin and gold as safe-haven hedgers for risk, encouraging investors to choose them instead of investments in the currently volatile capital market. The second study specifies the technology stock's profile within a global capital market, proxied by the MSCI ACWI index, and the similarities between investing in fin-tech companies, green bonds, and gold, also defining them as good risk hedgers. But both pieces of research focus on the risk, rather than reward, that SRIs could provide to their investors.

Around the beginning of the 21st century, scientific thought had mixed views regarding ESG investment returns. Guerard [26], Goldreyer and Diltz [24], Bauer et al. [2], and Brzeszczynski et al. [8] documented that ESG returns are either greater or at least equal to overall market performance. On the opposite point of view, Renneboog et al. [41] defined a so-called ethical penalty that investors must take into account. According to their research in capital markets in North America, Europa, Asia, and the Pacific ESG in-vestments underperformed, compared to traditional investments.

Hong and Kacperczyk [27] suggest that even if there is a contradiction between academics regarding the trade-off between returns, significant implications may result from the attributes of those returns. Riedl and Smeets [42] highlight that a major factor for investing in SRIs is social reference and responsibility, implying a long-term effect on stock pricing and return. Nilsson [36] believes the existence of a specific type of investor, that is socially responsible and thus invests in SRIs not only for the above-average return but also for altruistic motives.

The financial sector and more specifically the investment industry are increasingly seizing the opportunity of investing the capital within the CE. In the last years, there's been a steep increase in the issuance of equity and debt instruments related in some way to the CE. This includes ETFs, corporate bonds, sovereign bonds, venture capital, private stock, and debt, but also bank lending, insurance policies, and project finance. As of 30 June 2021, an analysis by the Ellen MacArthur Foundation shows the following numbers (Fig. 1).

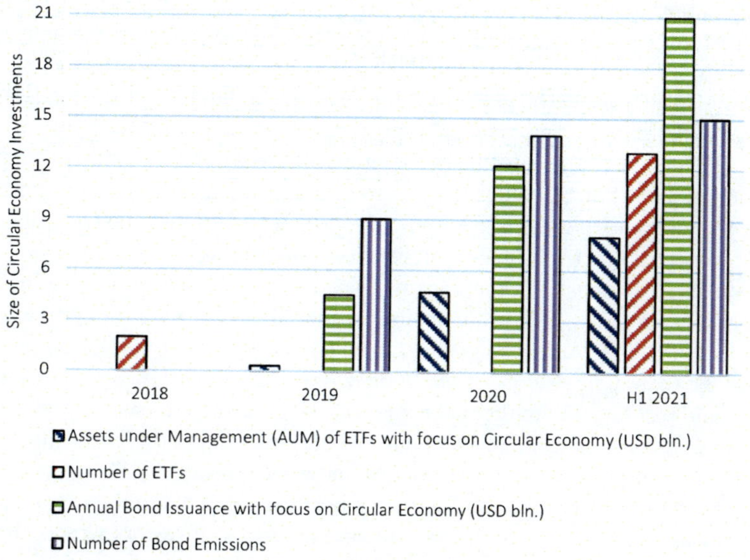


**Fig. 1.** Circular Economy—Assets under Management of ETFs and Emissions of Bonds. *Source* Ellen MacArthur Foundation [6].

As stated in the study providing these numbers, the size of the CE market grows from including just 2 funds in 2018 to 13 funds in 2021, issued by leading investment institutions like BlackRock, BNP Paribas, Credit Suisse, and Goldman Sachs. The growth of this market is over 26-fold increase and demonstrates a great potential for attracting capital. On the other side, are the bond emissions, underwritten and supported by financial institutions like Barclays, BNP Paribas, HSBC, ING, and Morgan Stanley.

Additionally, Intesa Sanpaolo launched a credit facility in the size of EUR 6 billion, Morgan Stanley launched a Plastic Waste Resolution for the removal of 50 million tons of plastic waste by 2030, the European Investment Bank partnered with 5 of the largest European countries' national financial institutions for the launch of EUR 10 billion loans and investment initiative dedicated to the CE.

The research, partnered by Bocconi University and Intesa Sanpaolo [6], samples a list of publicly traded companies, searching for a link between the degree of circularity and probability of default on a debt for companies with external credit rating on one hand and risk-adjusted return of public equity on the other. The results suggest that the more "circular" a given company is, the lower the risk of default on debt, in terms of one-year and five-year time horizons. Translating this into the investor language—the more "circular" a company is, the better investment it becomes.

Looking at Zara [50], the authors find they specify the effects of CE application on risk, return, and reputation of Financial Institutions, more specifically banks. That being said, a major part of the research findings is also applicable to companies of all sectors and industries. For the author the implementation of CE brings forth the "decoupling" (also see Peter et. al. [38]) of economic growth from virgin raw materials use, reducing company exposure to the growing trends in resource prices and their volatility. Finite materials are replaced by renewable, regenerative and, at minimum, second-hand ones, which entails a reduction in both systematic and idiosyncratic risks (further described in Zara and Shyham [50]). This reflects on the company's issued stock, by influencing long-term investment return stability, and by reducing the speculative component of return volatility (Zara et al. [53]).

Zara [50] elaborates further by intermingling the Sustainability Development Goals, adopted by the UN with the Environmental, Social, and Governmental framework [44] (previously mentioned within the principles of CE as ESG). Relating both allows investors to make the necessary step from simply classifying part of investments as ESG into exactly measuring the impact of sustainable investment within CE.

Additionally, the author interprets the CE's influence on companies' reputations. His attention is on the Financial Companies' opportunity to "clean their name" as one of the main perpetrators of the 2008–2009 financial crisis. By embracing the CE, they would reconnect with the real economy and contribute to societal well-being. From the point-of-view of companies' clients, a better reputation translates into higher profit, not just in terms of market share, but also in terms of premium clients are willing to pay for products and services that generate a positive impact on the environment and society. This also translates into the investment process; as investors would also prefer (ceteris paribus) investing in the stock of companies that care for the environment.

Maltais and Nykvist [32] embrace the idea of financial markets, or investors, in particular, as the main driver behind adopting CE. For McGlade and Ekins [34] investors are concerned about unsustainable business practices and how they lead to stranded assets. For Bloomberg et al. [5] investors are concerned about climate policy and climate impact risks. Galaz et al. [22], Folke et al. [21], and Gardner et al. [23] all comment on investors and their concerns about the transparency of perpetrators of financing unsustainable economic activity. Gardner et al. [23] and Buschet al. [10] specify how the ESG framework impacts the financial performance of assets.

Therefore, based on the above discussion, the work investigates the possibility to invest in a socially responsible manner, in a way that doesn't imply above-average return as the only motivation but also adds the altruistic motive of abiding by the requirements of the CE paradigm. The attention is not limited to only finding the socially responsible investments that meet these requirements but to also try and approbate a methodology for the diversification of these investments in terms of portfolio creation and combination of individual investments' competitive advantages with their existing relationship, for the realization of additional return at a lower level of risk.

Investment portfolio construction allows investors to combine the competitive advantages of individual assets with the relationship that exists between them, to realize additional returns at lower levels of risk. Following the linear economic model paradigm, a portfolio could be any combination of assets that only seeks to satisfy the investors' preferences, profit expectations, and risk tolerance. By the same logic, a portfolio would follow the CE paradigm if the assets combined are issued by companies that encompass the alternative model of production and consumption (as defined by EP).

Investment portfolio construction is often associated with the term Modern Portfolio Theory and the works of Markowitz [33], for whom a portfolio is the grouping of individual asset investments that enables risk-averse investors to maximize their return for a given level of risk, by ameliorating risk characteristics through this grouping, at least partially. Or to put it in simple terms—return stacks additively, bit risk not so much. To explain this, Markowitz uses a concept previously studied by de Finetti [13]—the variance-covariance matrix- to transcribe the risk split between a group of individual asset investments.

Figge et al. [20] find that modern portfolio theory (or MPT for short) continues to be not only popular in the financial world but also to be robust even today, as contemporary research continues to find its fresh nuances, extensions and applications, in works like Eaton et al. [16], Bajaj [1] and Ünlü and Xanthopoulos [45]. MPT's fundamental principles remain relatively intact. Brealey et al. [9] define the return and risk of a portfolio through the use of a two-asset example, where:

$$R_P = w_a * r_a + w_b * r_b \tag{1}$$

with: $R_P$ is portfolio return,

$w_a, w_b$—percentages invested in assets A and B, respectively,

$r_a, r_b$—returns of assets A and B, respectively; and

$$\sigma_P^2 = w_a^2 * \sigma_a^2 + w_b^2 * \sigma_b^2 + 2 * w_a * w_b * \sigma_a * \sigma_b * \rho_{ab} \tag{2}$$

with:

$\sigma_P$ is portfolio risk, measured via standard deviation or $\sigma_P^2$—variance,

$\sigma_a, \sigma_b$—the risk of assets A and B, respectively, measured via standard deviation,

$\rho_{ab}$—correlation of assets A and B.

These equations can be further expanded for any number of assets $n$, where:

$$R_P = \sum_{i=1}^{n} w_i * r_i \tag{3}$$

$$\sigma_P^2 = \sum_{i=1}^{n} w_i^2 * \sigma_i^2 + \sum_{i=1}^{n} \sum_{j<i}^{n} w_i * w_j * \sigma_i * \sigma_j * \rho_{ij} \tag{4}$$

with $i$ and $j$ being assets i and j within portfolio P, all other notations are the same as previously.

From Eqs. (3) and (4) one can see that while the relationship of return is additive, the relationship of risk is not. There exist an additional component $\rho$ denoting correlation or the mathematical expression of the relationship. From here, to diversify risk, the investor must only choose a combination of assets with negative correlation, which in turn will bring a negative sign of the second part of the equation, and thus reduce overall portfolio risk.

In a perfect world, there exists such a combination of assets that would allow for the diversification of the second part of the equation, also known as unsystematic risk, to a value of zero, reducing the equation to only its first part—the systematic risk. This reduction is the theoretical idea behind modern portfolio theory, which Markowitz introduced [33]. In the next part of the study, the authors will expand on these formulas, to define the so-called optimal portfolio and benchmark portfolio, both of which are important for investors who practice investment portfolio diversification.

The scientific paper follows the structure of Materials and Methods and discusses details about the data used and the methodology applied. Next comes the results section, where one can get acquainted with the findings. The authors later discuss what the results imply and how they fit into the existing paradigm of Portfolio Management and Circular Economy. After this, a conclusion is drawn which summarizes everything done and found.

## 3 Materials and Methods

### 3.1 Data

The study uses a sample of the historical NAV of BlackRock Equity BGF Circular Economy Fund A2 Euro (BGF Circular Economy Fund) [4], covering the workday frequency of data spanning from the index's inception—01 October 2019 up to 30 September 2021. The motivation behind this period is to have two full years of trade data, deployable within the confines of traditionally used reporting periods (like in this case from the beginning of Q4 of 2019 up to the end of Q3 of 2021 or eight quarters or two years). The reasoning behind using BlackRock's ETF is given in the next subsection—Methodology.

Additionally, the historical price performance of stocks from Investor's Business Daily's list of the 100 Best ESG Companies of 2021 is used (Investor's Business Daily [28]). The list consists of the highest-scoring companies, by a measure of ESG score, and is chosen by Dow Jones after screening over 6 000 global publicly traded companies. The data follows the same structure—workday frequency from October 20219 up to and including September 2021. For key financial and reporting statistics, the authors use publicly available data from YCharts [48]. To assess the Index using Sustainability Characteristics and Business Involvement Exposures, data from MSCI is used, as the main source of both methodological backgrounds behind different indicator construction, but also the data used for their calculation.

### 3.2 Methodology

Upon starting the research, the team finds the following investment opportunities, created around the notion of circular economy:

- BlackRock Equity BGF Circular Economy Funds;
- RobecoSAM Circular Economy Equities;
- BNP Paribas Easy ECPI Circular Economy Leaders UCITS ETFs.

After getting acquainted with their available prospectuses, the authors have found that the strategies behind the construction of each of the portfolios are similar in both the nature of pursued goals and the principles that are followed, both in terms of investment and circular economy. Additionally, the research team looked at the dynamics of each of the three funds' market performance, as presented in the following Fig. 2:

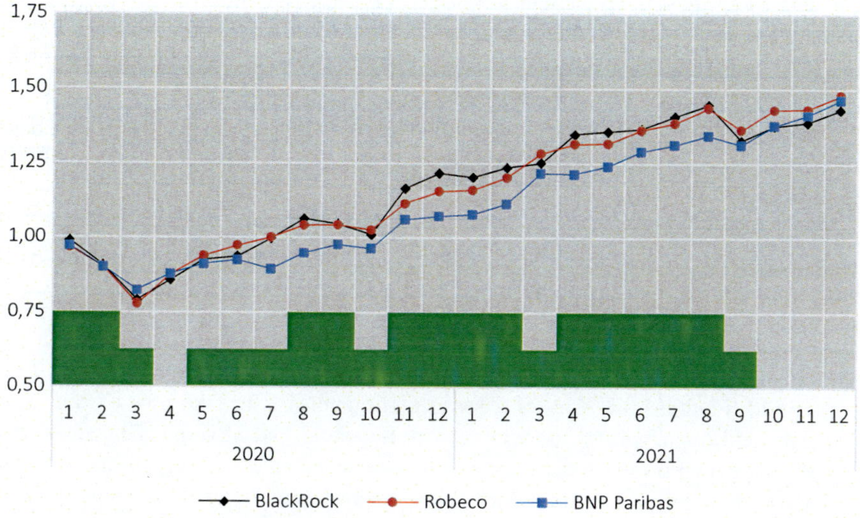


**Fig. 2.** Circular economy ETFs performance—2020 to 2021. *Source* BlackRock.com, Robeco.com, markets.ft.com (for BNP Paribas).

As seen in the figure, the performances of the three ETFs follow similar dynamics, where the only differences are due to short-term changes between each month. But no matter these changes, there is a trend of BlackRock outperforming the other two ETFs continuously. This can be seen by simply counting how often BlackRock is on top. BlackRock outperforms the others in 13 out of the 24 months and outperforms at least one additional 7 months. This gives us only 5 months during which the ETF underperforms. The team also looked at the same data but on a working day basis and have found that for the period from January 2020 to December 2021, BlackRock outperforms the other two in 51.08% of cases and outperforms at least one of the two in 31.76% of cases. This leaves BlackRock to underperform in only 17.16% of cases. Using this argument, the research team decided to use the BlackRock ETF as the object of study.

In giving a reason why BlackRock is a representative of the Circular Economy Paradigm, the team cites the Prospectus, given by the issuer. Issued at the beginning of October 2019, by BlackRock Inc. The ETF provides a return on investment through a combination of capital growth and income on its constituents. The fund invests globally at least 80% of its total asset value within stocks of companies, that benefit from and contribute to the advance of CE.

The main idea is to invest in a portfolio of stocks of companies of any market capitalization or industry sector, as long as they can be classified into one of the following four categories:

- Companies that adopt "circularity" in their business operations by either being involved in the sustainable fashion of production and resource use or that commit to using recycled resources within their production cycle;
- Companies that provide innovative solutions, aimed directly at solving the problem of inefficient use of resources and pollution, by recycling, reducing inputs like water or energy, and enabling sustainable logistics;
- Companies providing alternative materials to those that cannot be recycled and shift towards more easily recyclable resources, like for example natural or plant-based circular alternatives to non-recyclable and non-biodegradable resources;
- Companies aiming to facilitate efficiency and responsible consumption via their business model, including but not limited to digitalization and virtualization in one example or repair and resale to prolong the use of goods, in another.

To assess a company's benefit from or contribution to advancing the Circular Economy paradigm, the Investment Adviser uses the previously mentioned ESG concepts. The idea behind this way of assessing companies is to take into account not just the risks and opportunities associated with investing, but also the risks and opportunities associated with the Circular Economy—environmental concerns like the climate crisis and sustainability; social concerns like diversity, human rights, consumer protection, and animal welfare; corporate governance concerns like management structure, employee relations, executive and employee compensation, etc.

Further reading from the prospectus the research team finds that the portfolio adopts a "best in class" approach to investing. Initially, the Investment Adviser creates an investment universe containing the stocks of all companies to be potentially viable for inclusion within the portfolio. In this case, the investment universe is represented by the MSCI All Countries World Index (MSCI ACWI) which as a fair representation of an investment universe can also be used as a benchmark to compare the performance of the fund. Next, the Investment Adviser chooses the best issuers, from an ESG perspective, from each relevant industry sector, without exclusion. Although a big part of the portfolio will likely be invested in stocks of companies from the developed markets, it is possible to also include emerging markets.

The initial analysis of the index performance focuses on the investing point-of-view, both in terms of overall performance, as measured through price return, but also against a properly chosen benchmark, that is defined as the investment universe. This second comparative analysis is done for the purpose to assess how well the index performs in comparison to the market average. This comparative analysis is important as it tries to objectively answer the question is investment portfolio construction sustainable in the Circular Economy Paradigm. But before being able to do so, one must also answer another question, that very well can be interpreted as an opposite one—if a portfolio is assumed to comply with the principles of Circular Economy, is it even viable from an investment point-of-view? To answer this, one must first only assess the index's investment profile, before turning attention toward all other factors, that encompass the notion behind Circular Economy.

Next, the authors try to take into account said factors, as to find a way to measure the Index's performance from the point-of-view of the Circular Economy. For this a second benchmark must be defined, this one not just representing the investment world, but that part of it, that best encompasses the principles of Circular Economy. The research team has chosen to use Investor's Business Daily's list of the best ESG companies of 2021 as a starting point. The list includes companies with strong stocks and growth opportunities that don't sacrifice for environmental, social, or governance values.

The companies are screened by Dow Jones from a list of 6 000 globally traded equities. From there chosen are only the stocks trading over $10 or 2 360 companies. Next, Dow Jones looks at each stock's Composite Rating, (Investor's Business Daily [28]) taking into account only companies with a value of 85 or better. The Composite Rating is a measure that ranges from 1 to 99 where one means worst, while 99 means best possible investment. The rating takes into account:

- EPS rating—a product of comparing companies' performances by measures of quarterly earnings growth and annual earnings growth;
- SMR rating, Sales growth, Profit Margins, and Return on equity, ranked among all companies;
- Relative Strength (RS) rating—a product of comparing companies' last 52 weekly price returns.

Screening is done using the key indicator of ESG score, for which Dow Jones uses its data. All companies are ranked using ESG scores and the top 100 are chosen as the best ESG stocks. These hundred stocks will be the basis of the second benchmark, used as a proxy of an investment opportunity, taking into account the notion of a Circular Economy. But as these 100 companies are individual investments, the next step is to transform them into a single entity—a portfolio that represents them all, and thus also encompasses the Circular Economy paradigm.

While using portfolio construction as a basis for testing a hypothesis, this study's focus isn't on assessing how well a given portfolio construction methodology compares to others in terms of performance and levels of expected and realized return and risk. This is why the methodology used in this study is as close to the basic modern portfolio theory in terms of assumptions and biases. The authors assess either already constructed portfolios, like the BlackRock Equity BGF Circular Economy Fund A2 Euro index, or try to create their portfolios from a given investment universe, in this case, the top 100 best ESG companies of 2021.

Three different methods for portfolio creation are used:

- Naïve diversification—a heuristic method of essentially assuming that splitting the capital between many different assets is the best possible strategy to reduce overall risk.

The model doesn't calculate individual stock weights but instead gives equal weights to all assets included in the investment universe. In this case, this means that to construct a portfolio from the top 100 best ESG companies, it is sufficient to just give a weight of 1% to each of the hundred stocks. Somewhat primitive and thus called naïve, diversification still is a good way to project the characteristics of multiple individual objects into a single entity, in these cases the objects being stocks and the entity—a portfolio.

This assumption is made, following the work of DeMiguel et al. [14] who test how well portfolios perform against a portfolio constructed via naïve diversification. Their study assumes that there are two good reasons to use a naïve diversified portfolio as a benchmark for the performance of others. The first reason is the easier implementation due to the lack of need for estimation of moments of asset returns or the need for optimization calculations. The second reason is that, despite the existing sophisticated theoretical models for the estimation of parameters, investors tend to use simple allocation rules for investing their wealth in assets—Benartzi and Thaler [3].

A strange finding in their empirical tests of different portfolios is that naïve diversified ones tend to outperform portfolios built using sophisticated models. That being said, even if the paper doesn't study the performance of portfolios only from a return perspective, it is good to use a stable benchmark that could outperform the majority of portfolios. This allows one to escape a biased opinion on the performance of these portfolios, where one mistakenly attributes their additional return over the benchmark to the influence of circular economy factors, instead of, quite possibly, the low performance of the chosen benchmark.

- Quantitative assessment—a quantitative indicator method to determine the individual stock weights within the portfolio.

These indicators can be of any nature, but often are somehow connected to the investment logic and are either based on the market performance of the stock or the financial performance of the company, or even both. In this case, one can use any of the mentioned indicators, used for screening and ranking the top 100 ESG companies. Theoretically sound, this would somewhat distort the results further in the direction of the top companies at the expense of the bottom ones. The reason for this is that no matter what exact mathematical calculation one uses, still, the stocks at the top of the list will have higher weights than those at the bottom. To fight this, one can use any other theoretically sound logic, like for example weighing using market performance or weighing using the company's financial performance.

The use of indicators as a way of scoring stock weights within a portfolio emanates from the efficient-market hypothesis, which asserts that financial markets do reflect all relevant information, and thus stocks always trade at their fair value. This hypothesis assumes no possibility of purchasing undervalued or sell of overvalued stocks. In reality, markets are often semi-efficient (Kong et al. [30], Wen et al. [46], Daniel et al. [11]) but still allow for the partial use of financial indicators as predictors of expected return (Rasekhschaffe and Jones [39], Dai and Zhou [12], Wu et al. [47], Gu et al. [25]).

The research team follows the logic of Yeh and Liu [49] who analyze how well different metrics perform when used for the weighting of stocks in portfolio construction. For them one of the main indicators to be used is the "growth" factors, indicating that companies with better financial performance in terms of profit have stocks with greater returns. Similar results can be found in the works of Kong et al. [30] and Yeh and Hsu [54].

- Portfolio optimization—a purely mathematical method of finding what is the best portfolio.

As previously mentioned, the paper's focus isn't on assessing how good a portfolio construction methodology is, but instead on how well will portfolio construction measures against the Circular Economy Paradigm. Thus, the authors must try and find an alternative way to incorporate a purely mathematical way of portfolio construction. In this case, the authors choose to use Lagrangian Multipliers over the Variance-Covariance Matrix of the top 100 ESG companies. The main advantage of this is that the research team doesn't have to use an optimization model or computational power, but can just do a couple of matrix multiplications before reaching the results.

In terms of methodology to calculate the weights using Lagrange Multipliers, the model (Offiong et al. [37]) one can use the following. Let

$$w = (w_{i=1} \ldots w_{i=n})' \tag{5}$$

be a global minimum variance portfolio for *assets*. By solving a constrained minimization problem, given as:

$$\min_{w_{i=1} \ldots w_{i=n}} \sigma_P^2 = \sum_{i=1}^{n} w_i^2 * \sigma_i^2 + \sum_{i=1}^{n} \sum_{j<i}^{n} w_i * w_j * \sigma_i * \sigma_j * \rho_{ij} \tag{6}$$

$$s.t \sum_{i=1}^{n} w_i = 1 \tag{7}$$

the Lagrangian is

$$L(w_{i=1} \ldots w_{i=n}, \lambda) = \sigma_P^2 + \lambda * \left( \sum_{i=1}^{n} w_i - 1 \right) \tag{8}$$

with the following first-order conditions

$$\frac{\partial \mathrm{L}}{\partial w_{i=1}} = 2 * w_{i=1} * \sigma_{i=1}^2 + \sum_{j}^{n} 2 * w_{i=1} * COV_{i=1,j} + \lambda = 0 \tag{9}$$

$$\frac{\partial \mathrm{L}}{\partial w_{i=n}} = 2 * w_{i=n} * \sigma_{i=n}^2 + \sum_{j}^{n} 2 * w_{i=n} * COV_{i=n,j} + \lambda = 0 \tag{10}$$

$$\frac{\partial \mathrm{L}}{\partial \lambda} = \sum_{i=1}^{n} w_i - 1 = 0 \tag{11}$$

The above equations are for an order of $n + 1$ equation where the first $n$ are for the weights and by solving them, would return the global minimum variance portfolio weight values. The equations can be expressed by a matrix:

$$\begin{pmatrix} 2 * \sigma_{i=1}^2 & \cdots & COV_{i=1,j=n} & 1 \\ \vdots & \ddots & \vdots & \vdots \\ COV_{i=n,j=1} & \cdots & 2 * \sigma_{i=n}^2 & 1 \\ 1 & \cdots & 1 & 0 \end{pmatrix} * \begin{pmatrix} w_{i=1} \\ \vdots \\ w_{i=n} \\ \lambda \end{pmatrix} = \begin{pmatrix} 0 \\ \vdots \\ 0 \\ 1 \end{pmatrix} \tag{12}$$

which can be reduced to

$$\begin{pmatrix} 2\boldsymbol{\Sigma} & \boldsymbol{I} \\ \boldsymbol{I}' & 0 \end{pmatrix} * \begin{pmatrix} \boldsymbol{w} \\ \lambda \end{pmatrix} = \begin{pmatrix} 0 \\ 1 \end{pmatrix} \tag{13}$$

where $\boldsymbol{\Sigma}$ is the covariance matrix of size $n * n$,
$\boldsymbol{I}$ is an identity matrix-vector of size $n * 1$,
$\boldsymbol{I}'$ is the transpose of an identity matrix-vector of size $1 * n$,
$\boldsymbol{w}$ is a vector of size $n * 1$ containing the global minimum variance portfolio weights.
The solution of the matrix of equations is

$$\begin{pmatrix} \boldsymbol{w} \\ \lambda \end{pmatrix} = \begin{pmatrix} 2\boldsymbol{\Sigma} & \boldsymbol{I} \\ \boldsymbol{I}' & 0 \end{pmatrix}^{-1} * \begin{pmatrix} 0 \\ 1 \end{pmatrix} \tag{14}$$

Returning attention to the quantitative assessment method, the authors have decided to follow an alternative logic. The idea is to use key portfolio characteristics of the index, to be tested (BlackRock Equity BGF Circular Economy Fund A2 Euro index) for the construction of the benchmark against which the index will be tested. In this case, this key portfolio characteristic, to be used, is going to be the industrial concentration profile of the index, which would later be replicated when trying to quantitatively assess how to distribute the weight between hundred stocks. The research team believes that the industrial concentration profile of the index is of key importance for evaluating it from the point-of-view of the Circular Economy Paradigm, as its main concern is how well are resources used for production, how well is energy utilized, how much waste is created and how much the environment suffers from all of it.

In this sense, the choice of particular industries to include within a portfolio (no matter if we're talking about the index or any other portfolio) is important, as different industries tend to benefit from and contribute to environmental issues on a different scale. It is for example, quite easy to understand that a company within the Information Technology sector has quite a low environmental footprint compared with, for example, a company from the Materials sector.

Next, a list of Sustainability Characteristics is defined, such that would assess the Index from the point-of-view of how well it trades-off returns for social reference, responsibility, and long-term effect on both the market and the environment, thus creating an ethical penalty for the socially responsible investors. These sustainability characteristics can provide potential investors with information from non-traditional metrics enabling the evaluation of potential investments on environmental, social, and governance characteristics (ESG). These measurements do not make an estimation of the future performance of given investments and do not represent the potential risk or reward behind it, but just provide an indication of the current profile in terms of listed characteristics. They must not be considered in isolation, but instead as part of all available information that describes the investments' "environmental footprint".

Defined by MSCI the ESG rating is calculated to directly map the ESG quality score of companies to a letter rating category. Each investment—no matter single stock of a company or a whole portfolio, is scored with a rating on a scale from triple C to triple A. Investments with the lowest ratings, from triple C up to B, are laggard investments, followed by ratings from double B up to A which are classified as average investments. The upper group, of ratings double A or triple A, are leader investments.

To further quantify, the MSCI ESG rating can be transformed to a quality score on a scale from 0 to 10, calculated using a weighted average from the ESG scores of all investments within the scored one. This characteristic makes sense for portfolios but can be applied also to stocks of companies, that are holdings of multiple businesses. Here the methodology requires finding the ESG rating of subsidiaries which are later weighted to a single value. The score can consider the trend of the rating, the exposure of the investments to holdings to industry-specific risk, or, as mentioned above, against investments of the laggard category.

When looking at a sample of investment ratings and quality scores, it is a good idea to use relative interpretations like the MSCI ESG percent coverage and the MSCI ESG peer quality score. The first, again mostly for portfolio, but still usable for stocks of holding, measures what percent of the investment constituents have information about ESG fund rating and quality score. From here larger value means bigger transparency regarding ESG assessment. The second finds its roots in percentile ranking and comparison between a sample of potential investments in the so-called Lipper peer group, in which the assessed investment falls.

Another good measurement is the MSCI weighted average carbon intensity ratio which compares the total value of CO2 emissions against the total value of sales of a given company. The carbon emissions are measured in tons while the sales are in USD millions. To compare the ratio between peers within a sample group, a percentage coverage can be calculated where all values in the samples are normalized within a range of 0 to 100, where values closer to 100 would mean better ratios, or in other words, lower carbon emission corrected for sales volume.

The indicators mentioned allow for the quantification and comparison of the ESG performance of stocks of companies and portfolios. An alternative group of indicators is the exposure to different types of risks. Here risks aren't used in terms of investment volatility and the possibility of loss but instead risks for enforcement of the principles of Circular Economy. To measure the exposure to these risks one can, look at the investment's overall exposure to portfolios with specific profiles. Such portfolios, in this case, can be:

- MSCI controversial weapons—contains the stocks of companies that were identified (by MSCI ESG research) as having involvement with controversial weapons like cluster munitions, landmines, uranium-based weapons, biological or chemical weapons,

lasers with blinding or burning properties, non-detectable fragmentary or incendiary weapons;
- MSCI nuclear weapons—contains the stocks of companies identified as manufacturers of nuclear warheads, intercontinental ballistic missiles, submarines that carry ballistic missiles, and submarines capable of delivery of nuclear warheads;
- MSCI civilian firearms—contains the stocks of producers of firearms, small arms, and ammunitions for civilian markets;
- MSCI tobacco—contains the stocks of producers of tobacco and tobacco-related products;
- MSCI UN global compact violators—contains the stocks of companies that fail to comply with United Nations Global Compact Principles;
- MSCI thermal coal—contains the stocks of companies earning more than 5 percent of their total revenues from thermal coal mining;
- MSCI oil sands—contains the stocks of companies earning more than 5 percent of their total revenues from oil sand extraction.

## 4 Results

In Fig. 3 one can see the performance of BlackRock Equity BGF Circular Economy Fund A2 Euro, labeled as Index, and the MSCI ACWI global index, labeled as Benchmark. The figure also includes the spread between the Index and Benchmark, differentiating positive from negative values. At the beginning of the period, the Index underperforms compared to the Benchmark, resulting in a negative spread of average value of -1.01% for Q4 of 2019, resulting from an Index return of 7.70% versus a Benchmark return of 8.70%. A short volatility peak during March, followed by outperformance from august allows the Index to beat the Benchmark by 7.53% for the whole of 2020, resulting from an Index return of 22.00% against 14.47% for the Benchmark. Finally, for the first three quarters of 2021 the Index is still over the Benchmark, but with constant shifts in volatility around March and near the end of the period—during September. Here it can be seen that the Index closes the gap with the Benchmark while realizing a return of 9.13% versus 10.14% for the Benchmark or a negative spread of −1.01%.

If one looks at the overall performance of both investment alternatives, the Investment horizon of the index allows for the realization of a total Return of 43.38% against 37.05% total Return for the Benchmark. This difference is mainly due to outperformance during 2020 and high inertia and momentum in 2021, allowing for compensation of the descent in value near the end of the period. From an investment point-of-view, the Index is fully capable of outperforming the market and thus can be classified as a good overall investment, while not taking into account other factors like how well does in encompass the Circular Economy paradigm.

After analyzing the Index's performance from an investment point-of-view, the authors turn their attention toward the Circular Economy. As mentioned in the Methodology section, the research team would proxy this through the use of portfolios, constructed from the top 100 best ESG as listed by Investor's Business Daily [28]. As the list includes 100 companies, the research team chose to visualize their GICS Sector structure as a representative sample, both in terms of the number of companies in each industry and

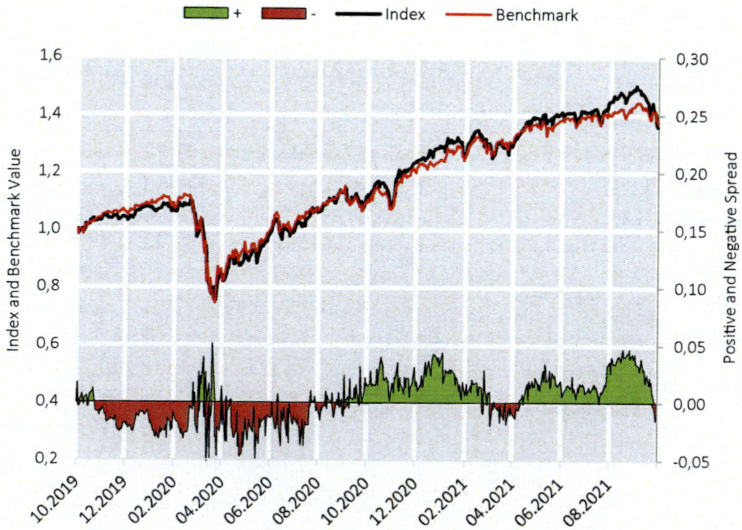


**Fig. 3.** Index and Benchmark performance for the period 30.09.2019–30.09.2021. *Source* BlackRock and Yahoo Finance.

the size of said industry, where the size of the companies' market capitalizations as of 30 September 2019 is used.

**Table 1.** Top 100 best ESG companies—Industry sector structure.

| Sector | Number | Mkt Cap (USD) | Mkt Cap % |
|---|---|---|---|
| Information Technology | 30 | 154 151 254 | 48.51% |
| Communication Services | 4 | 56 082 030 | 17.65% |
| Consumer Discretionary | 11 | 25 616 752 | 8.06% |
| Materials | 12 | 19 973 543 | 6.29% |
| Health Care | 8 | 19 752 798 | 6.22% |
| Industrials | 17 | 15 346 501 | 4.83% |
| Financials | 9 | 10 611 165 | 3.34% |
| Real Estate | 6 | 8 801 803 | 2.77% |
| Energy | 1 | 4 173 090 | 1.31% |
| Consumer Staples | 2 | 3 285 062 | 1.03% |
| Utilities | 0 | 0 | 0.00% |
| **Total** | **100** | **317 793 998** | **100.00%** |

*Source* Investor's Business Daily & YCharts. Authors' calculations, based on Market Cap data

Table 1 shows us that on the top of the list both in terms of the number of companies and the size of their market capitalization stands the Information Technology sector, responsible for almost a third of the number of companies and half of the market capitalization. On the bottom, one can first find a sector that isn't even part of the list—the Utility sector, with zero representatives. Next to it stand the Energy and Consumer Staples sectors, with one and two companies respectively, taking 1.31% and 1.03% of market capitalization.

Having chosen the 100 stocks of companies to be included in the potential portfolios, the research team constructs them, using the previously described methodologies, and analyzes their performances in time. Said performances are shown in the following Fig. 4.

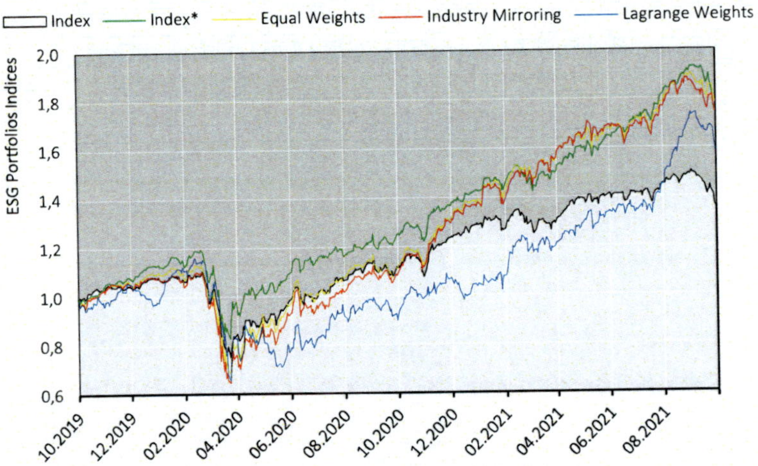


**Fig. 4.** ESG Portfolio Performance for the period 30.09.2019–30.09.2021. *Source* BlackRock.

Here one can find an overall positive performance of all portfolios, even if the dynamics behind them, and the reasons for these dynamics, can be interpreted quite differently. First, there is the index performance, which is the same as in the previous figure. Next to it, there is another portfolio, named Index* which is yet to be discussed. The third portfolio is the portfolio, a product of equal weighting. The fourth is a product of quantitative assessment, where the price-to-earnings ratios and industrial concentration profile of the index are used as a reflection of the portfolio weight distribution.

The price-to-earnings ratios are used to correct the weights of stocks, while the industry structure is used to limit the portfolio's concentration to industry exposure. The product of this is the performance named Industry Mirroring. Lastly, a fifth portfolio is visualized—a product of the mathematical construction of an optimal portfolio through the use of the matrix of variance and covariance relationship between the hundred stocks. To better assess the performances against the index, the following figure displays each portfolio's performance netted from the performance of the index, by use of the return spreads (Fig. 5).

The Equal Weights and Industry Mirroring Portfolios' spreads are very low, up to just 0.29 and 0.36% respectively for Q4 of 2019. Through the Q1 of 2020, they are quite volatile in negative terms, and continue this trend up until the end of Q3 of 2020, after which their spreads become positive. This shows an annual spread against the index

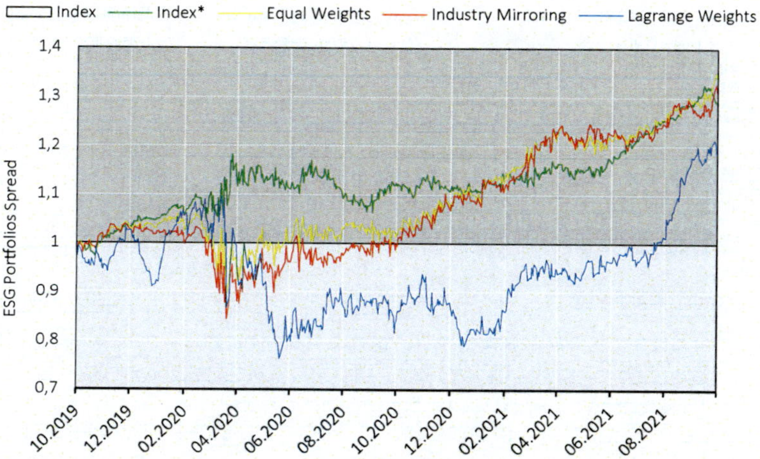


**Fig. 5.** ESG Portfolio spreads for the period 30.09.2019–30.09.2021. *Source* BlackRock.

of 7.03 and 7.12% respectively. The positive trend continues in 2021 and for the nine months, the portfolio's spread comes to 19.64 and 19.93% respectively. One can look at both at once, due to their almost identical volatility dynamics during the whole studied period.

The dynamic behind the spread of the Lagrange Weights Portfolio against the Index follows a mostly negative trend except for the months of August and September of 2021. Through 2019 the portfolio realizes a negative spread of 8.94%. For the year 2020, the negative spread is 10.60%. The monumental changes during 2021 bring the overall spread to a value of 43.38%.

In overall terms, the authors find that the Industry Mirroring portfolio realizes an overall return of 74.29% or 28.43% more than the index. For the Lagrange Weights Portfolio, the picture is 58.48% overall return or 16.77% more than the index.

The authors finally turn their attention toward the other portfolio shown on the two graphs, the one called Index*. While this is also a portfolio it consists not of the hundred stocks from the Top 100 best ESG companies (well at least from just a sample from them), but of just ten stocks of the companies that are part of the Index. These ten companies are the only ones made publicly available from BlackRock, whose names and weights within the Index are known. They are shown in the following Table 2:

The five stocks NKE, MSFT, ASML, A, and AVY are five of the hundred companies from the top 100 best ESG companies, which allows for the Index* to at least partially resemble the other portfolios. In terms of performance, the Index* outperforms the other portfolios. Its overall return is 74.86% spread over the Index's return of 35.73% with 39.13%. This difference while giving us information about the investment opportunities behind the Index* portfolio, consisting of just ten stocks, allows us to make one major discovery about the overall profile of the BlackRock Index—at its core, the investment return isn't the only priority.

This is a big argument in support of some of the studies, quoted above—Guerard [26], Goldreyer and Diltz [24], Bauer et al. [2], and Brzeszczynski et al. [8] in the sense

**Table 2.** BlackRock index known constituents.

| Company name | Sector | Industry | Symbol | Weight (%) |
|---|---|---|---|---|
| Nike Inc. Class B | Consumer discretionary | Textiles, apparel and luxury goods | NKE | 3.66 |
| Microsoft Corp | IT | IT services | MSFT | 3.66 |
| Asml Holding Nv | IT | Semiconductors and equipment | ASML | 3.50 |
| Agilent Technologies Inc | Health Care | Health care equipment and supplies | A | 3.30 |
| Avery Dennison Corp | Industrials | Commercial services and supplies | AVY | 3.01 |
| Veolia Environ Sa | Industrials | Waste management | VEOEY | 4.05 |
| Bureau Veritas Sa | Industrials | Consulting services | BVRDF | 3.37 |
| Sika Ag | Materials | Specialty chemicals | SKFO | 3.21 |
| L'Oréal Sa | Consumer discretionary | Household and personal products | LRLCY | 3.20 |
| Keyence Corp | IT | Scientific and technical instruments | KYCCF | 2.80 |
| **Total** | | | | **33.76** |

that Socially Responsible Investments can beat the market (as it was seen in the spread against MSCI ACWI) but also Renneboog et al. [41], Hong and Kacperczyk [27] and Riedl and Smeets [42] for whom there's a trade-off between the return from stocks of companies closely following the notion of Circular Economy. Next, the attention turns toward the benefits that the Index has, for making these trade-offs. First are the so-called Sustainability characteristics, shown in the following Table 3:

In terms of ESG Fund Rating, the Index is ranked quite high—double A is the second group after the best-performing peers of triple-A. In terms of score, this is quantified to a value of 8.42 from 10. The ESG percentage coverage is almost 100, so there is available information for almost all of the constituents behind the index—99.33. ESG Peer Quality Score has a value of 76.95 which translates into the Index being better than almost 77% of other investments, for which information about ESG performance is recorded. The number of these investments is 4 407.

In terms of Weighted Average Carbon Intensity, a value of 150.67 gives us little information, but when looked at in terms of comparison, the percentage coverage is 97.93, so the Index is performing better than almost 98% of all other 4 407 investments in the investment universe. Turning attention towards business involvement one can find no exposure towards any of the groups of stocks, that in some way be a violation of the principles of Circular Economy.

**Table 3.** Index sustainability characteristics and business involvement.

| Indicator | Value |
|---|---|
| MSCI ESG fund rating (AAA-CCC) | AA |
| MSCI ESG quality score (0–10) | 8.42 |
| MSCI ESG % coverage | 99.33 |
| MSCI ESG peer quality score | 76.95 |
| Investments in Peer Group | 4 41 |
| MSCI weighted average carbon intensity | 150.67 |
| MSCI weighted average carbon intensity percentage coverage | 97.93 |
| MSCI controversial weapons exposure | 0.00 |
| MSCI nuclear weapons exposure | 0.00 |
| MSCI civilian firearms exposure | 0.00 |
| MSCI tobacco exposure | 0.00 |
| MSCI UN global compact violators | 0.00 |
| MSCI thermal coal | 0.00 |
| MSCI oil sands | 0.00 |

## 5 Analysis and Discussion

Initially looking at the values and principles of Circular Economy, the authors have found that one of its principles is a key notion for the scientific study, namely the principle of policy and system reassessment, for translation from a short-term way of thought towards a long-term way of measuring the success of an economic system. These long-term measurements are expected to encompass the values and principles of Circular Economy, not only in the broader economic sense but more specifically in terms of investment theory and practice. This is found to already be happening, as given by two intertwined examples.

On one hand is the ESG concept, which already encompasses the Circular Economy Paradigm, and thus can be used as its proxy both in terms of measuring the companies' attitude towards implementing the core values and principles of Circular Economy, but also identifying said companies as issuers of investment opportunities that allow for investors to further support the "cause" of Circular Economy implementation.

On the other hand, stays the notion the authors have found while revising the current scientific paper, seen in Morea et al. [35]—for them the ESG concept has not just bridged the gap between investments and Circular Economy, but it already encompasses the Circular Economy fully, in its "E" element. Here the major implication for potential investors (that the research team not only agrees on but fully supports) is that if one wants to find which investments best represent the Circular Economy Paradigm, one must look with priority at ESG investments (as also mentioned in Nguyen et al. [37], Reboredo and Ugolini [40] and Saeed et al. [43]). By giving examples of existing ETFs one can invest in, and also showing the growth of the market, an implication can be made, for everybody whom it may concern, that ESG investments are not just a short-term fad, but grow their momentum towards becoming a key part of the financial markets. There exists the opportunity to invest not just in ETFs, but also in green bonds, credit facilities, loan and investment initiatives, etc. While the paper's scope is centered around the ETF part of the market, as it studies portfolios, a topic for future research can be defined—how well do Green Bonds have established themselves as a good investment, in terms of return and risk performance, but also in terms of incorporating Circular Economy values and principles.

Giving no investment recommendation the research team finds that the BlackRock ETF is better performing in terms of return, when put against two other ETFs, all three of which are built around the Circular Economy Notion, by being constructed mainly with ESG stocks. BlackRock outperforms the others in 82.84% of cases, as measured on a daily workday basis for the period from 2020 to 2021. In the case of BlackRock, the share of ESG stocks is around 80% of the whole portfolio, all of which the Investment Adviser has classified in one of four categories: adopter, enablers, beneficiaries, and business model winners, in terms of Circular Economy Paradigm implementation.

The authors don't limit themselves to existing portfolios but decide to build and test their own, using three different methodologies. In one case the authors use the concept of naïve diversification. The implication, that portfolio managers can be concerned about, here is that naïve diversified portfolios are still relevant, both in terms of their performance, but also in terms of being a good benchmark for the performance of others. The reasons for this are initially given in works like Benartzi and Thaler [3] and DeMiguel et al. [14] the late of whom perform an empirical test on these implications and find them to be true. This gives us the necessary argument to use a naïve constructed and diversified portfolio as a measuring stick for the performance of the chosen object of study—the BlackRock ETF.

Additionally, the research team looks at constructing a portfolio, using the methodology of weight-scoring, where the weights of individual stocks within the portfolio are somehow related to a specific financial and performance indicator of the companies that have issued them. This way of thought implies the existence of predicting power within said financial and performance indicators, built around historical and current information, but being able to foresee the future (at least partially). Here the chosen prediction is to use the historical price-to-earnings ratio of each company, as a correction of the individual company's stock weight within the portfolio, thus betting on the idea that companies with better historical performance must have a higher share within the portfolio.

Finally, a third portfolio is built, using a simple variance-covariance relationship between individual stocks, to optimize their weights within the portfolio. This method is the fundamental basis of Modern Portfolio Theory, requiring the calculation of such a combination of stock weights, as to maximize, minimize or optimize a given portfolio performance indicator. In this case, the authors choose not to do this, as this would create two caveats. First, an optimization would require computational power and simulation capabilities. Second, the resulting portfolio will only consist of some of the stocks that are used for portfolio construction but not all. To combat these, an implication is made—of using a mathematical transformation via Lagrangians, to artificially quantify the optimal weights within a portfolio consisting of all stocks available. This implication best works regarding portfolio construction and thus concerns potential investors.

Looking at the empirical results, the first implication for potential investors is related to the notion, mentioned by a list of authors, cited within the Literature Review—the notion of Return and the question is it greater than the overall return of the market. The authors tend to agree with Guerard [26], Goldreyer and Diltz [24], Bauer et al. [2], and Brzeszczynski et al. [8] for whom all the return is either greater or at least equal to the market. The authors see that in the provided empirical evidence—the return of BlackRock ETF against the return of MSCI ACWI is a measurement of the whole investment world's overall performance. Initially underperforming, the BlackRock ETF manages to beat the benchmark in September 2020, continuing to do so constantly until the end of the studied period. The total return of the ETF is 43.38% against 37.05% of the Benchmark.

The attention turns toward the list of 100 best ESG companies, that are studied to later construct their portfolios. After looking at it, the research team finds an interesting implication (for whomever it may concern)—the number and size of this list are largely influenced by two industries: the IT and the Telecom Sectors, measuring in a third (34) of the number of companies and two thirds (66.16%) of its whole market capitalization. These 100 companies are to be included in different portfolios, constructed following the mentioned methodologies.

The key findings related to the portfolio performance are as follows:

- The BlackRock ETF is easily beatable by using simple portfolios construction strategies like Naïve Diversification or Weight Scoring using Price-to-Earnings Ratio as weight correctors and the Industry Profile as Concentration Risk Measure;
- Weight Scoring doesn't add much additional even if able to beat the ETF, fails at delivering additional return over naïve diversification;
- Lagrangian optimal weight calculation results in underperformance through almost the whole period, but surprisingly manages to beat the ETF at the end of the period;
- It's enough to just invest in the top third of stocks of the BlackRock ETF, made publicly available, and one can outperform the whole ETF.

In terms of measuring how well the BlackRock ETF encompasses the Circular Economy's values and principles, one can only guess, using proxied existing indicators, in some way related to the ESG concept and Circular Economy. Fortunately, such indicators do exist and are calculated by MSCI, which allows us to quote that the BlackRock ETF has a triple-A rating, making it a leader investment in terms of following the values and principles of the Circular Economy. The same can be seen in ESG Quality Score, where BlackRock values 8.42 from 10.00.

Another measurement—the Weighted Average Carbon Intensity Ratio, allows us to measure that BlackRock ETF has a value of 150.67. Here the benchmark value is 100 so the close a stock or portfolio is toward it, the better the company (companies) is (are) at limiting their CO2 emissions, corrected for sales volume.

Finally, the authors look at exposures against a list of controversial businesses, each one of which is as far from the Circular Economy, as possible. These include the Production of Controversial Weapons, Nuclear Weapons, Civilian Firearms, Tobacco, but also UN Global Compact Violators, Thermal Coal, and Oil Sands. Against the whole list, the BlackRock ETF has 0.00 exposure.

## 6 Conclusions

Through the study, the authors tried to analyze and assess how well a publicly traded index blends the possibility of being an investment while promoting the values and principles of the Circular Economy, as presented in the Introduction. The research team first looked at existing examples of scientific studies neighboring the studied topic. This allowed us to describe the general idea and the notions that existed before the research. Next, the authors examined how to methodologically test the ideas through existing theories and models. This helped us choose the best way to empirically analyze for existence or lack of a causal relationship between an asset is both an investment and a way of supporting CE. Finally, the team looked at data, as to prove or disprove what was assumed to be true. It was found both what was expected, but also there were surprises to find—something else in support of the general idea.

Finding its roots around political discourse, the topic of Circular Economy doesn't allow for an isolated and independent analysis. This is mainly due to its principles which concern both economy and the ecosystem. The scientific world is talking about the improvement and replacement with an alternative of a process for gathering, production, and use of resources, developed throughout millennia. This is mainly due to the common belief that a linear economic model follows its "linear" nature to the tee, in the sense that, by continuing to use nature's finite resources linearly, humans as a species inhabiting this planet will inevitably come to a moment of lacking what is required to continue to exist and develop.

To fight this, society must enforce the discourse of the alternative CE, where finite resources are used better, energy is saved instead of wasted, and waste is treated and reused instead of ejected from product production and lifespan. But all must not end in the discussion, but in taking the necessary actions and precautions in the short and long term for strategic and forward-looking survival. And while companies using resources for production are the main agents of change towards CE, many other economic agents can give their contributions. A good example is investors who, even if positioned late far from resource use and the production cycle, can still enforce the principles of CE by influencing companies by choosing or refusing to invest within them.

This choice, somewhat controversial from an investment point-of-view, has a cost, and this cost is what makes it controversial. This cost is the price that investors must pay, by choosing to invest only in companies that respect and follow Circular Economy Principles. This cost is the real possibility to be able to realize a lower return than the overall market performance. But this real possibility, the team, as many more researchers before them, have found to be not so common. In the example given, the investment within a publicly traded index that encompasses the principles of responsibility through ESG concerns allowed for a return that is greater than the market. This greater return is significant enough, as to allow investors to choose the investment possibility of this ESG index instead of the market.

In terms of how well this index encompasses the principles around Circular Economy, the authors find themselves trying to assess this through the use of both the instruments of investing analysis and also through the use of a couple of new metrics and indicators. The authors find themselves asking the question "what are the best investments in terms of advocating and putting into practice the principles of the circular economy? The answer was given to us by a list of the hundred best companies in terms of ESG performance. These hundred companies while good individual benchmarks were most useful for the study as a single entity or a portfolio, that aggregates their characteristics. The research team discussed different ways of constructing these portfolios. Finally, the team was able to say which portfolio compared to the index performed in what way.

As a happy accident, the authors have found that a portfolio constructed from such stocks was able to beat the analyzed index, with additional returns larger than expected. Here the team wondered and continues to wonder what exactly this means. One way of thinking is to confirm the notion that yes, there is a cost that investors must pay when investing in the stocks of companies that encompass the principles of Circular Economy. But cost in comparison to what, because the portfolios constructed by us also consisted of stocks that encompass the same principles. Not only that, these stocks were in the top 100 in terms of encompassing exactly these principles. So, in the end, the authors only proved that there is an easy way to beat a given portfolio of ESG-responsible stocks with another portfolio of other ESG-responsible stocks. So, to put it simply, it's not just possible for portfolio construction to be sustainable in the Circular Economy Paradigm.

**Funding.** This work was supported by the D. A. Tsenov Academy of Economics, Svishtov, Bulgaria.

## References


1. Bajaj, S.: The interplay of equity shares: allocating investments via Markowitz model. IUP J. Appl. Financ. **27**(1), 25–54 (2021). https://ssrn.com/abstract=3820303
2. Bauer, R., Koedijk, K., Otten, R.: International evidence on ethical mutual fund performance and investment style. J. Banking Financ. **29**(7), 1751–1767 (2005). https://doi.org/10.1016/j.jbankfin.2004.06.035
3. Benartzi, S., Thaler, R.H.: Naive diversification strategies in defined contribution saving plans. Am. Econ. Rev. **91**(1), 79–98 (2001). https://doi.org/10.1257/aer.91.1.79
4. BGF Circular Economy Fund. https://www.BlackRock.com/sg/en/products/310164/BlackRock-circular-economy-fund. Accessed 04 Jan 2022
5. Bloomberg, M.R., Pavarina, D., Pitkethly, G., Thimann, C., Sim, Y.L.: Final report: recommendations of the task force on climate-related financial disclosures. Task Force on Climate-Related Financial Disclosures. Switzerland (2017). https://www.bloomberg.com/professional/blog/a-guide-to-the-task-force-on-climate-related-financial-disclosures/. Accessed Oct 2022
6. Bocconi University, Ellen MacArthur Foundation, Intesa Sanpaolo: The circular economy as a de-risking strategy and driver of superior risk-adjusted returns. (2021). http://www.ellenmacarthurfoundation.org/publications. Accessed Oct 2022
7. Braun, A.: ESG and circular economy (CE): Do they know each other? (2019) http://avkaksi.arctic-values.com/?page_id=601. Accessed Oct 2022
8. Brzeszczyński, J., McIntosh, G.: Performance of portfolios composed of British SRI stocks. J. Bus. Ethics **120**(3), 335–362 (2013). https://doi.org/10.1007/s10551-012-1541-x
9. Brealey, R.A., Myers, S.C., Allen, F., Mohanty, P.: Principles of corporate finance. Tata McGraw-Hill Education (2012)

10. Busch, T., Bauer, R., Orlitzky, M.: Sustainable development and financial markets: old paths and new avenues. Bus. Soc. **55**(3), 303–329 (2016). https://doi.org/10.1177/0007650315570701
11. Daniel, K., Mota, L., Rottke, S., Santos, T.: The cross-section of risk and returns. Rev. Financ. Stud. **33**(5), 1927–1979 (2020). https://doi.org/10.1093/rfs/hhaa021
12. Dai J., Zhou J.: A novel quantitative stock selection model based on support vector regression. In: 2019 ICEMME, IEEE, pp. 437–445 (2019). https://doi.org/10.1109/ICEMME49371.2019.00094
13. de Finetti B (1940) "Il problema dei pieni", Giornale Istituto Italiano Attuari, 9, 1–88. (English translation by L. Barone available as "The problem of "Full-risk insurances", Ch. 1 'The problem in a single accounting period', Journal of Investment Management, 4, 19–43, 2006)
14. DeMiguel, V., Garlappi, L., Uppal, R.: Optimal versus naive diversification: how inefficient is the 1/N portfolio strategy? Rev. Financ. Stud. **22**, 1915–1953 (2009) https://doi.org/10.1093/rfs/hhm075
15. Dutta, A., Das, D., Jana, R.K., Vo, X.V.: COVID-19 and oil market crash: revisiting the safe haven property of gold and bitcoin. Renewab. Energy **69**(April), 101816 (2020). https://doi.org/10.1016/j.renene.2016.03.078
16. Eaton, M.J., Yurek, S., Haider, Z., Martin, J., Johnson, F.A., Udell, B.J., Charkhgard, H., Kwon, C.: Spatial conservation planning under uncertainty: adapting to climate change risks using modern portfolio theory. Ecol. Appl. **29**(7), 1–19 (2019). https://www.jstor.org/stable/26797128
17. EP (2015) Circular Economy Action Plan of 2 December 2015 entitled 'Closing the loop-an EU action plan for the Circular Economy' (COM(2015)0614). https://eur-lex.europa.eu/resource.html?uri=cellar:8a8ef5e8-99a0-11e5-b3b7-01aa75ed71a1.0012.02/DOC_1&format=PDF. Accessed 08 Mar 2022
18. EC (2019) Communication from the Commission of 11 December 2019 entitled 'The European Green Deal' (COM(2019) 640 final). https://eur-lex.europa.eu/legal-content/EN/TXT/?qid=1588580774040&uri=CELEX:52019DC0640. Accessed 08 Mar 2022
19. EP (2021) European Parliament resolution of 10 February 2021 on the New Circular Economy Action Plan (2020/2077(INI)). https://www.europarl.europa.eu/doceo/document/TA-9-2021-0040_EN.html. Accessed 07 Jan 2022
20. EP (2021). Circular economy: definition, importance, and benefits. https://www.europarl.europa.eu/news/en/headlines/economy/20151201STO05603/circular-economy-definition-importance-and-benefits, Accessed 07 Jan 2022
21. Figge, F., Thorpe, A.S., Manzhynski, S.: Between you and I: a portfolio theory of the circular economy. Ecol. Econ. **190**(C) (2021). https://doi.org/10.1016/j.ecolecon.2021.107190. (Elsevier)
22. Folke, C., Österblom H., Jouffray, J., Lambin, E., Adger, W., Scheffer, M., Crona, B., Nyström, M., Levin, S., Carpenter, S.: Transnational corporations and the challenge of biosphere stewardship. Nat. Ecol. Evol. **3**, 1396–1403 (2019). https://doi.org/10.1038/s41559-019-0978-z
23. Galaz, V., Crona, B., Dauriach, A., Scholtens, B., Steffen. W.: Finance and the earth system–exploring the links between financial actors and non-linear changes in the climate system. Global Environ. Change **53**, 296–302 (2018). https://doi.org/10.1016/j.gloenvcha.2018.09.008
24. Gardner, T.A., Benzie, M., Börner, J., Dawkins, E., Fick, S., Garrett, R., Godar, J. et al.: Transparency and sustainability in global commodity supply chains. World Dev. **121** (September):163–177 (2019). https://doi.org/10.1016/j.worlddev.2018.05.025
25. Goldreyer, E.F., Diltz, J.D.: The performance of socially responsible mutual funds: incorporating sociopolitical information in portfolio selection. Manag. Financ. **25**, 23–36 (1999). https://doi.org/10.1108/03074359910765830

26. Gu, S., Kelly, B., Xiu, D.: Empirical asset pricing via machine learning. Rev. Financ. Stud. **33**(5), 2223–2273 (2020). https://doi.org/10.1093/rfs/hhaa009
27. Guerard, J.B.: Is there a cost to being socially responsible in investing? J. Forecast. **16**(7), 475–490 (1998). https://doi.org/10.1002/(SICI)1099-131X(199712)16:7%3c475::AID-FOR668%3e3.0.CO;2-X
28. Hong, H., Kacperczyk, M.: The price of sin: the effects of social norms on markets. J. Financ. Econ. **93**(1), 15–36 (2009). https://doi.org/10.1016/j.jfineco.2008.09.001
29. Investor's Business Daily. https://www.investors.com/news/esg-companies-list-best-esg-stocks-environmental-social-governance-values/. Accessed Jan 2022
30. Kirchherr, J., Reike, D., Hekkert, M.: Conceptualizing the circular economy: an analysis of 114 definitions. Resour. Conserv. Recycl. **127**(April), 221–232 (2017). https://doi.org/10.1016/j.resconrec.2017.09.005
31. Kong, D., Lin, C.P., Yeh, I.C., Chang, W.: Building growth and value hybrid valuation model with errors-in-variables regression. Appl. Econ. Lett. **26**(5), 370–386 (2019). https://doi.org/10.1080/13504851.2018.1486005
32. Le, T.-L., Abakah, E.J.A., Tiwari, A.K.: Time and frequency domain connectedness and spill-over among Fintech, green Bonds and cryptocurrencies in the age of the fourth industrial revolution. Technol. Forecast. Soc. Change **162**, 120382 (2020). https://doi.org/10.1016/j.techfore.2020.120382
33. Maltais, A., Nykvist, B.: Understanding the role of green bonds in advancing sustainability. J. Sustain. Financ. Invest. (2020). https://doi.org/10.1080/20430795.2020.1724864
34. Markowitz, H.M.: Portfolio selection. J. Financ. **7**, 77–91 (1952). https://doi.org/10.1111/j.1540-6261.1952.tb01525.x
35. McGlade, C., Ekins, P.: The geographical distribution of Fossil fuels unused when limiting global warming to 2 C. Nature **517**(7533), 187–190 (2015). https://doi.org/10.1038/nature14016
36. Morea, D.; Mango, F.; Cardi, M.; Paccione, C.; Bittucci, L. Circular economy impact analysis on stock performances: an empirical comparison with the Euro Stoxx 50® ESG index. Sustainability **14**, 843 (2022). https://doi.org/10.3390/su14020843
37. Nguyen, T.T.H., Naeem, M.A., Balli, F., Balli, H.O., Vo, X.V.: Time-frequency comovement among green bonds, stocks, commodities, clean energy, and conventional bonds Financ. Res. Lett. 40 101739 (2020). https://doi.org/10.1016/j.frl.2020.101739
38. Nilsson, J.: Segmenting socially responsible mutual fund investors. Int. J. Bank Market. **27**(1), 5–31 (2009). https://doi.org/10.1108/02652320910928218
39. Offiong, A.I., Riman, H.B., Eyo, E.E.: Determining optimal portfolio in a three-asset portfolio mix in Nigeria. J. Math. Financ. **6**, 524–540 (2016). https://doi.org/10.4236/jmf.2016.64041
40. Peter, L.; Keeble, J.; McNamara, R.: Circular Advantage–Innovative Business Models and Technologies to Create Value in a World without Limits to Growth. Accenture Strategy https://www.accenture.com/t20150523t053139__w__/us-en/_acnmedia/accenture/conversion-assets/dotcom/documents/global/pdf/strategy_6/accenture-circular-advantage-innovative-business-models-technologies-value-growth.pdf Accessed Oct 2022
41. Rasekhschaffe, K.C., Jones, R.C.: Machine learning for stock selection. Financ. Anal. J. **75**(3), 70–88 (2019)
42. Reboredo, J. C., Ugolini A.: Price connectedness between the green bond and financial markets. Econ. Model. **88**(C), 25–38 (2020). https://doi.org/10.1016/j.econmod.2019.09.004. (Elsevier)
43. Renneboog, L., Ter Horst, J., Zhang, C.: Socially responsible investments: institutional aspects, performance, and investor behavior. J. Banking Financ. **32**(9), 1723–1742 (2008). https://doi.org/10.1016/j.jbankfin.2007.12.039
44. Riedl, A., Smeets, P.: Why do investors hold socially responsible mutual funds? J. Financ. **72**(6), 2505–2550 (2017). https://doi.org/10.1111/jofi.12547

45. Saeed, F., Schleussner, C.F., Almazroui M.: From Paris to Makkah: Heat stress risks for Muslim pilgrims at 1.5°C and 2°C. Environ. Res. Lett. **16**, 024037 (2020). https://doi.org/10.1088/1748-9326/abd067
46. United Nations (2015) Resolution adopted by the General Assembly on 25 September 2015, Transforming our world: the 2030 Agenda for Sustainable Development https://www.un.org/sustainabledevelopment/sustainable-development-goals/. Accessed Oct 2022
47. Ünlü, R., Xanthopoulos, P.: A reduced variance unsupervised ensemble learning algorithm based on modern portfolio theory. Expert Syst. Appl. **180**, 115085 (2021). https://doi.org/10.1016/j.eswa.2021.115085
48. Wen, F., Xu, L., Ouyang, G., Kou, G.: Retail investor attention and stock price crash risk: evidence from China. Int. Rev. Financ. Anal. **65**, 101376 (2019). https://doi.org/10.47494/pbft.2021.1.25
49. Wu, X., Ye, Q., Hong, H., Li, Y.: Stock selection model based on machine learning with the wisdom of experts and crowds. IEEE Intell. Syst. **35**(2), 54–64 (2020)
50. YCharts: https://ycharts.com/. Accessed 04 Jan 2022
51. Yeh, I.-C., Hsu, T.-K.: Building real estate valuation models with comparative approach through case-based reasoning. Appl. Soft Comput. **65**, 260–271 (2018). https://doi.org/10.1016/j.asoc.2018.01.029
52. Yeh, I.-C., Liu, Y.-C.: Discovering optimal weights in weighted-scoring stock-picking models: a mixture design approach. Financ. Innov. **6**(1), 1–28 (2020). https://doi.org/10.1186/s40854-020-00209-x
53. Zara, C.: The transition towards a circular economy foresees offering an array of opportunities to banks and other financial actors. In: Transformative Economies. From the Circular Economy to the New Deal. FeltrinelliCamp 2020 (2020). https://fondazionefeltrinelli.it/app/uploads/2020/12/Transformative_Economies_.pdf
54. Zara, C., Iannuzzi, M., Shyham, R.: The impact of Circular Economy on Public Equity in Europe. Understanding de-risking effect and risk-adjusted performance. In: Presented at the 19th International Conference on Credit Risk Valuation, 24–25 September 2020. Università di Venezia (2020)

# Trade in Recyclable Raw Materials in EU: Structural Dynamics Study

Margarita Shopova[1], Mariana Petrova[2(✉)], and Lyubomir Todorov[1]

[1] D. A. Tsenov Academy of Economics, 2 Em. Chakarov Str, Svishtov, Bulgaria
lubomir.ivanov@uni-svishtov.bg

[2] St.Cyril and St, Methodius University of Veliko Tarnovo, Veliko Tarnovo, Bulgaria
mm_p@abv.bg

**Abstract.** The circular economy is a model of economic development in which, in accordance with the principles of sustainable development, resources are used efficiently, keeping them in economic circulation as long as possible, while waste and harmful effects on the environment are minimized. The key role of materials circularity is realized both by closing the supply chains in the companies and through the international trade, with the import of recyclable raw materials having an important role for countries with limited domestic supply. The aim of the paper was to study both the structure and the intensity of the structural changes in the import of recyclable raw materials in the European Union. The analysis covered the main characteristics of the dynamics in the structure of imports of recyclable raw materials in the EU for 2004–2021. By analyzing the dynamics of imports and its geographical structure, it was established that there is a unit root with drift in the dynamics of the indicator, and the integral coefficient of structural differences between imports and GDP for the period increased from 0.353 to 0.389. The continuous increase in imported volumes is accompanied by an increase in the differences between the related structures of imports of recyclable raw materials and GDP. In this way, the general movement towards a circular economy of the EU is realized at different speeds in the member states. In order to reduce the differences and to achieve a balanced accomplishment of the circular economy objectives, it is necessary to conduct a decisive and country-oriented policy at the EU level.

**Keywords:** Circular economy · Secondary raw materials · Dynamics · Structure · Structural changes

## 1 Introduction

The transformation of the economy from linear to circular one was adopted as main policy of the European Union on the way to a sustainable future in a fair and prosperous society with a competitive and resource-efficient economy [1, 2]. With its inherent features as neutrality towards the climate, effective waste management and saving of energy and materials, the circular economy makes substantial contribution for the achievement of the goals for sustainable development by promoting sustainable economic growth,


V. Koval et al. (Eds.): ISCMEE 2022, Lect. Notes in Managem., Industrial Engineering, pp. 43–64, 2023.
https://doi.org/10.1007/978-3-031-23463-7_3

creating new jobs, and reducing resource consumption. The circular economy seeks to create sustainable models for production and consumption by changing the behavior of producers and consumers [3, 4]. In this way, the value of goods, materials and resources is preserved in the economy as long as possible, and the generation of waste is minimized. Natural resources are utilized in the production of goods, which after use are re-integrated either into the natural chain as biodegradable materials or are introduced into a new economic cycle through repair, reconstruction, renovation, reuse and recycling [5, 6].

Among the main principles of the circular economy is the retention of the materials in economic circulation as long as possible [7], by extending the life cycle of the materials and the products produced from them. The movement of materials in the economy, which takes place both within individual countries and through international trade, is of key importance for reducing resource dependence and waste. The dynamics of the import of recyclable raw materials characterizes, on the one hand, the activity of countries in the international markets of secondary materials, and on the other hand, the readiness of the industry to process volumes of recycled raw materials exceeding the domestic supply. That's why "Trade of recyclable raw materials between the EU Member States and with the rest of the world" is one of the key indicators of Eurostat in the monitoring framework on the circular economy. The European Union actively participates in this trade, but within the Union there are certain differences caused by the specifics of individual countries in terms of their economic development, geographical, demographic, etc. factors. Therefore, the analysis of the dynamics and structure of the import of recycled raw materials is important for the formation and monitoring of the implemented policy regarding the movement of the EU countries towards a circular economy.

## 2 Literature Review

### 2.1 Theoretical and Conceptual Works

The trade in recyclable raw materials and the waste from which these secondary raw materials are derived was a subject of scientific interest from many different perspectives. They included technological processes and solutions, considered organizational problems, explored macroeconomic aspects, challenges to the economy and implications for the society.

The generation of waste, and in particular of waste that can be recycled, is directly related to the technology for the production of goods and the materials used in the process. With the implementation of the Sustainable Development Goals, the share of products with greater recyclability is expected to increase continuously.

Naturally, the economic benefit for the producers of such products may be less [8], but implementing marketing strategies based on public support for green policies and combining durability and recycling provided manufacturers with competitive advantages in the market [9]. In that way, the development of "*green products*" was stimulated, and the two systems—of production and waste management—were conceptually merged. Emphasis was placed on product design [10], which allowed both to account for material needs and their substitution with renewable or recyclable raw materials, and to increase energy efficiency. An example in this regard was building design and green construction

[11]. The process of integration was accelerated by the active application of the capabilities of information technologies, through the introduction of multi-criterial systems for the design of products that were safe for the environment, economical in terms of raw materials and with ever-increasing opportunities for recycling [12]. The consistent application of these principles, incl. the introduction of additional circles in the cities (the concept of "*circular cities*"), which are the main producers of waste [13], led to an increase in the volume of recycled materials—both as a demand from the producers and as a supply, due to the increasingly large number of green products. The intrinsic combination of the principles of the circular economy and sustainable development with the products' design determined the presence of a stable tendency towards the growth of the materials used and, respectively, of the trade with them both within individual countries and internationally.

The processes of waste collection, separation and processing were accompanied by a number of difficulties of technological and organizational nature, and the consequences for the participants were often unfavorable [14–17]. The researchers noted that the global pandemic was making the situation worse [18]. To improve working conditions and logistics in waste collection, partner networks were created and developed along the entire chain with the joint participation of central and local authorities [19]. An important element of achieving good conditions for efficient waste collection was the cooperation of those engaged in this activity, which allows them to contract directly with the recycling companies [20]. Adequate government assistance through certain programs should not be overlooked [21], which supported the production of goods from recycled materials. A result of cooperation at different levels was the sharing of benefits of trade in recyclable waste [22]. Solving the main problems related to logistics allowed to extract more and more recyclable raw materials from waste, as well as to efficiently process new and more valuable materials. This stimulated the development of trade in recyclable raw materials.

The processing of waste through a gradual transition from incineration to separation and extraction of the valuable components contained in it was a positive element of modern waste management. Separate collection significantly increased the amount of valuable materials recovered and had a beneficial effect in terms of global warming compared to incineration. Composting and biogas production from organic waste reinforced the economic and environmental benefits [23]. Key issues related to recycling and obtaining energy from waste could often be solved with existing infrastructure or its modifications, resulting in an effective reduction of harmful gas emissions [24]. At the same time, using waste as energy sources directly benefited fossil fuel substitution, but had limited effects on the environment and landfill capacity [25].

The accumulation of experience in waste management and the dissemination of good practices led to continuous improvement of processes, reduction of harmful effects on nature, lowering of costs and improvement of the financial status of companies and organizations [26–28]. Many examples could be cited in this regard. Leaders succeed in achieving full material recycling by closing the manufacturing and recycling loop within the corporation [29]. The introduction of special plans in the management of medical waste allowed to reduce the volume of non-recyclable materials and increase the volume of recyclable materials [30]. Management planning for the growing photovoltaic waste segment allowed for an appropriate monitoring, collection and storage system to be in place before the volume became large enough to make recycling economically viable [31]. The introduction of new technologies made it possible to process the increasing amounts of e-waste, which had a high potential for extracting valuable components and materials, including critical, strategic and scarce raw materials [32]. This increased the importance of trade in e-waste and derived materials, as well as environmental impact assessment [33].

The application of modern information and communication technologies allowed to increase the efficiency of waste collection, to neutralize the possibility of errors and adverse effects on people by automatically separating the various recyclable materials into separate containers with the help of sensors and intelligent systems [34]. Geographic information systems were implemented in measuring, reporting, analyzing and visualizing the location of stocks of recyclable resources or materials dispersed in an urban environment [35] as well as a routing tool (door-to-door model) in programs for the selective collection of recyclable materials. The results were expressed in better productivity in terms of distance traveled, duration of transportation and volume of materials collected [36].

## 2.2 Empirical Studies

Empirical studies of international trade in recyclables covered interesting aspects of its links with the circular economy, economic growth and recycling rates. The international division in terms of recycling depends on the economic development and geographical factors, but also on the recycling system used. Van Beukering and Bouman [37] found that for paper and lead waste, active trade in recycled materials was associated with higher recycling rates compared to countries with closed systems. Developed industrial countries are characterized by a higher degree of recycling than the utilization of the recycled raw materials, while the opposite is true for developing countries [38]. The volume and structure of the scrap trade was affected by both supply and demand and the availability or lack of relevant scrap processing infrastructure. For example, the USA exported low-value scrap to developing countries and imported high-value scrap from Europe [39]. The use of so-called sustainable logistics is also important, especially evident in Hungary [40]. To the extent that imported recycled raw materials were used for production purposes, their volume was logically related to the level of economic development. Empirical studies confirmed a significant relationship between wages, per capita gross domestic product, country population, and imports of recycled materials [41].

Regarding the formation of "*waste havens*", the results were contradictory. Higashida and Managi [41] did not confirm such a hypothesis based on a gravity model for a small number of products, while Okubo, Watabe & Furuyama [42] found evidence when analyzing the movement of waste streams from countries with high economic development and strict regulations to less developed countries, where nature conservation is not sufficiently regulated. According to the latter, in relation to Japan, the countries of the Asian continent represented a kind of "*waste haven*" when it comes to plastic, paper and metal waste. Cotta [43] considered waste trade concentrated along the North-South axis, where existing waste streams were expressed in the concentration of risks and burdens on the countries of the South, increasing inequality, unfair trade relations, lack of protection of fundamental rights of the countries. For example, it was stated that the trade in e-waste between the European Union and Africa exacerbated existing environmental risks and social problems in African countries, while at the same time creating new ones.

Empirical studies of the relations between trade in recycled materials and circular economy indicators were performed in many different directions. Tantau et al. [44] found that development costs, trade in recyclables, environmental taxes, productivity and domestic demand for materials had a significant effect on the level of municipal waste recycling. Torasa and Mekhum [45] registered a negative relationship between trade in recyclable raw materials and green activities in the transport and communications sector. Shpak et al. [46] analyzed waste recycling rates, showing that they have a significant impact on the trade of recyclable raw materials in the European Union. Pao and Chen [47] studied the relationship between carbon dioxide emissions and indicators characterizing the circular economy for European Union countries. They found that regarding the trade in recyclable raw materials, a positive but weakly elastic dependence was observed in the long-term aspect, while in the short-term aspect, the increase in the trade in recyclable raw materials led to a slight decrease in the amount of generated waste.

Analyzes of the dependencies between circular economy indicators and economic growth yielded conflicting results. Hysa et al. [48] identified the main elements of the circular economy that supported sustainable development and demonstrated a strong positive correlation between the circular economy and economic growth. Piecuch and Szarek [49] examined the relationship between bioeconomy indicators and economic growth in the 27 countries of the European Union. The results showed negative effect in terms of the economic growth of the export of recycled raw materials outside the European Union, the shares of recycled waste and e-waste. Chen and Pao [50] studied the causal relationships between circular economy indicators and economic growth in the European Union. In the short term, the increase in material recycling led to a reduction in generated waste, and as the volume of waste increases, an increase in investments related to the circular economy was observed. With economic growth, there was also growth in the circular economy, but the opposite remained unproven—i.e. an increase in circular economy activities is not found to lead to economic growth.

## 2.3 Paper Aim and Hypotheses

In general, the development of the concepts of sustainable development and circular economy, their implementation in the practice of more and more companies and countries, the improvement of collection technologies and organization, the implementation of modern tools of communication and information resources led to a continuous increase in the potential for trade in recyclables waste and the raw materials extracted in individual countries and in international trade. Empirical research showed the importance of trade in recyclable raw materials in terms of recycling, greenhouse gas emissions and economic growth. The specificity of individual countries, the level of their economic development and their potential for processing and utilizing recyclable raw materials determined the presence of stable streams and axes of movement of international flows of recyclable waste. This affected the structure of trade relations and led to its imbalance both for regions worldwide and for the countries of the European Union. Although reflections on structure of trade in recyclable raw materials can be found in some studies, research activity on this issue is weak.

Therefore, the aim of the paper was to analyze the structure and intensity of structural changes in the import of recyclable raw materials in the European Union. The interest was focused on its uniformity, stability and/or variability. In accordant with the main conclusions from the theoretical research and empirical studies two hypotheses were formulated for testing in the paper:

**Hypothesis One**: The volume of imports of recyclable raw materials in the EU is increasing, due to the adoption and implementation of the principles of the circular economy in relation to the materials circularity.

**Hypothesis Two**: The geographical structure of imports is not uniform, due to existing differences between individual countries such as the size of the economy, degree of circular economy development, access to raw materials, technological differences, specialization of production activities, etc.

# 3 Materials and Methods

## 3.1 Data

The analysis in the present study was carried out through several indicators. The main one among them, which was one of the indicators for characterizing the circular economy, was "*Trade in recyclable raw materials*". It was part of the indicator group "*Secondary raw materials*". The importance of this indicator for circular economy monitoring has several aspects. On the one hand, the implementation of one of the objectives set in the Plans [1, 2] can be followed—namely reducing the amount of waste. On the other hand, recycling waste, and extracting from it materials that can be used as secondary raw materials, reduced the raw material dependence of the economies of individual countries and the Union as a whole.

The indicator "*Trade in recyclable raw materials*" measures the quantities of recyclable waste that are transported between EU member states and across EU borders—both imports and exports. In our analysis a summary indicator had been used for the amount of recyclable waste that each EU member state imports—it was called "*Imports of recyclable raw materials*". The data were obtained by summation of the relevant data from the two main indicators "*Imports intra-EU27*" and "*Imports extra-EU27*". The source of the dataset for these two composite indicators was *International Trade in Goods Statistics* (*ITGS*), which was published by *Eurostat*. Time series were available since 2004, so the period of 2004–2021 was assumed for the study. The statistical data were presented in in tons and in euros. In order to eliminate the influence of prices and inflation in the analysis and to ensure comparability in both dynamics and statics, the analysis of the import of recyclable raw materials was performed in tons only.

To control for the relationship between the amount of imported recyclable raw materials and the economic power of countries, the indicator "*Gross Domestic Product*" was used. As known from economic theory, it accounts for the actual amount of goods produced in the economy over a certain period. Sources of statistical data for this indicator were various statistical studies, annual accounting reports of economic units and administrative data. They were collected exhaustively, for all statistical units (enterprises, households, state institutions).

### 3.2 Methods

The dynamics of the indicators with which the analysis of the import of recyclable raw materials was performed was studied through the average rate of development and the resulting average growth rate. They made it possible to objectively assess the speed (fast or slow) and the direction of change (increasing or decreasing), because their meanings do not depend on the scales of the investigated quantities. The average rate of development was calculated as an exponential average, because all members of the time series participated in the calculation procedure, which was why all values of the studied indicator were considered, not only part of them.

Trend detection is performed using the formal Dickey-Fuller unit root statistical tests [51, 52]. The test of the hypothesis of first-order or higher integration was based on an auxiliary regression equation:

$$\Delta y_t = D_t + \beta y_{t-1} + \sum_{j=1}^{p} \varphi_j \Delta y_{t-j} + \varepsilon_t, \tag{1}$$

where: $y_t$ was the time series;

$\varepsilon_t$—random components (white noise);

$\beta$, $\varphi_j$—regression parameters;

$D_t$—deterministic component.

Depending on the type of the deterministic component, the test was applied in three variants:

No constant or time trend—$D_t = 0$;

With constant only—$D_t = c$;

With constant and linear trend—$D_t = c + \delta t$.

When the null hypothesis of a unit root is rejected, the series is stationary when the linear trend parameter ($\delta$) is not statistically significant, and trending when the parameter is significant. If the null hypothesis is not rejected—the series is non-stationary, a statistically significant constant ($c$) is a signal of a deterministic linear trend, while a significant parameter in the linear trend ($\delta$) is evidence of a parabolic trend.

In the statistical analysis of the import of recyclable raw materials, the concept of statistical structure was used. It was defined as a property of the statistical population that characterizes its internal structure through the relative shares of its individual components.

Changes in the territorial structure of the import of recyclable raw materials have been studied through dynamic analysis. Its constituent parts were the relative shares of the quantities of recyclable raw materials imported into individual EU countries. The changes that had occurred had been evaluated by the integral coefficient of structural changes ($K_S$) [53]. Since it was normalized between 0 and 1, it allowed to evaluate the intensity of the changes. As a summary measure, it covered both absolute and relative structural changes. Its advantage over other indicators was that it reflected not only the absolute changes in the relative shares of the parts of the aggregate, but also their sizes. The integral coefficient of structural changes was calculated by the formula:

$$K_S = \sqrt{\frac{\sum (v_{i1} - v_{i0})^2}{\sum v_{i0}^2 + \sum v_{i1}^2}}, \tag{2}$$

where: $v_{i0}$ and $v_{i1}$ were the corresponding relative shares of the structure for the two periods.

When there were no structural changes $K_S = 0$. At $K_S = 1$, the changes led to a completely opposite structure.

Characterization of the differences between the relative shares of the imported quantities of recyclable raw materials in the EU countries gave the absolute inequality of the statistical structure. Its measurement was carried out by comparing the studied structure with a conditionally accepted uniform structure, which consists of parts with equal relative shares and in number as much as they were in the real structure [54].

To measure the absolute inequality, the following formula was used:

$$K_R = \sqrt{1 - \frac{20000}{10000 + k \sum v_i^2}}, \tag{3}$$

where: $v_i$ were the relative shares of the structure in percentages.

$K_R$ values could be between 0 and 1. The closer $K_R$ was to 0, the closer the actual structure was to the uniform one.

The differences between countries in terms of the quantities of imported recyclable raw materials was an objective characteristic that was caused by the power of their economies. This gave rise to the need to compare the import and GDP structures of countries, which was possible by applying a statistical analysis of structural differences. As a universal summary measure of differences, it was appropriate to use the integral coefficient of structural differences [55]:

$$K_D = \sqrt{\frac{\sum (v_{i2} - v_{i1})^2}{\sum v_{i1}^2 + \sum v_{i2}^2}}, \tag{4}$$

where: $v_{i1}$ and $v_{i2}$ were the relative shares of the two comparing structures.

The coefficient $K_D$ was normalized in the range from 0 to 1. The differences between the compared structures increased as the values of the coefficient increased from 0 to 1.

In the statistical analysis of structural changes and structural differences, a specific approach could be applied that allowed studying the influence of two factors–the degree and the direction of the inequality [55–58].

The methodological basis for the determination of factor influences was the assessment of absolute inequality. Through a special calculation procedure, the total structural change or structural difference was decomposed into two factor influences. One, the degree of inequality and denoted by $d_3$, characterized the difference, "*approaching*" or "*moving away*", of the studied structure from the uniform one. The other, direction of inequality and denoted by $d_4$, reflected the "*rearrangement*" in size of the relative shares of the constituent parts of the structure, i.e. characterized the restructuring. A positive value of the factor's influence $d_3$ indicated that the studied structure moved away from the uniform structure, and a negative value was associated with an approach to the uniform structure. The values of factor's influence $d_4$ were always positive. The ratio between factor's influences $d_3$ (its absolute value) and $d_4$ was a measure of their importance for the overall structural change of a structure, or for the difference between two static structures.

# 4 Results

## 4.1 Dynamics of Import of Recyclable Raw Materials

The import of recyclable raw materials into the EU for 2004–2021 could be seen in Fig. 1. The graphic image showed an upward trend in the quantities of imported recyclable raw materials by 33% in 2021 compared to 2004. The dynamics is characterized by about 6% yearly relative growth with minor fluctuations during the period.

**Fig. 1.** Import of recyclable raw materials in EU 2004–2021.

Checking for unit root with Dickey-Fuller tests gave the following results (Table 1). The values of the test statistics did not allow rejection of the hypothesis of the presence of a single root in the studied indicator, since they were smaller in absolute value than the critical limits at 5%. When testing the second unit root hypothesis, the test statistics became statistically significant, exceeded the theoretical limits, and allowed the null hypothesis to be rejected. Therefore, the time series of imported recyclable raw materials contained unit root.

Based on the Dickey-Fuller test equation (Table 2), it could be observed that the parameters characterizing the deterministic trend—the constant and the time trend—were statistically significant at 5%. This was evidence of the presence of a stable linear or parabolic trend in the series. Their positive values indicated that the trend was upward.

How the scale of the economy affected the quantities of imported recyclable raw materials can be established by analyzing the relationship between these quantities, measured in tons, and GDP, measured in million euros. The dynamics of this relation for the EU was presented in Fig. 2.

The comparison between the absolute magnitude plots depicted in Fig. 1, and the relative magnitude depicted in Fig. 2, showed similarities in their patterns, and upward trending as well, but the magnitude of the fluctuations was larger for the relative indicator.

**Table 1.** Dickey-fuller test for unit root in import of recyclable raw materials in EU.

| Hypotheses and specification | Test statistics | Critical values at 5%[a] |
|---|---|---|
| *Integrated of order 1* | | |
| No constant | 2.674 | −1.968 |
| Constant | 0.806 | −3.099 |
| Constant and time trend | −3.338 | −3.760 |
| *Integrated of order 2* | | |
| No constant | −3.907 | −1.964 |
| Constant | −4.361 | −3.099 |
| Constant and time trend | −4.541 | −3.791 |

[a] Critical values were computed by MacKinnon [59], as one-sided for 20 observations. Lag length was chosen according to SIC

**Table 2.** Results from ADF test for import of recyclable raw materials in EU.

| Parameter | Value | t-stat | p-value |
|---|---|---|---|
| $\beta$ | −1.708 | −3.338 | 0.0075 |
| $\varphi_1$ | 0.715 | 1.924 | 0.0832 |
| $\varphi_2$ | 0.454 | 1.467 | 0.1732 |
| $c$ | 181000000 | 3.344 | 0.0074 |
| $\delta$ | 3240565 | 3.451 | 0.0062 |

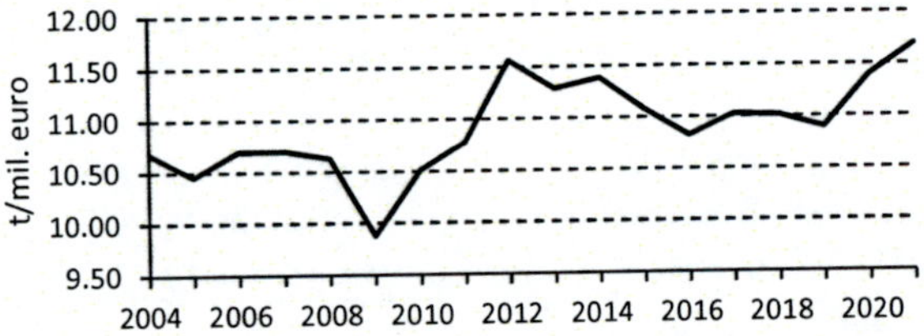


**Fig. 2.** Import of recyclable raw materials in EU per mil. euro GDP.

The check for the presence of a trend the Dickey-Fuller tests was applied. The obtained results were similar to the dynamics of the absolute volumes of imports (Table 3). The hypothesis of a unit root could not be rejected for any of the variants, and the hypothesis of a second unit root was rejected for all three variants at 5%. The conclusion was again that the import of recyclable raw materials in relation to the GDP for the whole EU was integrated of first order. The test equation parameters allowed to establish the presence of a trend also of a linear or parabolic type (Table 4).

**Table 3.** Dickey-fuller test for unit root in import of recyclable raw materials in EU relative to GDP.

| Hypotheses and specification | Test statistics | Critical values at 5%[a] |
|---|---|---|
| *Integrated of order 1* | | |
| No constant | 0.592 | −1.963 |
| Constant | −1.440 | −3.052 |
| Constant and time trend | −3.126 | −3.760 |
| *Integrated of order 2* | | |
| No constant | −4.280 | −1.964 |
| Constant | −4.305 | −3.066 |
| Constant and time trend | −4.186 | −3.733 |

[a] Critical values were computed by MacKinnon [59], as one-sided for 20 observations. Lag length was chosen according to SIC

**Table 4.** Results from ADF test for import of recyclable raw materials in EU relative to GDP.

| Parameter | Value | t-stat | p-value |
|---|---|---|---|
| $\beta$ | −1.097 | −3.126 | 0.011 |
| $\varphi_1$ | 0.530 | 1.739 | 0.113 |
| $\varphi_2$ | 0.528 | 1.896 | 0.087 |
| $c$ | 11.308 | 3.103 | 0.011 |
| $\delta$ | 0.068 | 2.602 | 0.026 |

While imports in general for the EU were growing both in absolute terms and in relation to the size of the GDP, certain differences appeared in relation to individual countries. The trend in Germany, the Netherlands, Italy and Belgium could be defined as upward. A downward trend was observed in Spain until 2009, after which the imported quantities fluctuated around the same level. On the contrary, in Italy the fluctuations were around the same level until 2008, and after that the tendency was upward. After falling by 36.9% in 2005, Luxembourg's imported quantities have been hovering around the same level. For the rest of the countries, there was an increase in imported quantities.

In Slovenia, Latvia, Belgium, the Netherlands, Lithuania, Croatia, Denmark, the ratio between the import of recyclable raw materials and GDP was slowly increasing, and in Luxembourg it was decreasing. For Estonia, Poland, the Czech Republic, Ireland, Spain and Cyprus, two periods could be distinguished in the patters of the relative indicator—in one period its values were below the EU average, and in the other they were above.

## 4.2 Structural Changes in Import of Recyclable Raw Materials

The structure of the import of recyclable raw materials in the EU countries for selected years from the period 2004–2021 was presented in Table 5. The selection of the intermediate years of the period was related to the world economic crisis (2009) and the adoption of the first plan for circular economy in the EU (2015). The results showed that Germany (on average 15.93%), the Netherlands (on average 14.02%), Belgium (10.06%), Spain, France and Italy (on average between 9.55 and 9.84%) had the largest relative shares in imports, i.e. they had a dominant role in the structure. Imports in these six countries formed more than 2/3 of EU imports, and at the beginning of the period their total share was over 70%. The relative share of imports in ten of the remaining countries was below 1% on average. The variations in the relative shares around the average level for individual countries were within wide limits—from 4.48% for Germany to 35.90% for Romania. It was noteworthy that in the countries with relative shares determining the export structure—Germany, Belgium and the Netherlands—the variations around the average level were small, between 4.28 and 7.45%. Quite the opposite—although the changes in the very small relative shares (below 1%) were insignificant, the fluctuations around their average level were bigger and reached up to 34.57% (for Latvia).

As could be seen from the results in the last column of the table, the changes in the relative shares were very small, usually around 2–3% on average per year. Exceptions were a few countries where larger changes were noted—for example, Romania with an average annual increase of 8.08%, Lithuania with 5.86%, Latvia with 4.85%, Poland with 4.48%, Denmark with 4.19% and Malta with an average annual decrease of 4.09%. It was noteworthy that for four of the six countries with dominant relative shares, an average annual decrease of these relative shares is established—for Germany, the Netherlands, France and Spain. In addition, half of the countries with the smallest relative shares in the structure had the largest average annual changes—Lithuania, Latvia, Malta, Romania, Estonia.

During the studied period, the structure of the import of recyclable raw materials in the EU countries was characterized by a strong inequality, estimated by the integral coefficient of the absolute inequality. The graphical representation in Fig. 3 showed that the above-described changes in the individual relative shares in the studied structure were reflected in slight reduction of inequality. The values of the integral coefficient decreased with negligible fluctuations during the period—from 0.68 to 0.63. Therefore, at the end of the period the structure of imports was somewhat closer to the uniform structure than at the beginning.

A generalized assessment of the intensity of the structural changes of the import of recyclable raw materials in the EU countries was obtained through the values of the integral coefficient of the structural changes. The results of the calculations on a chain basis and on a constant basis (2004) were presented in Fig. 4. It was found that in each year of the studied period there were minor changes in the import structure compared to the previous year, as the values of the integral coefficient were between 0.03 in 2017 and 0.07 in 2009. Th changes were in the same direction though, and as result of their accumulation, it could be concluded that moderate changes have occurred in the structure of imports for the entire 18-year period.

**Table 5.** Structure of import of recyclable raw materials in EU.

| Countries | 2004 | 2009 | 2015 | 2021 | Average for 2004–2021 | Relative variance (%) | Average growth rate (%) |
|---|---|---|---|---|---|---|---|
| Belgium | 9.41 | 10.81 | 10.63 | 9.53 | 10.06 | 5.06 | 100.38 |
| Bulgaria | 0.30 | 0.48 | 0.46 | 0.54 | 0.47 | 16.27 | 102.15 |
| Czechia | 1.36 | 1.28 | 1.56 | 1.93 | 1.47 | 12.30 | 102.02 |
| Denmark | 2.89 | 2.86 | 4.24 | 4.90 | 3.83 | 23.51 | 104.19 |
| Germany | 15.42 | 15.56 | 16.13 | 14.90 | 15.93 | 4.28 | 99.81 |
| Estonia | 0.19 | 0.12 | 0.16 | 0.22 | 0.20 | 31.01 | 103.19 |
| Ireland | 1.14 | 2.05 | 1.77 | 1.95 | 1.82 | 17.87 | 102.56 |
| Greece | 1.65 | 1.65 | 0.94 | 1.45 | 1.43 | 27.13 | 97.43 |
| Spain | 11.98 | 10.29 | 9.71 | 8.95 | 9.84 | 15.87 | 97.56 |
| France | 11.81 | 10.25 | 9.55 | 8.04 | 9.63 | 10.84 | 98.06 |
| Croatia | 0.80 | 0.59 | 0.69 | 1.14 | 0.75 | 19.67 | 101.97 |
| Italy | 10.41 | 8.05 | 9.58 | 9.68 | 9.55 | 5.71 | 100.22 |
| Cyprus | 0.13 | 0.14 | 0.14 | 0.19 | 0.15 | 15.11 | 100.26 |
| Latvia | 0.23 | 0.51 | 0.54 | 0.97 | 0.59 | 34.57 | 104.85 |
| Lithuania | 0.35 | 0.40 | 0.54 | 0.95 | 0.56 | 33.67 | 105.86 |
| Luxembourg | 4.64 | 2.55 | 2.39 | 2.59 | 2.83 | 19.71 | 97.56 |
| Hungary | 1.08 | 0.94 | 1.15 | 1.41 | 1.20 | 12.01 | 101.49 |
| Malta | 0.02 | 0.03 | 0.02 | 0.01 | 0.02 | 26.15 | 95.91 |
| Netherlands | 14.94 | 16.11 | 12.40 | 14.41 | 14.02 | 7.45 | 99.20 |
| Austria | 2.97 | 3.32 | 4.05 | 4.08 | 3.53 | 11.28 | 101.92 |
| Poland | 2.18 | 3.63 | 4.66 | 4.47 | 3.88 | 22.49 | 104.48 |
| Portugal | 1.09 | 2.07 | 2.19 | 1.54 | 1.97 | 14.45 | 101.19 |
| Romania | 0.26 | 0.75 | 0.83 | 1.13 | 0.77 | 35.90 | 108.08 |
| Slovenia | 0.98 | 1.96 | 2.01 | 1.80 | 1.82 | 15.49 | 102.28 |
| Slovakia | 0.57 | 0.59 | 0.70 | 0.96 | 0.76 | 14.40 | 102.12 |
| Finland | 1.15 | 0.95 | 0.72 | 0.78 | 0.87 | 23.61 | 97.02 |
| Sweden | 2.01 | 2.06 | 2.24 | 1.49 | 2.05 | 13.65 | 100.00 |

The structural dynamics of the import of recyclable raw materials in the EU countries had been studied in more detail by decomposing the general structural change into factor influences. The results of the analysis were presented in Table 6. They were the basis for the following conclusions:

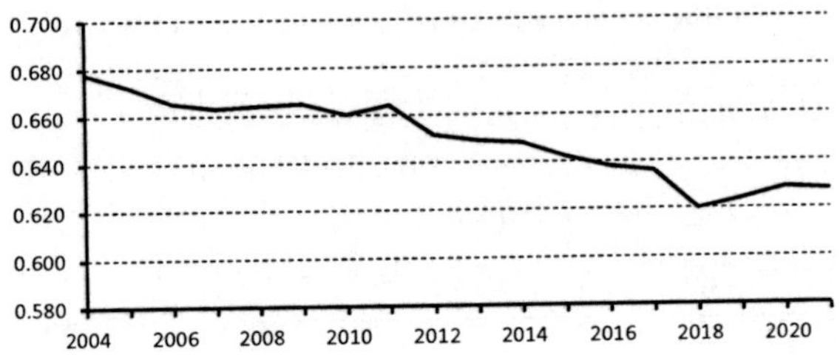


**Fig. 3.** Integral coefficient of absolute structural inequality in imports of recyclable raw materials in EU.

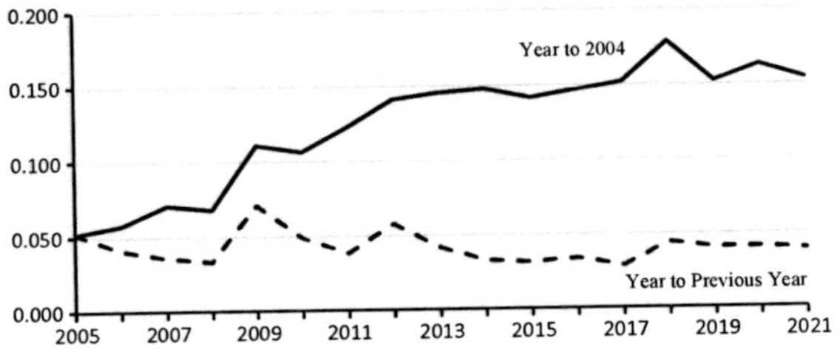


**Fig. 4.** Integral coefficient of structural change.

- As a result of the smooth increase in the values of the quantity $d_0$, changes in the structure of imports tended to move away from the basic structure of 2004.
- With minor deviations, the tendency to decrease the inequality of the studied structure was manifested, as the values of the quantity $d_2$ decrease. For the entire studied period, the value of the corresponding metrics decreased by 11.26%.
- The influence of the factors on the overall structural change was increasing, with the size of the relative shares ($d_3$) changing its importance by 12.8% on average per year, and the rearrangement of relative shares ($d_4$) increasing by 5.6% on average per year.
- The structural changes were determined by both factors, where the influence of the rearrangement of the relative shares of the groups in the structure was stronger than the changes in their size, because the values of the $d_4$ were greater than the absolute value of the $d_3$. These differences decreased during the considered period, which was proven by the ratio of the values of the respective measures—at the beginning it is 6.20, and at the end it is 1.72 in favor of the factor measuring the direction of changes.

Additional aspect of the present analysis was the study of the differences between the territorial structures for the EU countries of the import of recyclable raw materials and the structure of GDP. The evaluation of the structural differences was performed by calculating the integral coefficient values presented in Fig. 5. They slightly increase from 0.353 to 0.389 and were rated as significant. The graphical representation shows greater variation at the beginning of the period than at the end.

The general structural difference between the structures of imports of recyclable raw materials and GDP was decomposed into factor influences. The values of the quantities

**Table 6.** Decomposition of structural changes to factors.

| Year | *d0* | *d2* | *d3* | *d4* |
|---|---|---|---|---|
| 2005 | 2,33 | 24,73 | −0,37 | 2,28 |
| 2006 | 2,56 | 24,26 | −0,84 | 2,38 |
| 2007 | 3,14 | 24,12 | −0,98 | 2,93 |
| 2008 | 3,02 | 24,18 | −0,92 | 2,82 |
| 2009 | 4,90 | 24,24 | −0,86 | 4,74 |
| 2010 | 4,69 | 23,94 | −1,16 | 4,44 |
| 2011 | 5,42 | 24,18 | −0,92 | 5,25 |
| 2012 | 6,16 | 23,38 | −1,72 | 5,71 |
| 2013 | 6,34 | 23,24 | −1,86 | 5,83 |
| 2014 | 6,43 | 23,17 | −1,93 | 5,89 |
| 2015 | 6,13 | 22,81 | −2,29 | 5,42 |
| 2016 | 6,35 | 22,55 | −2,55 | 5,51 |
| 2017 | 6,56 | 22,44 | −2,66 | 5,67 |
| 2018 | 7,63 | 21,51 | −3,59 | 6,23 |
| 2019 | 6,52 | 21,74 | −3,36 | 5,20 |
| 2020 | 7,00 | 22,01 | −3,09 | 5,88 |
| 2021 | 6,60 | 21,95 | −3,15 | 5,42 |

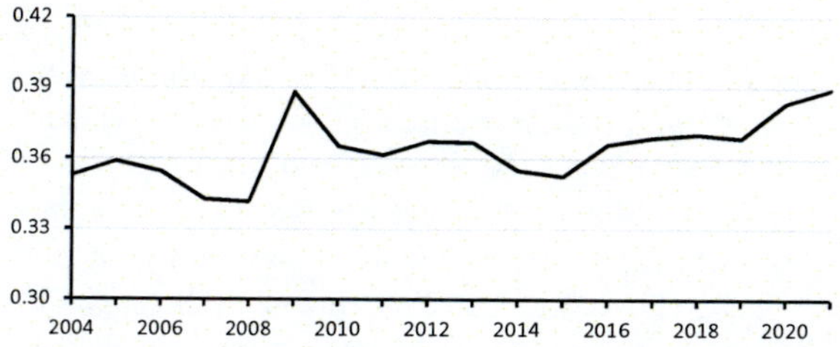


**Fig. 5.** Integral coefficient of structural differences.

$d_3$ and $d_4$, characterizing the degree and direction of the differences, respectively, were presented in Fig. 6. The graphic images were a reason to conclude that:

- The influence of the factors on the general structural difference between the structures of the import of recyclable raw materials and the GDP was 3–4 times stronger than the influence of the respective factors on the structural dynamics of the import of recyclable raw materials in the EU countries.
- The influence of the factors on the total structural difference was uneven. The role of changes in the size of the relative shares ($d_3$) was increasing due to an increase in the

values of the factor by 2.2% on average per year. The rearrangement of the groups according to the size of the relative shares ($d_4$) had less and less importance for the structural differences, with the influence decreasing by 0.6% on average per year.

- The structural differences were determined by both factors, where the influence of the rearrangement of the relative shares of the groups in the structures was stronger than the changes in the size of the relative shares, because the values of $d_4$ were greater than $d_3$.

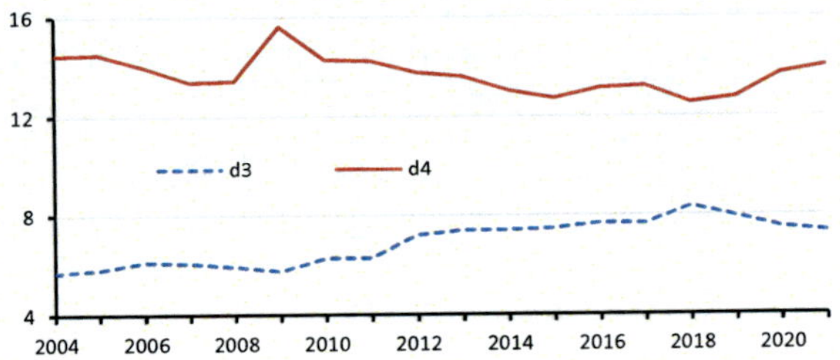


**Fig. 6.** Factor influences on the structural differences between import of recyclable raw materials and GDP.

## 5 Discussion

The presence of a tendency towards an increase in the volume of imports of recyclable raw materials in the EU for the period 2004–2021 showed the continuous involvement in the production processes of increasingly large amounts of secondary raw materials. This was a clear sign of the movement of the Union members on the path of transition to a circular economy. The parabolic nature of the trend was evidence of an acceleration of these processes, although the acceleration was relatively weak, and the movement itself was not without problems. Two anomalies of high intensity were noticeable in the established trend. The first was in 2009 and could be explained by the global economic crisis, which led to a contraction of production, as a result of which the recyclable materials subject to trade logically decreased. The decline was temporary amounting 11.1% compared to 2008, after which imports returned to their previous level. The second exception is in 2021, which saw an increase of 8.14% compared to 2020. Here the explanation was related to the resumption of manufacturing and economic activities as many of the restrictions were lifted in 2020 as a result of subsiding Covid-19 pandemic.

The results of the analysis of the structure of the absolute country data showed that they could be divided into three groups according to the amount of imported recyclable raw materials. The first group included Germany, the Netherlands, Belgium, Spain, France and Italy, which stand out from the rest as the largest importers. In each of them, an average of over 9.5 million tons were imported annually during the researched period. Luxembourg, Austria, Denmark and Poland belonged to the second group of countries with imports between 3 and 7 million tons of recyclable raw materials. In the third group were the remaining 17 EU countries with less than 3 million tons of recyclable raw materials imported annually on average. The smallest annual quantities were in Estonia and Cyprus—below 400 thousand tons, and Malta—below 35 thousand tons.

According to the meanings of the relationship between the import of recyclable raw materials and GDP, groups of countries could again be distinguished, but the composition of these groups significantly differed from the composition of the groups formed according to the quantities of imported recyclable raw materials. Luxembourg, Slovenia, Latvia, Belgium, the Netherlands, Lithuania, Croatia, Denmark were in the group of countries that imported the largest quantities of recyclable raw materials per 1 million euros of GDP. Other countries in which the relative value was above the EU average were Portugal, Austria, Bulgaria, Hungary, Slovakia, Estonia. The import of recyclable raw materials per 1 million euro of GDP in the Czech Republic, Ireland, Spain, Cyprus, Poland fluctuated around the average European level. The last group consisted of Greece, Finland, Germany, Italy, France, Sweden, Malta and Romania. Here, the ratios were below the EU average.

The high values of the integral coefficient of structural inequality and the formation of clusters of countries indicated the presence of heterogeneity in relation to the import of recyclable raw materials. This was true for both absolute and relative figures (relative to the size of their economies). The presence of heterogeneity arose from various factors. Chief among them were specialization, access to resources, level of technological development, capital accumulation. Reasons for the differences could be found in the specifics of the countries regarding their movement towards a circular economy and the implementation of the goals of sustainable development.

In the dynamics of the structure of the import of recyclable raw materials, very weak changes were observed in individual years. However, the changes were in the same direction and accumulated throughout the period, leading to a reduction in inequality. There was a move away from the base structure for 2004, indicating a move towards an even structure. The overall structural change was due to the simultaneous and increasing influence of two factors. One of them—the changes in the sizes of the relative shares of individual countries in the structure of imports had a weaker role than the second—the rearrangement in size of the relative shares.

The increase in structural differences between imports and GDP of individual countries was evidence of deepening inequalities between them in terms of the use of imported recyclable raw materials. This could also be interpreted as a signal that the movement towards a circular economy does not proceed in the same way, but at a different speed. The overall structural difference was due to the joint influence of two factors. One of them, depending on the differences in the sizes of the relative shares between the two structures, had a growing role. The importance of the second factor, caused by the different arrangement in the two structures of the respective relative shares in size, decreased during the studied period.

On this basis the first hypothesis could be assumed to hold. There was indeed substantial increase in the import of recyclable raw materials in the EU. The second hypothesis was also supported by the data. There was heterogeneity among the EU members concerning import of recyclable raw materials. Even more important was the fact that the differences increased over time suggesting further research in this area in order to identify the main factors and their influences.

## 6 Conclusions

The current study was focused on the import of recyclable raw materials in the EU for the period 2004–2021. Based on data on the absolute volumes of imports both from countries outside the EU and internal, a number of regularities in the dynamics and structure in a territorial aspect have been established.

Unit root tests showed that there was a clear linear or parabolic upward trend in the dynamics of imports in absolute terms and relative to the size of GDP. This meant that the volume of secondary raw materials used in production processes in the Union was constantly increasing, at a faster rate than the GDP. It was a clear signal of the EU's steady movement towards a circular economy.

In relation to EU countries, heterogeneity was found both in imports and when comparing it with the territorial structure of GDP. While the absolute structural changes for the period 2004–2021 led to a weak but distinct movement towards more even structure, when comparing with GDP, there was an increase in differences. This could be interpreted as a deepening of disparity between countries in their move towards a circular economy, at least in terms of the import (and possibly use) of recyclable raw materials.

The managerial implications of the study are twofold. First, the entire process of transition towards circular economy, as linked with the use and trade of recyclable raw materials, requires constant monitoring and extension of the already establish indicators with additional ones, that will take into consideration the country specifics in economic development. Second, to correct the multi-speed transition, additional measures were needed to stimulate both the import of recyclable raw materials and their utilization in the various spheres of industry and services. Measures should be specific and oriented to those countries where there were problems with the transition to a circular economy and the use of traditional raw materials was dominant.

## References

1. European Commission: Closing the loop-an EU action plan for the Circular Economy. COM (2015) 614 final, Brussels (2015). https://eur-lex.europa.eu/resource.html?uri=cellar:8a8ef5e8-99a0-11e5-b3b7-01aa75ed71a1.0012.02/DOC_1&format=PDF. Accessed 21 July 2022
2. European Commission: a new Circular Economy Action Plan "For a cleaner and more competitive Europe". COM (2020) 98 final, Brussels (2020). https://eur-lex.europa.eu/resource.html?uri=cellar:9903b325-6388-11ea-b735-01aa75ed71a1.0017.02/DOC_1&format=PDF. Accessed 21 July 2022
3. Ünal, E., Urbinati, A., Chiaroni, D., Manzini, R.: Value creation in circular business models: the case of a US small medium enterprise in the building sector. Resour. Conserv. Recycl. **146**, 291–307 (2019)
4. Philippidis, G., Sartori, M., Ferari, E., M'Barek, R.: Waste not, want not: a bio-economic impact assessment of household food waste reductions in the EU. Resour. Conserv. Recycl. **146**, 514–522 (2019)

5. UNIDO: Development of recycling industries within the UNIDO circular economy approach (2019). https://www.unido.org/sites/default/files/files/2019-07/Development%20of%20recycling%20industries%20within%20the%20UNIDO%20circular%20economy%20approach.pdf. Accessed 21 July 2022
6. UNIDO: Addressing the challenge of Marine Plastic Litter using Circular Economy methods (2019). https://www.unido.org/sites/default/files/files/2019-06/UNIDO_Addressing_the_cha-llenge_of_Marine_Plastic_Litter_Using_Circular_Economy.pdf. Accessed 21 July 2022
7. Bocken, N.M.P., de Pauw, I., Bakker, C., van der Grinten, B.: Product design and business model strategies for a circular economy (2016). https://www.tandfonline.com. https://doi.org/10.1080/21681015.2016.1172124. Accessed 21 July 2022
8. Rizea, M.: General considerations on waste management in Romania. In: 16th International Conference the Knowledge-Based Organization Management and Military Science, pp. 610–617 (2010)
9. Brouillat, E.: An evolutionary model of recycling and product lifetime extension. Technol. Forecast. Soc. Chang. **76**(4), 471–486 (2009). https://doi.org/10.1016/j.tech-fore.2008.03.023
10. Djuric Ilic, D., Eriksson, O., Ödlund (former Trygg), L., Åberg, M.: No zero burden assumption in a circular economy. J. Clean. Prod. **182**, 352–362 (2018). https://doi.org/10.1016/j.jclepro.2018.02.031
11. Sun, S., Fang, Z., Chen, J.: Using renewable and recyclable materials in planning green building. Adv. Mater. Res. **280**, 89–92 (2011). https://doi.org/10.4028/www.scientific.net/AMR.280.89
12. Kinoshita, Y., Yamada, T., Gupta, S.M., Ishigaki, A., Inoue, M.: Decision support model of environmentally friendly and economical material strategy for life cycle cost and recyclable weight. Int. J. Prod. Econ. **224**, art. no. 107545 (2020). https://doi.org/10.1016/j.ijpe.2019.107545
13. Domenech, T., Borrion, A.: Embedding Circular Economy Principles into Urban Regeneration and Waste Management: Framework and Metrics. Sustainability (Switzerland) **14**(3), art. no. 1293 (2022). https://doi.org/10.3390/su14031293
14. Arantes, B., Borges, L.: Recyclable Materials Collector: productive chain and precarious conditions. Arquivos brasilieros de psicologia. **65**(3), 319–337 (2013)
15. de Souza, J., Martins, M.: Map of risks in cooperatives of collectors of recyclable materials in the municipality of Campina grande-PB. Sistemas & Gestao. **13**(2), 232–245 (2018). https://doi.org/10.20985/1980-5160.2018.v13n2.1385
16. Zolnikov, T., da Silva, R., Tuesta, A., Marques, C., Cruvinel, V.: Ineffective waste site closures in Brazil: a systematic review on continuing health conditions and occupational hazards of waste collectors. Waste Manag. **80**, 26–39 (2018). https://doi.org/10.1016/j.wasman.2018.08.047
17. Ribeiro, N., Cantoia, S.: The Cuiaba dump and the generation of socio-environmental impacts. Geosaberes **11**, 100–116 (2020). https://doi.org/10.26895/geosaberes.v11i0.778
18. Maia, C., Feitosa, A., Galvao, A., de Araujo, D., Andrade, J.: Reflections on the impact of the pandemic coronavirus on the performance of the recyclable materials collector. Revista pegada **21**(3), 416–432 (2020)
19. Cruz, J., Quandt, C., Kato, H., Martins, R., Martins, T.: How does the structure of social networks affect the performance of its actors?-a case study of recyclable materials collectors in the Brazilian context. Resour. Conserv. Recycl. **78**, 36–46 (2013). https://doi.org/10.1016/j.resconrec.2013.06.002
20. Fuzzi, F., Leal, A.: Organized networks of cooperatives and associations of recyclable materials collectors: rede cataoeste, Sao Paulo, Brazil. Revista formacao online. **25**, 123–155 (2018)

21. Zambon, P., Lima, J.: The challenge of solid waste management in Brazilian municipalities: study of the Parana ecocidadao program. Revista de direito da cidade-city law. **11**(2), 830–848 (2019). https://doi.org/10.12957/rdc.2019.37844
22. Suthar, S., Rayal, P., Ahada, C.: Role of different stakeholders in trading of reusable/recyclable urban solid waste materials: a case study. Sustain. Cities Soc. **22**, 104–115 (2016). https://doi.org/10.1016/j.scs.2016.01.013
23. Cimpan, C., Rothmann, M., Hamelin, L., Wenzel, H.: Towards increased recycling of household waste: Documenting cascading effects and material efficiency of commingled recyclables and biowaste collection. J. Environ. Manag. **157**, 69–83 (2015). https://doi.org/10.1016/j.jenvman.2015.04.008
24. Stanisavljevic, N., Levis, J.W., Barlaz, M.A.: Application of a life cycle model for European union policy-driven waste management decision making in emerging economies. J. Ind. Ecol. **22**(2), 341–355 (2018). https://doi.org/10.1111/jiec.12564
25. de Lorena Diniz Chaves, G., Siman, R.R., Chang, N.-B.: Policy analysis for sustainable refuse-derived fuel production in Espírito Santo, Brazil. J. Clean. Prod. **294**, art. no. 126344 (2021). https://doi.org/10.1016/j.jclepro.2021.126344
26. Yu, P., Leu, H., Lin, S.: Analysis of a municipal recyclable material recycling program. Resour. Conserv. Recycl. **17**(1), 47–56 (1996). https://doi.org/10.1016/0921-3449(96)01104-4
27. Leu, H., Lin, S.: Cost-benefit analysis of resource material recycling. Resour. Conserv. Recycl. **23**(3) (1998). https://doi.org/10.1016/S0921-3449(98)00020-2
28. Chung, S., Poon, C.: Characterization of municipal solid waste and its recyclable contents of Guangzhou. Waste Manag. Res. **19**(6), 473–485 (2001). https://doi.org/10.1177/0734242X0101900603
29. Sakai, T., Yamaguchi, S.: Establishment of complete treatment system of recyclable materials generating from home appliance recycling plants at Mitsubishi Materials Corporation. In: H. SM (ed.) EPD Congress, pp. 509–514 (2008)
30. Moreira, A., Gunther, W.: Assessment of medical waste management at a primary health-care center in Sao Paulo Brazil. Waste Manag. **33**(1), 162–167 (2013). https://doi.org/10.1016/j.wasman.2012.09.018
31. Kim, H., Park, H.P.V.: Waste management at the crossroads of circular economy and energy transition: the case of South Korea. Sustainability **10**(10) (2018). https://doi.org/10.3390/su10103565
32. Witkowska-Kita, B., Biel, K., Blaschke, W., Orlicka, A.: Critical, strategic and deficit raw materials in the waste of electrical and electronic equipment and in batteries and accumulators. Inzynieria mineralna-J. Polish Miner. Eng. Soc. **1**, 189–194 (2019). https://doi.org/10.29227/IM-2019-01-36
33. Prasad, M.N.V.: Waste electrical and electronic equipment in India: diversity, flows, and resource recovery approaches. In: Handbook of Electronic Waste Management: International Best Practices and Case Studies, pp. 207–241 (2019). https://doi.org/10.1016/B978-0-12-817030-4.00008-5
34. Senturk, S., Gulmez, H., Gul, M., Kirci, P.: Detection and separation of transparent objects from recyclable materials with sensors. Commun. Comput. Inf. Sci. **1534**, 73–81 (2022). https://doi.org/10.1007/978-3-030-96040-7_6
35. Zhu, X.: GIS and Urban mining. Resour. Basel **3**(1), 235–247 (2014). https://doi.org/10.3390/resources3010235
36. Oliveira, R., Lima, R., Lima, J.: Arc routing using a geographic information system: application in recyclable materials selective collection. Adv. Mater. Res. **838–841**, 2346–2353 (2014). https://doi.org/10.4028/www.scientific.net/AMR.838-841.2346
37. Van Beukering, P.J.H., Bouman, M.N.: Empirical evidence on recycling and trade of paper and lead in developed and developing countries. World Dev. **29**(10), 1717–1737 (2001). https://doi.org/10.1016/S0305-750X(01)00065-1

38. Van Beukering, P.J.H., Van Den Bergh, J.C.J.M.: Modelling and analysis of international recycling between developed and developing countries. Resour. Conserv. Recycl. **46**(1), 1–26 (2006). https://doi.org/10.1016/j.resconrec.2005.06.002
39. Lyons, D., Rice, M., Wachal, R.: Circuits of scrap: Closed loop industrial ecosystems and the geography of us international recyclable material flows 1995–2005. Geogr. J. **175**(4), 286–300 (2009). https://doi.org/10.1111/j.1475-4959.2009.00341.x
40. Mesjasz-Lech, A., Michelberger, P.: Sustainable waste logistics and the development of trade in recyclable raw materials in Poland and Hungary. Sustainability (Switzerland) **11**(15), art. no. 4159 (2019). https://doi.org/10.3390/su11154159
41. Higashida, K., Managi, S.: Determinants of trade in recyclable wastes: evidence from commodity-based trade of waste and scrap. Environ. Dev. Econ. **19**(2), 250–270 (2014). https://doi.org/10.1017/S1355770X13000533
42. Okubo, T., Watabe, Y., Furuyama, K.: Export of recyclable materials: evidence from Japan. Asian Econ. Pap. **15**(1), 134–148 (2016). https://doi.org/10.1162/ASEP_a_00414
43. Cotta, B.: What goes around, comes around? Access and allocation problems in Global North–South waste trade. Int. Environ. Agreem. Polit. Law Econ. **20**(2), 255–269 (2020). https://doi.org/10.1007/s10784-020-09479-3
44. Tantau, A.D., Maassen, M.A., Fratila, L.: Models for analyzing the dependencies between indicators for a circular economy in the European Union. Sustainability (Switzerland) **10**(7), art. no. 2141 (2018). https://doi.org/10.3390/su10072141
45. Torasa, C., Mekhum, W.: Impact of green logistics activities on circular economy: panel data evidence from ASEAN. Int. J. Supply Chain Manag. **9**(1), 239–245 (2020)
46. Shpak, N., Melnyk, O., Horbal, N., Ruda, M., Sroka, W.: Assessing the implementation of the circular economy in the EU countries. Forum Scientiae Oeconomia **9**(1), 25–39 (2021). https://doi.org/10.23762/FSO_VOL9_NO1_2
47. Pao, H.-T., Chen, C.-C.: The dynamic interaction between circular economy and the environment: evidence on EU countries. Waste Manag. Res. **40**(7), 969–979 (2022). https://doi.org/10.1177/0734242X211057015
48. Hysa, E., Kruja, A., Rehman, N.U., Laurenti, R.: Circular economy innovation and environmental sustainability impact on economic growth: an integrated model for sustainable development. Sustainability (Switzerland) **12**(12), art. no. 4831 (2020). https://doi.org/10.3390/SU12124831
49. Piecuch, J., Szarek, J.: Dynamic panel model in bioeconomy modelling. Agric. Econ. (Czech Republic) **68**(1), 20–27 (2022). https://doi.org/10.17221/156/2021-AGRICECON
50. Chen, C.-C., Pao, H.-T.: The causal link between circular economy and economic growth in EU-25. Environ. Sci. Pollut. Res. (2022). https://doi.org/10.1007/s11356-022-21010-6
51. Dickey, D.A., Fuller, W.A.: Distribution of the estimators for autoregressive time series with a unit root. J. Am. Stat. Assoc. **74**(366), 427–431 (1979). https://doi.org/10.1080/01621459.1979.10482531
52. Dickey, D.A., Fuller, W.A.: Likelihood ratio statistics for autoregressive time series with a unit root. Econometrica **49**(4), 1057–1072 (1981). https://doi.org/10.2307/1912517
53. Gatev, K.: Metodi za statisticheski analiz na ikonomicheski i socialni strukturi. Nauka i izkustvo, Sofia (1987)
54. Gatev, K.: Metodi za analiz na strukturi i strukturni efekti. Akademichno izdatelstvo Stopanstvo, Sofia (2007)
55. Yankova, N.: Statistichesko izsledvane na strukturni izmenenia. Akademichno izdatelstvo Prof. Marn Drinov, Sofia (2007)
56. Khristov, E., Yankova, N.: Strukturni razlichiya–faktorni vliyaniya i analiz v dinamika. Ikonomicheska missal. **8**, 55–68 (1986)

57. Bozhanova, V., Korenyuk, P., Lozovskyi, O., Belous-Sergeeva, S., Bielienkova, O., Koval, V.: Green enterprise logistics management system in circular economy. Int. J. Math. Eng. Manag. Sci. **7**, 350–363 (2022). https://doi.org/10.33889/ijmems.2022.7.3.024
58. Koval, V., Mikhno, I., Udovychenko, I., Gordiichuk, Y., Kalina, I.: Sustainable natural resource management to ensure strategic environmental development. TEM J. 1022–1030 (2021). https://doi.org/10.18421/tem103-03
59. MacKinnon, J.G.: Numerical distribution functions for unit root and cointegration tests. J. Appl. Economet. **11**, 601–618 (1996). https://doi.org/10.1002/(SICI)1099-1255(199611)11:6%3C601::AID-JAE417%3E3.0.CO;2-T

# Impacts of Sustainable Entrepreneurship and Income on Sustainable Food Consumption

Simeonka Petrova[1], Ivan Marinov[1], Zoya Ivanova[1], and Yerbol Akhmedyarov[2](✉)

[1] D. A. Tsenov Academy of Economics, 2 Em. Chakarov Str., Svishtov, Bulgaria
[2] Sh. Ualikhanov Kokshetau University, 76 Abaya Str., Kokshetau, Kazakhstan
ahmediarov_e_a@mail.ru

**Abstract.** Sustainable food consumption depends on complex and interrelated factors, including sustainable entrepreneurship business models and disposable income, as well as a number of other economic and demographic indicators. Reducing the impact on the environment, related to food consumption is one of the main challenges that requires efforts in all stages of the value chain. The purpose of this paper is to analyze the impacts of sustainable entrepreneurship and income on the development of sustainable food consumption and, on this basis, to draw relevant conclusions and generalizations. In order to measure these impacts, various economic concepts and theories have been studied. Respondents in this study are households from the 27 Member States of the European Union. Data on their income, expenditure and consumption refer to the period 2012–2021. The study is based on the simple linear regression and correlation method. This method is applied by using the statistical software product IBM SPSS Statistics. Eurostat is used as the main source of data for the survey of household budgets. The results reveal the impacts of sustainable entrepreneurship and household income on sustainable food consumption. In studying sustainability, the approach of focusing on food consumption is of particular importance. It is precisely this approach that determines the relevance of the study.

**Keywords:** Sustainable entrepreneurship · Household food expenditure · Household income · Sustainable food consumption

## 1 Introduction

In our modern world, sustainability is developing as one of the most influential areas in economic theory. The concept of sustainability is gaining importance for the development of economies. The increased interest in alternative approaches and solutions to achieve sustainability is becoming a global trend. This paper interprets the doctrine of sustainability as understanding the balance between economic and social interests while preserving the ecological balance. It also seeks to fulfil the requirement for unity of the economic, social and market environment and discusses important economic issues in terms of the fundamental principle of economics about the scarcity of resources and the need for their sustainable consumption.


V. Koval et al. (Eds.): ISCMEE 2022, Lect. Notes in Managem., Industrial Engineering, pp. 65–78, 2023.
https://doi.org/10.1007/978-3-031-23463-7_4

The main aim of the present study is to analyze the impacts of sustainable entrepreneurship and income on the development of sustainable food consumption and, on this basis, to draw relevant conclusions and generalizations. In order to achieve the aim, the following research tasks are set: interpreting concepts and empirical studies of sustainability issues in terms of sustainable consumption, including food consumption; determining basic factors influencing sustainable consumption – sustainable entrepreneurship and consumer income; highlighting empirical dimensions of the relationship between sustainable food consumption, income and sustainable entrepreneurship; drawing some more important conclusions and generalizations from the conducted theoretical-empirical study.

Respondents of the study are households from the 27 Member States of the European Union. The data on their income, expenditure and consumption refer to the period 2012–2021. The chosen ten-year period provides tracking of the dynamics and trends in the development of the studied phenomenon.

The study of European household budgets is based on the use of the simple linear regression and correlation method. This method is applied by using the statistical software product IBM SPSS. In studying European household budgets, Eurostat, the statistical office of the European Union, is used. The obtained results give grounds for identifying the impacts of sustainable entrepreneurship and household income on sustainable food consumption.

## 2 Literature Review

Globally, consumption patterns should be transformed almost radically to meet sustainability requirements [1]. According to documents of a number of world institutions, a large part of the developing countries need to build new infrastructure, and deal with the problems of poverty and health care. A number of negative side effects of food consumption are manifested. The various forms of production, entrepreneurship and consumption have an impact on the environment. According to a report of the European Commission, it is recommended that a low-carbon economy is reached and the levels of greenhouse gas emissions are reduced to 80% below 1990 levels by 2050 [2]. The relationship between the rates of economic growth and the environment is subject to a special reassessment and specifying.

Global companies manage consumption patterns through various entrepreneurial business models. in their papers devoted to studying the life cycle of products and approaches to cleaner production, researchers C. Bakker, F. Wang, J. Huisman give the definition that the so-called "linear economy" focuses on the model "take, make and waste", where the planned obsolescence of products usually does not lead to sales [3]. Unsustainable consumption patterns are developed and provided by unsustainable business practices and entrepreneurships.

Alternative business models should include longer product life as well as opportunities to promote slower forms of consumption and processing. In this area, specialized literature provides the research results of A. Kollmuss and J. Agyeman in the field of sustainability through the prism of the environmental activities of consumers and barriers to pro-environmental behaviour [4]. These scholars outline the factors affecting consumption patterns. According to them, there are three groups of factors as follows: individual factors (the personality, education and acquired knowledge of consumers), social factors (e.g. social environment), and broader contextual factors such as: political, economic, environmental, societal, technological and infrastructural.

In a broader aspect, it can be explained that the overall understanding of the socio-economic profile of a given economy requires that some corresponding factors and indicators are analyzed and assessed. Such indicators are the demographic and economic ones. In modern economic theories and studies, demographic factors and indicators are often cited as leading for the socio-economic development rates. Their group includes: number of the population, age and age structure of the population, distribution of the population according to various criteria – for example, by regions, districts, municipalities and settlements, ongoing demographic processes, etc. [5]. In the range of economic factors affecting sustainable food consumption, the following stand out: household disposable income, per capita income, household expenditure, household consumption, including food consumption. Thus, for example, a sufficiently high rate of increase in household income can mean a shift towards consumption of more sustainable products. Therefore, household budgets and, in particular, disposable income can be defined as one of the factors with a priority influence on consumption patterns, including sustainable food consumption.

As N. Bocken competently confirms, businesses are also in a position to manage the sustainability of consumption patterns through the approaches and ways in which business is done [6]. The above-cited author states that sustainable entrepreneurship is recognized as a key factor in creating and enforcing sustainable products, processes and business models. New entrepreneurships appear as a response to public needs. The periodic review of entrepreneurs' activities is carried out using the criteria for assessing the proximity and compliance of this activity with the concept of sustainability.

### 2.1 Sustainable Consumption

In this paper sustainable consumption is discussed and assessed in terms of the interpretations of sustainable entrepreneurship and consumer income. The terminology of sustainable consumption is relatively new to economic theory and practice. For this reason, it can be interpreted based on more distant in time and thorough judgments about consumption itself, its dependencies on disposable income and on the adopted entrepreneurial business models.

Adhering to the research aim set in the present paper we begin by clarifying the essential nature of consumption based on sustainability. In its nature, the term "sustainable development" was introduced in 1987 in the report of the World Commission on Environment and Development, entitled "Our Common Future". The basis of the document is the definition of sustainable development as "a global strategy for cooperation between the development of the current generation and the environment, but with care for the needs of the future generations" [7]. In general, the sought international synergy is found in the three key pillars – economic development, social solidarity and ecological balance.

Adopted in 2015, the United Nations 2030 Agenda for Sustainable Development and the 17 Sustainable Development Goals represent a modern reflection of the principles underlying the above-mentioned report, as well as a general model of the long-term efforts of world organizations and governments to build a road map for global sustainable development, applicable to all countries, regardless of their location and economic status [8].

Considered in its essence, the concept of "sustainable consumption" is defined as consumption based on sustainability, which is associated with the use of goods or services that can meet the basic needs and can bring a better quality of life, while minimizing the use of natural resources, toxic materials and emissions throughout the life cycle without jeopardizing the needs of the future generations [9].

In sustainability research, the focus on food consumption is relatively new, but increasingly important to economies and society. The review of the specialized literature shows that in their work researchers, led by A. Drewnowski, adopt the view that sustainable food consumption is brought to the consumption of nutrient-rich, accessible and culturally acceptable foods while respecting the environmentally-friendly principle [10]. This means that a multifaceted approach should be applied to considering the concept of sustainable food consumption. The following are determined as separate aspects: socio-economic, cultural, dietary and ecological.

Under the influence of the COVID-19 pandemic, most of the developing economies in Europe and the world are characterized by higher rates of inflation, ongoing processes of recession and a decline in economic activity. The highlighted processes lead to an increase in the prices of food products, including sustainable products. According to a GFK report, the price levels of organic products are much higher than those of their conventional equivalents [11]. The upward trends in food prices affect the generating of serious difficulties for those households that have lower annual income. Thus, the more vulnerable households have to spend a significant part of their income on food. According to data from the above cited GFK report, more than 2 billion people in the developing world spend nearly 70% of their disposable income on food [11]. As a rule, vulnerable groups of the population, in particular of the households, are those who have an income lower than the average level in different regions of the world.

The arguments presented so far confirm the statement and the belief that sustainable consumption patterns underlie sustainable development of economies and society. Of particular importance is the establishment of the role of both consumers and businesses in stimulating sustainable consumption, including food consumption.

## 2.2 Income and Sustainable Consumption

Understanding sustainable consumption patterns should be considered equivalent to understanding individual consumer behaviour. In the present study, the research respondents are selected households from the European Union and their income. For this reason, household consumption is seen as particularly important in explaining the changes in consumption patterns as a reflection of disposable income.

Basing our arguments on J. Keynes's scientific and practical research allows us to adopt the concept that consumption function is an important element of any macroeconomic model [12]. This importance is presupposed by the high relative share of individual household consumption in the overall structure of the Gross Domestic Product [5].

As confirmed by a number of market studies, modern consumption patterns in Europe are undergoing changes leading to differences from the patterns followed a decade or more ago [13]. The factors contributing to the changes can be sought in different aspects. In this regard, the following are identified as more important socio-economic factors determining consumption: number of households, age of household members, age structure of households, distribution of households by regions, districts, municipalities and settlements, demographic processes, gross domestic product, household income, minimum wage level, average income per capita, etc.

According to the aim of the study, household income can be highlighted among the above listed factors. Of particular interest is the measuring of the power of impact of individual household income on consumption by intended use. Essentially, disposable income and consumption are among the key components of a household's economic well-being. At the aggregate level, data on household income and consumption allow establishing the dependencies and interrelationships between them. The analysis and assessment of the results obtained can support the processes of developing appropriate economic and social policies, including in terms of consumption of more sustainable products.

### 2.3 Sustainable Entrepreneurship and Sustainable Consumption

Today's entrepreneurs are directing more and more strategies and investments to ensure a sustainable and responsible entrepreneurial business. Entrepreneurs are reassessing the sustainability indicators of the products offered, trying to meet the requirements of more and more consumers. The increased focus on the sustainability of entrepreneurial business allows us to identify and examine some of the driving factors in this area.

A number of papers in the specialized literature confirm that sustainable entrepreneurs can do a business with social and environmental effects. By indicating certain characteristics, B. Parrish arrives at the statement that sustainable entrepreneurs differ from conventional ones in that they focus on an organizational goal that maximizes the value of human and natural resources [14]. This type of entrepreneur seeks to create synergies and multiply benefits by focusing on meeting the needs of multiple stakeholders. Therefore, these entrepreneurs follow a marked value-based approach and intend to create positive changes in the environment and society.

Studying alternative forms of business models for sustainable innovation, F. Boons and F. Lüdeke-Freund determined different forms of sustainable entrepreneurship [15]. According to the authors, business models of sustainable entrepreneurship can take the following more significant forms: technical – for example, switching to renewable energy sources; social – such as those related to a shift from products to services, or organizational innovation – for example reorienting businesses towards stronger social impacts.

Implementing sustainability principles obliges entrepreneurs to reconsider their operating models [16–19]. Just before the COVID-19 pandemic, the entrepreneurial focus was placed on achieving sustainable development and dealing with the consequences of climate status worsening. In the post-pandemic reality, in addition to climate challenges, entrepreneurs are challenged to find new ways to survive under the conditions of highly disrupted supply chains and a higher degree of responsibility towards the health and safety of consumers.

There are also criticisms of the classic definitions of sustainable entrepreneurship. The understanding underlying them is that one should not deny the fact that, in the long run, it is normal for companies to aim at satisfying their own shareholders as best as possible and at maximizing their profit, while at the same time the company's strategy is not infrequently focused on social and environmental aspects with unclear economic benefits for the business [20]. It is an extremely difficult task to measure the degree of sustainability of an entrepreneurial business given the different viewpoints of shareholders and society. The influence of various factor conditions should be taken into account, including those related to market objectives and implementation [21, 22]. Companies adopt business models in order to more fully satisfy their consumers [23, 24].

## 3 Research Methodology

The theoretical formulations presented so far are the grounds for adopting the view that sustainable entrepreneurs have a clearly expressed potential to manage the transformation towards sustainable consumption patterns. Disposable income is also among the factors of priority importance for the sustainability of consumption, including food consumption.

Adhering to the main aim of the present paper gives grounds to analyze the impacts of sustainable entrepreneurship and income on the development of sustainable food consumption and to draw relevant conclusions and generalizations on this basis. Sustainability entrepreneur strategies, developed in accordance with the requirements of consumers and the environment, are specifically studied.

Households from the 27 Member States of the European Union were selected as respondents to the study. The data on their income, expenditure and consumption refer to the period 2012–2021. The chosen ten-year period provides tracking of the dynamics and trends in the development of the studied consumption. In its entirety, the study is based on Eurostat data on European household budgets.

The survey is representative and allows the application of two-stage nested selection. The total population from which the sample is formed is based on all households in Europe surveyed by Eurostat. The survey unit is any randomly selected ordinary European household regardless of the number of persons and their property and personal situation.

Methodologically, the study of European household budgets is based on the use of the simple linear regression and correlation method. This method is applied by using the statistical software product IBM SPSS statistics.

The application of the simple linear regression and correlation method using the given software product allows to analyze and assess the relationship between the annual income and the annual food expenditure of the households in the 27 countries of the European Union. On this basis, emerging trends can be outlined and opportunities for more sustainable consumption can be identified.

The survey of annual household food expenditure is based on the classification of individual consumption according to purpose – COICOP.

With a view to achieving the main aim, the present study adopts the hypothesis: household consumption expenditure is a function of disposable income – a fundamental concept on the basis of which the economic theory of consumption function is built.

## 4 Results

Today's consumers clearly state their view that they find sustainability extremely important. In this area, Deloitte's market research conducted in 2021 focused on tracking the global state of consumers worldwide [25]. The survey provides an opportunity to analyze the collected data relating to consumer purchases during four consecutive weeks of a certain month of 2021. According to the survey results, 55% of surveyed consumers answered that they purchased a sustainable product or service, 32% said that they paid significantly more for their sustainable purchase than the cost of an alternative product or that they are willing to wait longer to receive it. The results show that the largest category of sustainable purchases made by respondents was in food and beverages – with a relative share of 42%, followed by household goods – with a relative share of 25%.

The adoption of sustainable strategies is in response to the realization that entrepreneurial business has to meet the demands of consumers, staff and the environment regarding security and environmental compliance. According to another sustainability study by Deloitte, called Deloitte 2022 CxO Sustainability Report, 86% of consumer industry entrepreneurs agree that immediate action is needed that can limit the worst impacts of climate change and help the move towards a better future [26]. The aforementioned report places special emphasis on how entrepreneurs develop sustainability strategies in line with consumer demands. The following table visualizes how entrepreneurs develop their sustainability strategies in accordance with consumer and environmental demands (Table 1). As the data in Table 1 shows, a very large proportion of respondents (67%) indicated that their companies use more sustainable materials, such as recycled materials. According to the survey data, 66% of the surveyed companies strive to increase the efficiency of energy use, including energy efficiency in buildings.

A significant proportion of the surveyed companies – between 55% and 57% – report that they are adopting strategies to reduce the amount of air travel post-pandemic, to train their employees on climate change actions and impacts, and to use energy-efficient or climate-friendly machinery, technologies and equipment.

Regarding activities harder to implement aimed at sustainable entrepreneurial business, the respondents indicate the following: 49% of them express preferences for the development of new climate-friendly products or services, for 46% of the surveyed companies the requirement that suppliers and business partners meet specific sustainability criteria is of particular importance. The analysis of the data presented in this way allows us to assess that globally, entrepreneurs are moving towards developing their sustainability strategies through the prism of complying with consumer and environmental demands.

The above statements and empirical evidence reinforce the researchers' belief that entrepreneurs' decisions regarding sustainability should be considered as a significant factor in the relationship between entrepreneurial business and consumption sustainability. In connection with this, in order to reach economically efficient market solutions,

**Table 1.** Entrepreneurial sustainability strategies and activities.

| Key activities aimed at sustainable entrepreneurial business |
| --- |
| • 67% - using more sustainable materials (e.g. recycled materials, lower-emitting products)<br>• 66% - increasing the efficiency of energy use (e.g. energy efficiency in buildings)<br>• 57% - using energy-efficient or climate-friendly machinery, technologies and equipment<br>• 57% - training employees on climate change actions and impacts<br>• 55% - reducing the amount of air travel post-pandemic |
| Activities harder to implement aimed at sustainable entrepreneurial business |
| • 49% - developing new, climate-friendly products or services<br>• 46% - requiring suppliers and business partners to meet specific sustainability criteria<br>• 44% - updating/relocating facilities to make them more resistant to climate impacts<br>• 40% - incorporating climate considerations into lobbying/political donations<br>• 37% - tying senior leader compensation to environmental sustainability performance |

*Source* Deloitte 2022 CxO Sustainability Report

entrepreneurs must know the specific characteristics and peculiarities of consumers, as well as the consumption patterns they follow, including food consumption.

The analysis of food consumption patterns can be based on individual household budgets within a given country, region or territory. The research emphasis on household consumption is driven by the belief that this consumption is particularly important in explaining the changes in an economy, including those from a sustainability perspective. This importance is reinforced by the high relative share of individual household consumption in the overall structure of the Gross Domestic Product.

The study of household budgets in the 27 countries of the European Union allows an assessment to be made of the relationship between annual income and annual food expenditure. On this basis, emerging trends can be highlighted and opportunities for more sustainable consumption can be outlined.

Before analyzing household food expenditure from the 27 countries included in the European Union, a brief overview of the distribution of household expenditure by category can be made (Fig. 1) [27].

As shown by the data from Fig. 1 in 2020 household consumption in the European Union decreased by 8.10% compared to the previous year. Household consumption expenditure on products offered by the restaurant and hotel industry decreased at the highest rates – by 37.80%. During the studied period, the consumption of clothes and shoes shrank by 17.30%, followed by a reduction in the consumption expenditure of the products offered by the transport sector – by 16.80% and the products for recreation and culture – where the decrease compared to 2019 is by 16.70%. Only three groups of EU household final consumption expenditure showed an increase in 2020 compared to the previous year – expenditure on consumption of food and non-alcoholic beverages – an increase of 3.20%, expenditure on communications – by 2.40%, expenditure on the house, water, electricity, gas, etc. – by 0.30%.

The data presented above allow us to analyze the changes in the structure of final consumption expenditure of the surveyed households. The view that these changes have a

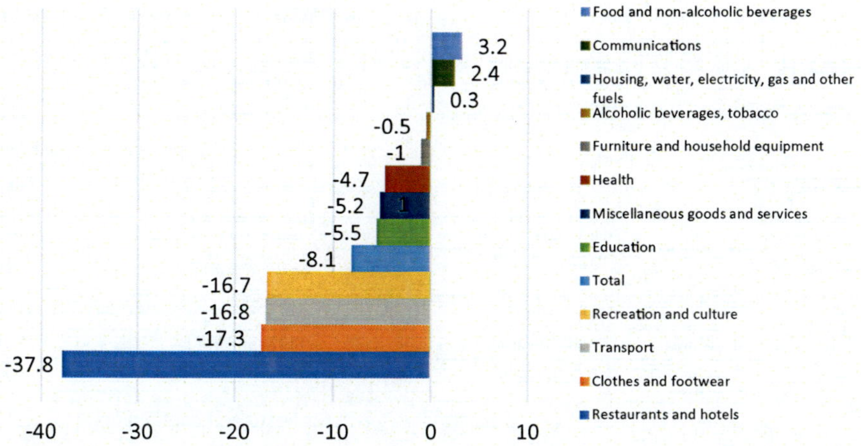


**Fig. 1.** Dynamics of household expenditure by category, European Union, in 2020 compared to the previous year. *Source* Eurostat. Available at: https://ec.europa.eu/eurostat/data/database.

significant impact on the overall economic well-being of the European Union consumers can be adopted.

The analysis of household food expenditure of the 27 countries of the European Union can be conducted through the prism of their dependence on disposable income. In this way, the economically related phenomenon – "food consumption expenditure - household disposable income" can be assessed. Such an assessment provides an opportunity to compare the economic well-being of consumers from the separate countries within the European Union. Based on the simple linear regression and correlation method using SPSS IBM, the results obtained are reliable, allowing the relationship between the disposable income of the EU-27 households and their food consumption expenditure to be analyzed and assessed. The main hypothesis adopted is that household consumption expenditure is a function of disposable income – a fundamental concept on which the economic theory of the consumption function is built. The results are illustrated in the following Tables 2, 3 and 4 and Fig. 2.

**Table 2.** Model summary.

| R | R square | Adjusted R square | Std. error of the estimate |
|---|---|---|---|
| 0.985 | 0.970 | 0.966 | 14357.538 |
| The independent variable is Disposable income | | | |

*Source* Eurostat. Available at: https://ec.europa.eu/eurostat/data/database

The variance estimation can be done according to the data in Table 3.

The data in Table 4 assist in presenting the coefficients calculated using the simple linear regression and correlation method.

**Table 3.** ANOVA.

| | Sum of squares | df | Mean square | F | Sig |
|---|---|---|---|---|---|
| Regression | 52461372449.162 | 1 | 52461372449.162 | 254.495 | 0.000 |
| Residual | 1649111123.614 | 8 | 206138890.452 | | |
| Total | 54110483572.776 | 9 | | | |
| The independent variable is Disposable income | | | | | |

*Source* Eurostat. Available at: https://ec.europa.eu/eurostat/data/database

**Table 4.** **Coefficients.**

| | Unstandardized coefficients | | **Standardized coefficients** | t | Sig. |
|---|---|---|---|---|---|
| | B | Std. error | Beta | | |
| Disposable income | 0.128 | 0.008 | 0.985 | 15.953 | 0.000 |
| (constant) | −89969.741 | 62551.464 | | −1.438 | 0.188 |

*Source* Eurostat. Available at: https://ec.europa.eu/eurostat/data/database

Figure 2 illustrates the relationship between the EU-27 household food consumption expenditure and disposable income.

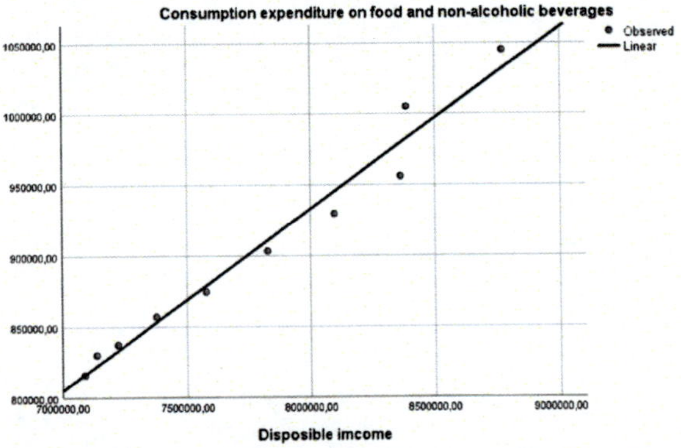


**Fig. 2.** Relationship between food consumption expenditure and disposable income of EU-27 households. *Source* Eurostat. Available at: https://ec.europa.eu/eurostat/data/database.

Conducting univariate linear regression analysis requires that the model adequacy hypothesis be checked. Thus, the adequacy check of the linear model constitutes a basis for its adoption and further use.

The main dependence measures that are important for the analysis are the regression coefficients, the correlation coefficient and the coefficient of determination. The regression coefficient measures the average change in the outcome variable for a unit increase in the factor variable. Since this coefficient takes the value of 0.128, it can be concluded that when the household disposable income increases by €1, there is an average increase of 0.128 in the food consumption expenditure. As the data in Table 2 show, the correlation coefficient is 0.985. This coefficient is a measure of the degree of dependence, which is also called the "strength of the relationship" between the variables. A review of the specialized literature shows that at $0.7 < R < 1.0$ there is a strong dependence. Therefore, the adopted value of the correlation coefficient allows establishing a high degree of dependence or a high correlation between disposable income and household food expenditure. In essence, the coefficient of determination measures the relative share of the variation in consumption expenditure that can be explained by the change in household disposable income. Thus, the coefficient measures the degree of success of the linear model in explaining differences between units of consumption expenditure related to the corresponding differences in household disposable income. According to the data in Table 2, the coefficient of determination analyzed is as follows: $R2 = 0.970$. This means that 97% of the differences between households in their food consumption expenditure is related to the differences between them in their disposable income.

As shown in Fig. 2 the values of the disposable household income are plotted on the abscissa, and the values of the food consumption expenditure – on the ordinate. Due to the existence of a dependence between the two quantities, the actual values of food consumption expenditure tend to be distributed around the regression line.

## 5 Discussion

The results obtained in the present study are a basis for interpreting the possible impacts of sustainable entrepreneurship and income on sustainable food consumption. In this aspect, the research interest is the grounds for comparing results of other researchers in their studies in the problem area.

Thus, for example, there are numerous studies in the field of entrepreneurship that offer specific methodologies for assessing sustainable development. According to a group of researchers, including Hockerts and Wüstenhagen [28], sustainable entrepreneurs balance between economic development, social justice for consumers and environmental sustainability. The adduced balance is achieved by means of the entrepreneurial behaviour followed. Developing their views in this direction, the two authors highlight that within the framework of sustainable entrepreneurship, economic opportunities are created for the transformation of businesses and markets towards an ecologically and socially more sustainable state. Thus, the building of sustainable business models for consumption is stimulated. There are a number of untapped opportunities and alternatives in this area [29].

Under modern conditions, the built food systems are characterized by a certain degree of unsustainability [30]. Unsustainability arises above all from the change in the applied consumption patterns. The consumption of highly processed foods is rising. The symmetries in ensuring food security are broken. The percentage of food waste is increasing.

In this area, part of the environmental impacts can be explained by food consumption. In the processes of production and market realization of food products, there are some negative effects resulting from technologies, means and methods of energy use, soil pollution, use of pesticides and failure to protect agricultural lands, introduction of physical, chemical and biological substances into the waters, generation of greenhouse gas emissions [31, 32].

Therefore, the assessment of the impact of sustainable entrepreneurship on consumption sustainability should be related to both positive and negative effects for the economic and socio-ecological system.

## 6 Conclusion

In its entirety, food consumption in the Member States of the European Union is a component of particular importance for the economic well-being of households. Both consumers themselves, represented by households, and business structures are relevant to this issue. Consumer perspectives related to sustainability require an almost radical transformation of consumption patterns, including food consumption patterns.

Resource scarcity requires that business, as an interested party, adopt sustainable entrepreneurial models. Sustainable entrepreneurs manage their businesses by converting to the principles containing ideas of: fruitful life of better quality in harmony with nature, efficient use of resources, environmental protection, preserving the integrity of ecosystems, eliminating poverty and food insecurity. In all the listed activities there is a vast untapped potential for future development. In this area, the answer to the question of how these sustainable entrepreneurships can contribute to a sustainable transformation and promotion of building sustainable consumption patterns becomes particularly important.

## References

1. IPCC: Climate Change 2014, Synthesis Report, Summary for Policy Makers. www.ipcc.ch/pdf/assessment-report/ar5/syr/AR5_SYR_FINAL_SPM.pdf. Accessed 28 July 2016
2. European Commission: 2050 Low-Carbon Economy. http://ec.europa.eu/clima/policies/strategies/2050/index_en.htm. Accessed 28 July 2016
3. Bakker, C., Wang, F., Huisman, J., Hollander, M.: Products that go round: exploring product life: exploring product life extension through design. J. Clean. Prod. **69**, 10–16 (2014)
4. Kollmuss, A., Agyeman, J.: Mind the gap: Why do people act environmentally and what are the barriers to pro-environmental behaviour? Environ. Educ. Res. **8**(3), 239–260 (2002)
5. NSI. National Statistical Institute. https://nsi.bg/bg. Accessed 6 May 2022
6. Bocken, N.: Business-led sustainable consumption initiatives: impacts and lessons learned. J. Manag. Dev. **36**(1), 81–96 (2017)
7. UN: Brundtland Report (1987). https://www.are.admin.ch/are/en/home/media/publications/sustainable-development/brundtland-report.html. Accessed 28 July 2016
8. UN: Transforming our world: the 2030 Agenda for Sustainable Development. https://www.un.org/ga/search/view_doc.asp?symbol=A/RES/70/1&Lang=E. Accessed 8 June 2022
9. SDC, 2016: Sustainable Development Commission, Challenge Paper: Measuring What Matters in Light of the Stiglitz Report. www.sdcommission.org.uk/data/files/publications/SDC_Indicators_Challenge_Paper_2011.pdf. Accessed 28 July 2016

10. Drewnowski, A.: Sustainable Healthy Diets: Models and Measures. Springer International Publishing (2017). https://doi.org/10.1007/978-3-319-55942-1-2
11. **GFK, 2007: Organic food products surge ahead: the findings of a GFK survey on consumption** of organic products. Press Release Nurmberg: GFK AG (2007)
12. Keynes, J.: The General Theory of Employment, Interest and Money. Hristo Botev Publishing House (1993)
13. McKinsey & Company: Understanding and shaping consumer behavior in the next normal. https://www.mckinsey.com/business-functions/. Accessed 28 July 2016
14. Parrish, B.: Sustainability-driven entrepreneurship: principles of organization design. J. Bus. Ventur. **25**, 510–523 (2010)
15. Boons, F., Ludeke-Freund, F.: Business models for sustainable innovation. state-of-the-art and steps towards a research agenda. J. Clean. Prod. **45**, 9–19 (2013)
16. Em, O., Georgiev, G., Radukanov, S., Petrova, M.: Assessing the market risk on the government debt of Kazakhstan and Bulgaria in conditions of turbulence. Risks **10**, 93 (2022). https://doi.org/10.3390/risks10050093
17. Seitzhanov, S., Kurmanov, N., Petrova, M., Aliyev, U., Aidargaliyeva, N.: Stimulation of entrepreneurs' innovative activity: evidence from Kazakhstan. Entrep. Sustain. Iss. **7**(4), 2615–2629 (2020). https://doi.org/10.9770/jesi.2020.7.4(4)
18. Mussapirov, K., Djalkibaev, J., Kurenkeyeva, G., Kadirbergenova, A., Petrova, M., Zhakypbek, L.: Business scaling through outsourcing and networking: selected case studies. Entrep. Sustain. Iss. **7**(2), 1480–1495 (2019). https://doi.org/10.9770/jesi.2019.7.2(48)
19. Odinokova, T., Bozhinova, M., Petrova, M.: Promotion of innovative entrepreneurship under sustainable development. E3S Web Conf. **41**, 04015 (2018). https://doi.org/10.1051/e3sconf/20184104015
20. Dvořáková, L., Zborková, J.: Integration of sustainable development at enterprise level. Procedia Eng. **69**, 686–695 (2014)
21. Filipova, T.: Trade Intermediation in Receivables Transactions (Economic and Managerial Aspects, Fragmentary Approbation), pp. 14–18. Academic Publishing House "Tsenov", Svishtov (2009)
22. Iliychovski, S.: Trade Crediting (Economic and Managerial Aspects), pp. 73–78. Academic Publishing House "Tsenov", Svishtov (2008)
23. Midova, P.: Management of Commercial Sites, pp. 26–38. Academic Publishing House "Tsenov", Svishtov (2007)
24. Perkov, V.: Commercial Modeling (Application of Mathematical Methods and Models in Commercial Activity), pp. 18–37. University Publishing House "V. Aprilov", Gabrovo (2018)
25. Deloitte. Deloitte. https://www2.deloitte.com/uk/en/pages/consumer-business/articles/sustainable-consumer.html. Accessed 6 June 2022
26. Deloitte: 2022 CxO sustainability report. https://www2.deloitte.com/content/dam/. Accessed 6 June 2022
27. Eurostat. https://ec.europa.eu/eurostat/data/database. Accessed 1 June 2022
28. Hockerts, K., Wüstenhagen, R.: Greening Goliaths versus emerging Davids: theorizing about the role of incumbents and new entrants in sustainable entrepreneurship. J. Bus. Ventur. **25**(5), 481–492 (2010)
29. Omarov, E.: Social entrepreneurship and what does it mean for management of consumer behavior. Access to science, business, innovation in digital economy. Access Press **1**(2), 86–102 (2020). https://doi.org/10.46656/access.2020.1.2(1)
30. Liashenko, O., Kravets, T., Prokopenko, M.: Consumer behavior clustering of food retail chains by machine learning algorithms. Access to science, business, innovation in digital economy. Access Press **2**(3), 234–251 (2021). https://doi.org/10.46656/access.2021.2.3(3)

31. Tukker, A., Huppes, G., Geerken, T., Nielsen, P.: Environmental impact of products (EIPRO). Draft report of the Institute for Prospective Technological Studies (IPTS) and the European Science and TO, Brussels (2005)
32. Koval, V., Mikhno, I., Hajduga, G., Gaska, K.: Economic efficiency of biogas generation from food product waste. E3S Web Conf. **100**, 00039 (2019). https://doi.org/10.1051/e3sconf/201910000039

# Clustering EU Countries—The Relationship Between Circular Economy, Resource Efficiency and Sustainable Development

Petranka Midova[1], Ventsislav Perkov[1(✉)], and Viktor Koval[2]

[1] D. A. Tsenov Academy of Economics, 2 Em. Chakarov Str., Svishtov, Bulgaria
v.perkov@uni-svishtov.bg

[2] Izmail State University of Humanities, 12 Repina Str., Izmail, Ukraine

**Abstract.** The study shows a current level of the European Union (EU) countries on their way to implement the principles of the circular economy. It is limited to a number of factors that are partially and simultaneously involved in several basic EU policy packages—Resource efficiency indicators, Circular economy indicators and Sustainable development indicators. The hierarchical and two-stage cluster analysis shows that the leading criteria by which the countries of the European Union differ are the level of energy efficiency, trade in recyclable raw materials and resource efficiency. As a result, three clusters have been identified: the first cluster of seven countries with high resource efficiency, highly developed trade in recyclable products and a medium level of energy efficiency; the second cluster—catching-up countries—low and medium level of the top three indicators and low level of waste generation and a third cluster consisting of three countries—energy examples—a high level of energy and product productivity due to a high level of waste generation and a low level of trade in recyclable waste. The analysis can be used to improve research in the field of the circular economy, in decision-making at various management levels and in building the reverse supply chain of recyclable raw materials.

**Keywords:** Circular economy · Resource efficiency · Sustainable development · European Union · Correlation analysis · Cluster analysis

## 1 Introduction

The growing concern about the environment at the political and public level is ex-pressed in the adoption of a number of normative acts and ideas aimed at reducing pollution and preserving ecosystems. At the same time, in order to limit harmful emissions and global warming, society must improve the resources use efficiency, especially energy, by implementing the principles of the circular economy. These actual problems generally reflect on the concept of sustainable development.

The aim of the paper is to classify the EU countries according to selected criteria, which partially and simultaneously are included in several main packages of EU policies—Resource efficiency indicators, Circular economy indicators and Sustainable


V. Koval et al. (Eds.): ISCMEE 2022, Lect. Notes in Managem., Industrial Engineering, pp. 79–94, 2023.
https://doi.org/10.1007/978-3-031-23463-7_5

development indicators. The data cover 27 countries and the period is 1 January 2020–31 December 2020. To achieve this aim, the following tasks are set: considering and summarizing the theoretical and empirical research works of the authors working on the problems of the circular economy and classification of various objects—mostly countries; developing methodological tools based on hierarchical and two-step cluster analysis; discussing the problematic moments arising from the literature sources and from the conduct of the empirical research, making recommendations regarding the application of the obtained results.

## 2 Literature Review

Pineiro-Villaverde and García-Álvarez [1] study the impact of the circular economy, sustainable consumption and sustainable production on resource productivity in order to establish the level of sustainability in the European Union (EU) countries. Using various statistical techniques, they found that recycling, as part of the circular economy, has a significant positive effect on the level of resource productivity. In addition, the sustainable use of resources is linked to sustainable production, especially the level of energy use.

A United Nations report [2] states that worldwide resource consumption has increased over a period of 17 years (2000–2017) by 70%. It notes that the circular economy is focused on the circulation of flows of resources, materials and products and this can have a positive impact on the development of society as a whole.

Zarębska et al. [3] link eco-innovation to sustainable development and the circular economy, focusing on examining the dynamics of indicators in specific Objective 9 for Poland. In the context of environmental sustainability, the circular economy model is expressed in a strategy which reduces the negative impact on the environment by offering an alternative to the traditional linear business model, in which the final product is a source of value-creation, reaching its highest point with its consumption [4]. Dura et al. [5] consider that the concept of a circular economy is a holistic approach to sustainable development designed to unite and strengthen the community, the business and the environment. They focus their research on wholesalers due to their specific position in the supply chain and their ability to implement the principles and practices of the circular economy in it.

Mazur-Wierzbicka [6] notes that the assessment of the indicators characterising the circular economy can be considered firstly in terms of progress and transformation into a circular economy, secondly in terms of the effectiveness of the implementation of the circular economy objectives, thirdly by comparing and revealing differences in the indicators of the circular and linear economy, fourthly through the use of reference points to report and monitor the level of development of the circular economy, fifthly assessment of the effectiveness of the implementation of the circular economy at national, regional and local level. She assesses the progress in the field of the circular economy for the period 2010–2018 by averaging 13 indicators for this period and creating 13 scales, which are subsequently standardised in one scale for the purpose of the analysis. As a result, problems have been reported to developed EU countries related to waste generation, pollution, overuse of resources, etc., and the countries of Central and Eastern Europe are moving more slowly towards a circular economy, due to the relatively smaller problems mentioned above. Hierarchical cluster analysis and the k-means clustering technique are used to classify the countries.

Polyakov et al. [7] use the cluster analysis technique (k-means) to assess progress towards a circular economy at EU level on the basis of three main groups—Waste production, Waste management and Trade-in secondary raw materials, covering 20 indicators. Based on this, they identify three clusters of uneven progress in the circular economy, defining countries as—leaders, followers and outsiders.

Moraga et al. [8] identify six groups of circular economy strategies that focus on: product and service functions, product life cycle, preservation of product components, preservation of materials, conservation of energy and scenario analysis. Their analysis shows that the circular economy indicators used by the EU focus mainly on strategy four—preservation of materials, with the first and fifth strategies suffering from a lack of progress indicators.

Stankovic et al. [9] identify four clusters at EU level, distinguishing four groups of production and consumption, waste management, secondary raw, materials, competitiveness and innovation, containing 11 indicators. They address the results of their research to the political leaders who need to take into account successful practices and the role of the circular economy in sustainable development.

Razminienė and Tvaronavičienė [10] emphasise the competitiveness at the level of an individual enterprise as a key factor for the implementation of the circular economy, as a result of which the efficiency of the resources used is perceived as a leading indicator for achieving the goal.

Gomonov et al. [11] analyse material flows and waste recycling as the main factors of the circular economy influencing the environment. They use the k-means algorithm to classify EU countries according to their level of development of the circular economy in four clusters, and the highest quality of classification is reported for the variable "Generation of municipal waste per capita".

The considered authors' views emphasise the significance of the problems related to the development of the circular economy in the EU countries, as most studies classify the EU countries according to the degree of progress in the field of the circular economy. As can be seen, the individual authors focus on a combination of various clustering and measurement techniques of the indicators, included in the different groups. The literature review shows that there is no consensus on the indicators to be used for clustering, and the existing indicators are used mainly to characterise the use and recycling of materials and the amount of pollution measured and related to various leading variables—GDP, the size of the population, etc. It should be noted that, in rare cases, the efficient use of resources and energy sources, which simultaneously correspond to the circular economy and the goals of sustainable development, are taken into account.

## 3 Methodology

The research methodology covers six stages. In the first stage, S*election of an information source and data collection*, the Eurostat database was studied, and three main directions were considered: Resource efficiency indicators, Circular economy indicators and Sustainable development indicators. On this basis, criteria for classifying countries have been selected on the three main guidelines of European policy.

The first two criteria are related to the use of energy and energy dependence—Share of renewable energy in gross final energy consumption [T2020_RD330], Energy dependence [T2020_RD320].

Secondly, four indicators for use, trade in recyclable raw materials and waste generation are included—Circular material use rate [SDG_12_41] [CEI_SRM030], Domestic material consumption per capita [T2020_RL110], Trade in recyclable raw materials [CEI_SRM020], Generation of municipal waste per capita [CEI_PC031].

Thirdly, productivity indicators are included—Resource productivity [T2020_RL100], Energy productivity [SDG_07_30]. After examining the data for individual countries for 2020, some of them are dropped from the survey—Switzerland, due to lack of data on the criteria Energy dependence [T2020_RD320], Circular material use rate [SDG_12_41]; Iceland and Norway due to lack of data on the criterion Circular material use rate [SDG_12_41]; The United Kingdom and Bosnia and Herzegovina due to lack of data on the criterion Share of renewable energy in gross final energy consumption [T2020_RD330]; Albania and Bosnia and Herzegovina due to lack of data on Domestic material consumption per capita [T2020_RL110]. Iceland and Liechtenstein due to lack of data on the criterion Trade in recyclable raw materials [CEI_SRM020]. Serbia, Turkey, Northern Macedonia, Montenegro and Kosovo due to lack of EU membership. As a result, EU-27 countries were included in the survey.

*The second* stage is related to the performance of correlation analysis. The criteria were tested for the presence of a very strong dependence (above 0.9) on each other by correlation analysis. Pearson's correlation coefficient was used in performing the correlation analysis [12]:

$$r_{x/y} = \frac{\sum(x_i - \overline{x}) \times (y_i - \overline{y})}{\sqrt{\sum(x_i - \overline{x})^2 \times \sum(y_i - \overline{y})^2}}, \tag{1}$$

*where*: $x_i$, $y_i$—variable values; $\overline{x}$, $\overline{y}$—average values of the variables.

*The third stage* is aimed at performing *cluster analysis*. The research work uses the so-called agglomerative clustering techniques, which are part of the hierarchical method of clustering. Euclidean squared distance is used as a measure of the similarity/difference (distance) between the individual objects. The Ward method was used to perform the clustering. Data are standardized using z-score. The procedure described by Mooi and Sarstedt [13] for cluster analysis with the software product SPSS was then performed.

*The fourth stage* requires *a decision on the number of clusters*. When performing the cluster analysis, it is necessary to provide information for determining the number of clusters, based on Fisher's F-test, obtained during the one-way analysis of variance (from 2 to 12 clusters).

The application of the F-criterion is used in the calculation of the VRC coefficient of Calinski and Harabasz. The coefficient is needed to calculate $w$, the smallest value of which determines the number of clusters [13].

*The fifth stage involves evaluating the validity and stability of the clusters.* After determining the number of clusters, it is necessary to assess the validity and stability of the derived clusters. This was achieved through two approaches—the application of different distance measures and different clustering methods and a double clustering method. When testing with Euclidean squared distances, the following methods were used: Average Linkage (Between Groups), Average Linkage (Within Groups) and Complete Linkage. Then the same methods were used but with a distance—Euclidean distance. In clustering with all methods, standardized values for the criteria were used.

*The sixth stage is aimed at interpreting the results obtained.* Interpretation of the results is performed using dendrograms and one-way analysis of variance. When performing a one-way analysis of variance, the F-criterion is significant for each of the clustering criteria, which means that there are significant differences between the individual clusters. In this case, the one-way analysis of variance cannot be used as a means of establishing the differences between the clusters. Then the post hog multiple comparison test—Tukey HSD—is used as an auxiliary tool.

## 4 Results Analysis

According to the specified methodological framework and in the same sequence, the results of the study for classification of EU countries according to the level of energy consumption, use and recycling of materials and waste, and productivity are presented. Table 1 shows the results of the correlation analysis between the selected eight indicators. The selected cells show the degree of correlation between the variables at significance levels of 0.05 and 0.01.

From the data in the table, it is evident that between the studied indicators there are no strong linear dependencies above 0.9, as the strength of the relationship between the variables SDG_12_41 and T2020_RL100 is positive and average. An average positive relationship between the variable SDG_07_30 and two other indicators T2020_RL110 and CEI_PC031 can also be reported. As a consequence of the results of the correlation matrix, it can be concluded that there are no significant strong positive dependencies between the indicators, due to which it will be necessary to drop any of the variables. Therefore, at the next stage of the analysis, the cluster analysis is performed with the selected eight criteria.

In the third stage of the study, a hierarchical cluster analysis was performed using the Ward method, the distance was measured by the Euclidean squared distance, and the values were standardized. As mentioned, the fourth stage is aimed at determining the number of clusters.

The cluster analysis was performed sequentially as the minimum number of tested clusters was two and the maximum—12. Subsequently, ANOVA analysis was performed, which is the basis for calculating the VRC coefficient of Calinski and Harabasz.

The results of the calculation of $w$ are given in Table 2.

The analysis of the obtained values of the coefficient $w$ shows that its lowest value is $\omega_3 = -17.504$, which is a reason to adopt a clustering strategy with three clusters.

**Table 1.** Correlation dependencies between the selected indicators.

| | SDG_12_41 | SDG_07_30 | T2020_RD330 | T2020_RD320 | T2020_RL110 | T2020_RL100 | CEI_SRM020 | CEI_PC031 |
|---|---|---|---|---|---|---|---|---|
| SDG_12_41 | 1 | 0.020 | −0.320 | 0.047 | **−0.415***  | **0.683**** | **0.512**** | 0.061 |
| | | 0.923 | 0.103 | 0.816 | **0.031** | **0.000** | **0.006** | 0.762 |
| SDG_07_30 | 0.020 | 1 | −0.027 | 0.260 | 0.108 | **0.583**** | 0.150 | **0.568**** |
| | 0.923 | | 0.896 | 0.191 | 0.592 | **0.001** | 0.455 | **0.002** |
| T2020_RD330 | −0.320 | −0.027 | 1 | **−0.506**** | 0.465* | −0.326 | −0.118 | −0.067 |
| | 0.103 | 0.896 | | **0.007** | 0.015 | 0.097 | 0.558 | 0.739 |
| T2020_RD320 | 0.047 | 0.260 | **−0.506**** | 1 | **−0.450*** | **0.471*** | 0.214 | **0.456*** |
| | 0.816 | 0.191 | **0.007** | | **0.018** | **0.013** | 0.284 | **0.017** |
| T2020_RL110 | **−0.415*** | 0.108 | **0.465*** | **−0.450*** | 1 | −0.368 | **−0.500**** | 0.063 |
| | **0.031** | 0.592 | **0.015** | **0.018** | | 0.059 | **0.008** | 0.755 |
| T2020_RL100 | **0.683**** | **0.583**** | −0.326 | **0.471*** | −0.368 | 1 | **0.559**** | **0.489**** |
| | **0.000** | **0.001** | 0.097 | **0.013** | 0.059 | | **0.002** | **0.010** |
| CEI_SRM020 | **0.512**** | 0.150 | −0.118 | 0.214 | **−0.500**** | **0.559**** | 1 | 0.070 |
| | **0.006** | 0.455 | 0.558 | 0.284 | **0.008** | **0.002** | | 0.730 |
| CEI_PC031 | 0.061 | **0.568**** | −0.067 | **0.456*** | 0.063 | **0.489**** | 0.070 | 12 |
| | 0.762 | **0.002** | 0.739 | **0.017** | 0.755 | **0.010** | 0.730 | |
| N | 27 | 27 | 27 | 27 | 27 | 27 | 27 | 27 |

*Pearson Correlation is significant at the 0.05 level (2-tailed)
**Pearson Correlation is significant at the 0.01 level (2-tailed)

**Table 2.** Auxiliary table for selecting the number of clusters

| $\omega_3$ | $\omega_4$ | $\omega_5$ | $\omega_6$ | $\omega_7$ | $\omega_8$ | $\omega_9$ | $\omega_{10}$ | $\omega_{11}$ |
|---|---|---|---|---|---|---|---|---|
| −17.504 | −13.51 | 10.111 | 3.427 | 3.906 | 4.293 | −0.777 | 5.05 | −5.198 |

After re-performing a cluster analysis with three clusters, it is proceeded to the fifth stage, in which the stability and validity of the formed three clusters must be analysed. Table 3 shows the results of the different clustering methods used, as well as different metrics. The first column contains the results of the cluster analysis, as each digit indicating which cluster the given country is included in. When using the Average Linkage (Between Groups) method with Euclidean squared distance, there is a transfer of two of the countries from the second cluster (marked cells) to the first cluster and respectively one country from the third to the first cluster.

When using the Average Linkage (Within Groups) method, only two countries move from the third cluster to the first and second clusters. When applying the Complete Linkage method, some countries move from the first and second clusters to the third. The next three columns show the results of the three methods used, but when the distance is changed—the Euclidean distance. As can be seen in the first case, there are no marked cells, i.e. there is a complete match between the formed clusters by the Ward method and the Average Linkage (Between Groups). When changing the metric and applying the other two methods, the results of columns 3 and 4 are repeated. The analysis shows that relatively stable clusters have actually been formed.

As a confirmation of the conclusion made about the stability of the clusters, the result is presented in Fig. 1. After performing a two-step cluster analysis with three clusters, the quality of the clusters is indicated. As can be seen, it is approximately 0.5, in the area where the quality can be said to be relatively good.

Once the level of quality of the formed clusters is established, the results are interpreted. The interpretation of the results successively goes through the analysis of the dendrogram (Fig. 2), analysis of ANOVA results (Table 4) and Tukey HSD results (Table 5).

The dendrogram clearly shows that there are significant differences between the individual EU countries in the selected classification criteria. The first cluster includes seven countries—Belgium, Germany, Greece, Italy, Spain, France and the Netherlands, the second cluster includes Bulgaria, Czechia, Estonia, Croatia, Cyprus, Latvia, Lithuania, Hungary, Malta, Austria, Poland, Portugal, Romania, Slovenia, Slovakia, Finland and Sweden. The third cluster includes only three countries—Denmark, Ireland and Luxembourg. The analysis of the dendrogram (see Fig. 2) shows that the countries of the first cluster are divided into two conditional convergence groups, the first group includes Belgium, France and Italy, and the second group includes Greece and Spain. Germany and the Netherlands are included in this cluster, but remain outside the other countries due to differences in some of the criteria. The second cluster covers a group of 17 countries, which according to the degree of proximity can be formed into five conditional groups, which, in addition to the criteria used, are close in geographical

**Table 3.** Cluster stability using different methods and metrics (3 clusters).

| Case | Ward linkage | Average linkage (between groups) | Average linkage (within groups) | Complete linkage | Average linkage (between groups) | Average linkage (within groups) | Complete linkage distance |
|---|---|---|---|---|---|---|---|
| 1. Belgium | 1 | 1 | 1 | 1 | 1 | 1 | 1 |
| 2. Bulgaria | 2 | 2 | 2 | 2 | 2 | 2 | 2 |
| 3. Czechia | 2 | 2 | 2 | 2 | 2 | 2 | 2 |
| 4. Denmark | 3 | 3 | **2** | 3 | 3 | **2** | 3 |
| 5. Germany | 1 | 1 | 1 | **3** | 1 | 1 | **3** |
| 6. Estonia | 2 | 2 | 2 | 2 | 2 | 2 | 2 |
| 7. Ireland | 3 | 3 | 3 | 3 | 3 | 3 | 3 |
| 8. Greece | 1 | 1 | 1 | **3** | 1 | 1 | **3** |
| 9. Spain | 1 | 1 | 1 | **3** | 1 | 1 | **3** |
| 10. France | 1 | 1 | 1 | 1 | 1 | 1 | 1 |
| 11. Croatia | 2 | 2 | 2 | 2 | 2 | 2 | 2 |
| 12. Italy | 1 | 1 | 1 | 1 | 1 | 1 | 1 |
| 13. Cyprus | 2 | **1** | 2 | **3** | 2 | 2 | **3** |
| 14. Latvia | 2 | 2 | 2 | 2 | 2 | 2 | 2 |
| 15. Lithuania | 2 | 2 | 2 | 2 | 2 | 2 | 2 |
| 16. Luxembourg | 3 | **1** | **1** | 3 | 3 | **1** | 3 |
| 17. Hungary | 2 | 2 | 2 | 2 | 2 | 2 | 2 |
| 18. Malta | 2 | **1** | 2 | **3** | 2 | 2 | **3** |
| 19. Netherlands | 1 | 1 | 1 | 1 | 1 | 1 | 1 |

*(continued)*

**Table 3.** (*continued*)

| Case | Ward linkage | Average linkage (between groups) | Average linkage (within groups) | Complete linkage | Average linkage (between groups) | Average linkage (within groups) | Complete linkage distance |
|---|---|---|---|---|---|---|---|
| 20. Austria | 2 | 2 | 2 | 2 | 2 | 2 | 2 |
| 21. Poland | 2 | 2 | 2 | 2 | 2 | 2 | 2 |
| 22. Portugal | 2 | 2 | 2 | 2 | 2 | 2 | 2 |
| 23. Romania | 2 | 2 | 2 | 2 | 2 | 2 | 2 |
| 24. Slovenia | 2 | 2 | 2 | 2 | 2 | 2 | 2 |
| 25. Slovakia | 2 | 2 | 2 | 2 | 2 | 2 | 2 |
| 26. Finland | 2 | 2 | 2 | 2 | 2 | 2 | 2 |
| 27. Sweden | 2 | 2 | 2 | 2 | 2 | 2 | 2 |

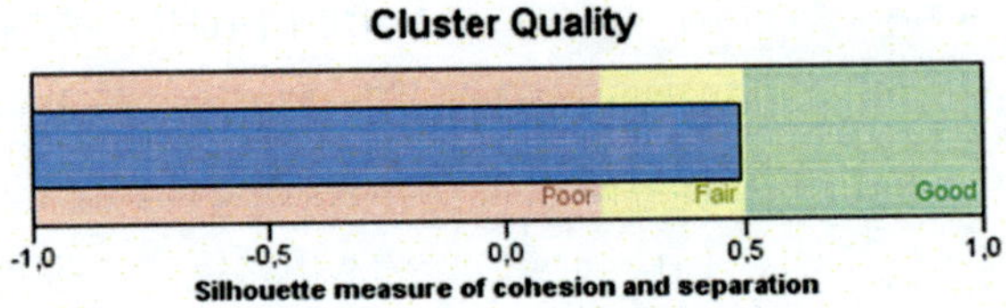


**Fig. 1.** Cluster quality.

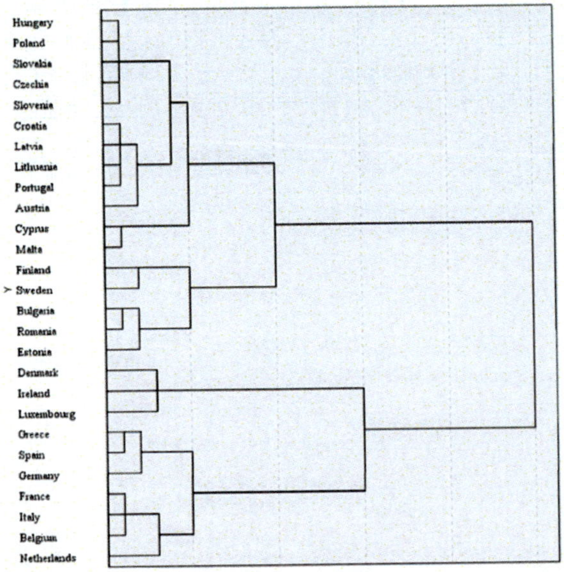


**Fig. 2.** Dendrogram of the EU countries.

terms. A group of countries with Central European positioning is being formed here—Hungary, Poland, Slovakia, Czechia и Slovenia, the Mediterranean group—Cyprus и Malta, Scandinavian group—Finland and Sweden, the South Eastern group—Bulgaria и Romania and the North Eastern group—Latvia, Lithuania. The remaining countries in cluster two, regardless of their geographical location, are also positioned here. Cluster 3 can be conditionally called North Western.

The dendrogram graphically shows the degree of difference/similarity between the individual clusters, but not the sources of these differences. One of the techniques for revealing the reasons for the differences is to perform a One-way Analysis of Variance. The result of the ANOVA analysis is shown in Table 4.

As shown in Table 4, the main differences between countries are due to 6 criteria that have a significance of less than 0.01. The other two criteria (marked cells) T2020_RD330 and T2020_RD320 are not statistically significant, therefore their influence in determining the similarities/differences between the clusters is not significant. On the basis of two-step cluster analysis, the criteria with the greatest significance for the differentiation of the clusters can be indicated (Fig. 3). The criterion comes first SDG_07_30—Energy

**Table 4.** Results of the conducted ANOVA

| | | Sum of squares | df | Mean square | f | Sig. |
|---|---|---|---|---|---|---|
| SDG_12_41 | Between groups | 612.7967 | 2 | 306.398 | 8.807 | 0.001354 |
| | Within groups | 834.9818 | 24 | 34.791 | | |
| | Total | 1447.7785 | 26 | | | |
| SDG_07_30 | Between groups | 363.4769 | 2 | 181.738 | 36.858 | 0.000000 |
| | Within groups | 118.3373 | 24 | 4.931 | | |
| | Total | 481.8141 | 26 | | | |
| T2020_RD330 | Between groups | 490.3997 | 2 | 245.200 | 2.008 | **0.156222** |
| | Within groups | 2930.9974 | 24 | 122.125 | | |
| | Total | 3421.3971 | 26 | | | |
| T2020_RD320 | Between groups | 1772.3786 | 2 | 886.189 | 2.163 | **0.136885** |
| | Within groups | 9833.4205 | 24 | 409.726 | | |
| | Total | 11605.7991 | 26 | | | |
| T2020_RL110 | Between groups | 478.8287 | 2 | 239.414 | 8.010 | 0.002164 |
| | Within groups | 717.3802 | 24 | 29.891 | | |
| | Total | 1196.2089 | 26 | | | |
| T2020_RL100 | Between Groups | 24.9046 | 2 | 12.452 | 21.713 | 0.000004 |
| | Within groups | 13.7642 | 24 | 0.57351 | | |
| | Total | 38.6687 | 26 | | | |
| CEI_SRM020 | Between groups | 3403699744487.47 | 2 | 1701849872243.73 | 28.126 | 0.000001 |
| | Within groups | 1452210220353.05 | 24 | 60508759181.38 | | |
| | Total | 4855909964840.52 | 26 | | | |

*(continued)*

productivity, second is the criterion CEI_SRM020—Trade in recyclable raw materials,

**Table 4.** (*continued*)

| | | Sum of squares | df | Mean square | f | Sig. |
|---|---|---|---|---|---|---|
| CEI_PC031 | Between groups | 171172.14 | 2 | 85586.07 | 8.872 | 0.001304 |
| | Within groups | 231513.71 | 24 | 9646.41 | | |
| | Total | 402685.85 | 26 | | | |

third is the criterion T2020_RL100—Resource productivity. Next are the other three criteria with a value below 0.40.

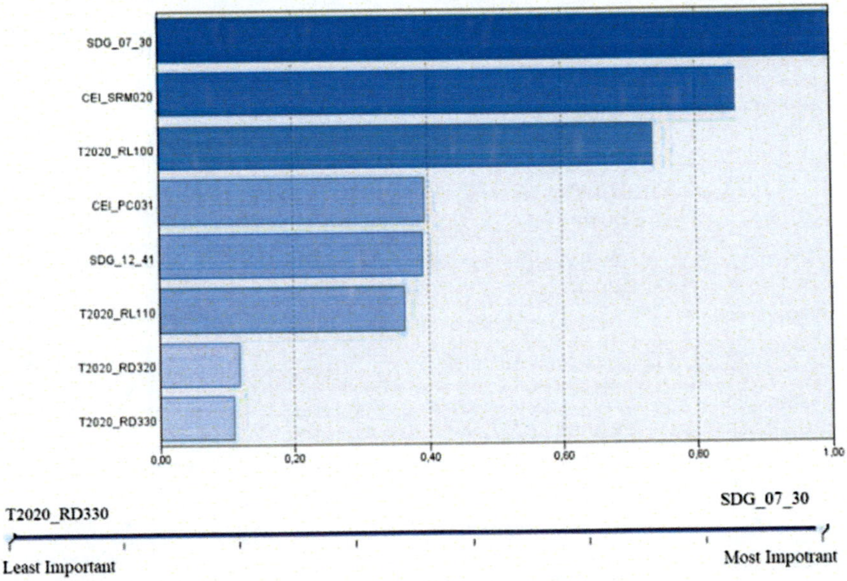


**Fig. 3.** Importance of clustering criteria.

In general, the differences between clusters can be interpreted through the following conditional levels—low, medium and high. The first cluster is characterised by a medium level of energy productivity, a high level of trade in recycled raw materials, a high level of resource productivity, a low to medium level of waste generated per capita, a high level of recycled and returned to the economy materials, low to medium level of own resources used in the economy per capita.

The second cluster is characterised by low energy productivity, low level of trade in recycled raw materials, low resource productivity, low to medium level of waste generated per capita, low to medium level of recycled and returned to the economy materials, medium to high level of the own materials used in the economy per capita.

The third cluster is characterised by a high level of energy productivity, low level of trade in recycled raw materials, high level of resource productivity, high level of waste generated per capita, from low to medium level of recycled and returned to the economy materials, medium to high level of own resources invested in the economy per capita.

The results of Tukey's post-hoc multiple comparison test are aimed at identifying the sources of differences between the clusters. The analysis shows (Table 5) that for the variable Circular material use rate SDG_12_41 there are significant differences between the countries in the first cluster and those in the second and third clusters, namely in the first cluster the level is high and in the other two—from low to medium.

This is confirmed by the insignificance of the coefficient for the second and third cluster. The differences between the clusters in terms of the degree of Energy productivity [SDG_07_30] are statistically significant, as the countries in the first cluster have a medium level of energy productivity, those in the second cluster—a low level and those in the third cluster—a high level of Energy productivity. For the variable Domestic material consumption per capita [T2020_RL110], the differences between the countries of the first cluster and the countries of the second and third clusters are statistically significant, namely the countries of the first cluster have low to medium level and those of the second and third cluster from medium to high level.

According to the Resource productivity criterion [T2020_RL100], there are statistically significant differences between the countries in the second cluster and those in the first and third clusters, which are expressed in low resource productivity in the second cluster and high resource productivity in the other clusters. The indicator Trade in recyclable raw materials [CEI_SRM020] shows statistically significant differences between the countries of the first cluster and those of the second and third, as the countries of the first cluster have a high level of trade, while the countries of the second and third cluster—low level.

According to the criterion Generation of municipal waste per capita (CEI_PC031), statistically significant differences were reported between the countries from the third cluster and those from the first and second clusters. The difference is expressed in the high level of waste generation per capita in the third cluster and the low to medium level in this indicator for the other two clusters.

## 5 Discussion

The study shows a snapshot of the EU countries on their way to implement the circular economy principles. It is limited to a number of factors that are partially and simultaneously involved in several major EU policy packages—resource efficiency indicators, circular Economy indicators and sustainable development indicators. The studied authors choose different approaches to the research of the indicators included in one package, through a separate group of indicators within the package to selected indicators, which may be included in other packages [14, 15]. The clear interdependence between EU policies is the basis for choosing several packages containing quantitative indicators, and it is not possible to give an unambiguous answer as to whether all indicators should be included in the research or only part of them, which are selected according to a study of theoretical and empirical studies, or those reduced by factor analysis techniques (the

**Table 5.** Results of the Tukey post hoc test for multiple comparisons

| Dependent variable | | (I) Ward method | (J) Ward method | Mean difference (I–J) | Std. error | Sig. |
|---|---|---|---|---|---|---|
| SDG_12_41 | Tukey HSD | 1 | 2 | 10.925210 | 2.648897 | **0.001** |
| | | | 3 | 10.542857 | 4.070270 | **0.041** |
| | | 2 | 1 | −10.925210 | 2.648897 | **0.001** |
| | | | 3 | −0.382353 | 3.693710 | 0.994 |
| | | 3 | 1 | −10.542857 | 4.070270 | **0.041** |
| | | | 2 | 0.382353 | 3.693710 | 0.994 |
| SDG_07_30 | Tukey HSD | 1 | 2 | 2.996303 | 0.997212 | **0.016** |
| | | | 3 | −8.814286 | 1.532306 | **0.000** |
| | | 2 | 1 | −2.996303 | 0.997212 | **0.016** |
| | | | 3 | −11.810588 | 1.390545 | **0.000** |
| | | 3 | 1 | 8.814286 | 1.532306 | **0.000** |
| | | | 2 | 11.810588 | 1.390545 | **0.000** |
| T2020_RL110 | Tukey HSD | 1 | 2 | −8.623471 | 2.455281 | **0.005** |
| | | | 3 | −12.534000 | 3.772762 | **0.008** |
| | | 2 | 1 | 8.623471 | 2.455281 | **0.005** |
| | | | 3 | −3.910529 | 3.423726 | 0.498 |
| | | 3 | 1 | 12.534000 | 3.772762 | **0.008** |
| | | | 2 | 3.910529 | 3.423726 | 0.498 |
| T2020_RL100 | Tukey HSD | 1 | 2 | 1.943851 | 0.340096 | **0.000** |
| | | | 3 | −0.144110 | 0.522588 | 0.959 |
| | | 2 | 1 | −1.943851 | 0.340096 | **0.000** |
| | | | 3 | −2.087961 | 0.474241 | **0.001** |
| | | 3 | 1 | 0.144110 | 0.522588 | 0.959 |
| | | | 2 | 2.087961 | 0.474241 | **0.001** |
| CEI_SRM020 | Tukey HSD | 1 | 2 | 800908.563025 | 110469.213946 | **0.000** |
| | | | 3 | 856590.190476 | 169745.971523 | **0.000** |

(*continued*)

main components method). In order to trace the dynamics of time, different authors use metrification of time series and their reduction to a value by which to classify countries, another possibility is to take a series of snapshots for a certain period of years by observing the shift of countries in individual groups. The problem with choosing the same indicators remains. In the present study, the leading factor causing the difference

**Table 5.** (*continued*)

| Dependent variable | | (I) Ward method | (J) Ward method | Mean difference (I–J) | Std. error | Sig. |
|---|---|---|---|---|---|---|
| | | 2 | 1 | −800908.563025 | 110469.213946 | **0.000** |
| | | | 3 | 55681.627451 | 154041.959119 | 0.931 |
| | | 3 | 1 | −856590.190476 | 169745.971523 | **0.000** |
| | | | 2 | −55681.627451 | 154041.959119 | 0.931 |
| CEI_PC031 | Tukey HSD | 1 | 2 | 43.571429 | 44.107748 | 0.591 |
| | | | 3 | −215.428571 | 67.775557 | **0.011** |
| | | 2 | 1 | −43.571429 | 44.107748 | 0.591 |
| | | | 3 | −259.000000 | 61.505315 | **0.001** |
| | | 3 | 1 | 215.428571 | 67.775557 | **0.011** |
| | | | 2 | 259.000000 | 61.505315 | **0.001** |

between clusters is energy productivity [SDG_07_30]. On the other hand, the criteria share of renewable energy in gross final energy consumption [T2020_RD330], and energy dependence [T2020_RD320] are of low importance (statistically insignificant F) and do not cause significant differences between clusters. It follows the question of the relationship between energy productivity and its sources of production, moreover, the entire EU is clearly dependent on external sources in the field of energy. The second most important factor is trade in recyclable raw materials [CEI_SRM020]. It shows uneven development in the field of trade in recyclable raw materials, which raises the question of how to collect, separate and deliver possible recyclable raw materials to traders and recycling companies. This is one of the big problems of Bulgaria and other EU countries.

## 6 Conclusion

The results of the study show the presence of three groups of countries dissimilar in the included eight criteria. The leading factor causing the difference between clusters is Energy productivity [SDG_07_30], the second most important factor is Trade in recyclable raw materials [CEI_SRM020], and the third important factor is criterion T2020_RL100—Resource productivity.

It should be noted that there are countries with a high level of energy and resource use efficiency, but also a high level of waste per capita and a low level of recyclability of materials—the third cluster. The second cluster are the countries positioned mainly at low and medium level by most criteria, and the first cluster are the so-called highly developed countries that need to improve their level of energy efficiency and reach a high level of resource efficiency and material recycling.

**The analysis can be used to develop specific geo-economic measures for neighbouring countries**, to create a common map to track the reverse supply chains and the development of trade in recyclable raw materials, to reduce the level of waste per capita by introducing prevention of this pollution.

## References


1. Pineiro-Villaverde, G., García-Álvarez, M.T.: Sustainable consumption and production: exploring the links with resources productivity in the EU-28. Sustainability **12**, 8760 (2020). https://doi.org/10.3390/su12218760
2. United Nations Economic Commission for Europe. Background paper on Measuring and monitoring the circular economy and use of data for policy-making. United Nations Development Account (2021). https://unece.org/sites/default/files/2021-10/Background%20report%20on%20measuring%20circular%20economy_ENG.pdf. Accessed 21 Mar 2022
3. Zarębska, J., Żabińska, I., Zarębski, A.: Eco-innovations in Poland—the extent of changes, development and barriers. Sci. Pap. Silesian Univ. Technol. Organ. Manag. **135**, 245–256 (2019). https://doi.org/10.29119/1641-3466.2018.135.19
4. Kowszyk, Y., Maher, R.: Case studies on circular economy models and integration of sustainable development goals in business strategies in the EU and LAC. Hamburg, Germany (2018). 10.12858.1018EN
5. Dura, C.C., Iordache, A., Ionescu, A., Isac, C., Breaz, T.O.: Analyzing performance in wholesale trade Romanian SMEs: framing circular economy business scenarios. Sustainability **5567**, 1–34 (2022). https://doi.org/10.3390/su14095567
6. Mazur-Wierzbicka, E.: Circular economy: advancement of European Union countries. Environ. Sci. Eur. **33**(1), 1–15 (2021). https://doi.org/10.1186/s12302-021-00549-0
7. Polyakov, M., Khanin, I., Bilozubenko, V., Korneyev, M., Shevchenko, G.: Factors of uneven progress of the European Union countries towards a circular economy. Probl. Perspect. Manag. **19**(3), 332–344 (2021). https://doi.org/10.21511/ppm.19(3).2021.27
8. Moraga, G., et al.: Circular economy indicators: What do they measure? Resour. Conserv. Recycl. **146**, 452–461 (2019). https://doi.org/10.1016/j.resconrec.2019.03.045
9. Stanković, J.J., Janković-Milić, V., Marjanović, I., Janjić, J.: An integrated approach of PCA and PROMETHEE in spatial assessment of circular economy indicators. Waste Manag. **128**, 154–166 (2021). https://doi.org/10.1016/j.wasman.2021.04.057
10. Razminienė, K., Tvaronavičienė, M.: Detecting the linkages between clusters and circular economy. Terra Econ. **16**(4), 50–65 (2018). https://doi.org/10.23683/2073-6606-2018-16-4-50-65
11. Gomonov, K., Ratner, S., Lazanyuk, I., Revinova, S.: Clustering of EU countries by the level of circular economy: an object-oriented approach. Sustainability **13**, 7158 (2021). https://doi.org/10.3390/su13137158
12. Koop, G.: Analysis of Economic Data, 2nd ed. Wiley Ltd. (2005)
13. Mooi, E., Sarstedt, M.: A Concise Guide to Market Research the Process, Data, and Methods Using IBM SPSS Statistics. Springer, Berlin, Heidelberg (2011)
14. Cudečka-Puriņa, N., Atstāja, D., Koval, V., Purviņš, M., Nesenenko, P., Tkach, O.: Achievement of sustainable development goals through the implementation of circular economy and developing regional cooperation. Energies **15**, 4072 (2022). https://doi.org/10.3390/en15114072
15. Atstaja, D., Cudecka-Purina, N., Hrinchenko, R., Koval, V., Grasis, J., Vesere, R.: Alignment of circular economy business models for framing national sustainable economic development. Acta Innov. 5–14 (2022). https://doi.org/10.32933/actainnovations.42.1

# Transformational Opportunities for Business Entities in the Circular Economy

Vladyslav Iakovenko(✉), Roman Pavlov, Tetiana Pavlova, and Oksana Levkovich

Oles Honchar Dnipro National University, Dnipro 49000, Ukraine
vladislavyakovenko@gmail.com

**Abstract.** The study based on the available key principles of the circular economy functioning proposes a conceptual scheme of the circular economy ecosystem, which includes key elements such as the country's population, investments, emissions, and GDP. The ecosystem also considers nine key principles of circular economy functioning, namely: refuse; reduce; reuse; repair; refurbish; remanufacture; repurpose; recycle; recover, which are integrated into functions and supported by them, namely: mining; production; consumption, service; reuse; processing; utilization, forming a chain of a closed cycle of usage of the raw materials, goods, and resources. The study also proposed a conceptual system-dynamic model of the circular economy ecosystem functioning, based precisely on its functions, presented in 7 blocks of the model. Each block covers the influence of circular economy principles on the model's indicator elements. According to each block of functions, the authors offer opportunities to form new organizational structures that will support the viability of the circular economy concept, provide waste minimization, additional service and processing of goods and waste, and allow the closing of the cycle from extraction to disposal. Further research established the simulation experiments' specifics and the opportunity to build the model blocks.

**Keywords:** Circular economy · Modeling · Business structures · Ecosystem · Ecological footprint

## 1 Introduction

The circular economy viability began to be studied in 1970 by the "ecological footprint" term [1]. The problems faced by society, the economy of countries, and the global environmental condition were dealt with long before the principles of the circular economy appeared. However, such trends are currently observed at the micro and meso levels of countries' economies; at the state level, there has been an understanding of the complexity and systemic nature of the problems of minimizing the "ecological footprint" and ensuring the development of the country's circular economy for a long time. According to the circularity gap report 2022 [2], currently, only 8.6% of the total world economy has fully implemented the circular economy principles. The circular economy complexity lies primarily in the interaction of mentality, technologies, the handling resources culture, the economic state profile, and its natural and ecological specifics of existence. All


V. Koval et al. (Eds.): ISCMEE 2022, Lect. Notes in Managem., Industrial Engineering, pp. 95–103, 2023.
https://doi.org/10.1007/978-3-031-23463-7_6

those components should be based on the principles of 9R [3] Refuse-Reduce-Reuse-Repair-Refurbish-Remanufacture-Repurpose-Recycle-Recover. Undoubtedly, compliance with these principles can create an ecosystem for monitoring the interaction of system elements of the circular economy at the state level.

The research will consist of the systematization of the elements of the circular economy ecosystem, the construction of a model of the circular economy ecosystem that will consider the 9R principles, and we will also conduct modeling of circular economy behavior and identify entrepreneurial and organizational opportunities in the newly created interdependencies of the circular economy ecosystem. The paper considers the meaningful systematization of the circular economy elements [4]. According to the new industrial design and entrepreneurial opportunities in the production sphere, the main principles of consumer goods production are clarified in the papers [5] Significant achievements in the development evaluation of the circular economy and the systematization of indicators are developed in the paper [6].

Notably, this research at the first stage is theoretical in nature and aims to establish key parameters of the circular economy ecosystem and establish entrepreneurial opportunities for economic entities. That stage represents research at the macro level, not a separate entity of entrepreneurial activity. Each existing entity can choose certain opportunities, which are established in the research, considering the specifics of the business entity's activity.

## 2 Material and Methods

The research methodological basis will be the elements of system analysis to establish the elements of the circular economy ecosystem. The analysis of scientific sources and expert reports proves the complexity of the ecosystem and the sufficient difference in understanding this system's elements.

The principles of cognitive modeling were used to establish the influence vectors among system elements. When building a conceptual model, positive or negative dependencies among indicators are primarily established; the combination of those dependencies in levels allows for a global vector of influence (positive or negative) on the ecosystem.

Simulation modeling, as a modeling subdivision of economic processes and the world balance, allows several approaches. The agent approach represents the behavior of system elements according to certain algorithms (behavior of agents). The discrete-event approach was used to model the system based on the principle of mass service systems and J. Forester's system dynamics [7], which allows for dynamic modeling of indicators of system elements according to established dependencies and predictive values considering feedback and a cybernetic approach. The study authors chose the system dynamics to build a conceptual model of the circular economy ecosystem existence. At this first stage of research, system dynamics, as a simulation modeling method, will be used for structuring the basic parameters of the circular economy ecosystem model and graphic and analytical representation between them in the circular economy ecosystem elements section. In the next research steps, the number of parameters and variables of each ecosystem element will be expanded, allowing for simulation experiments at the macro level of the circular economy. That will provide an opportunity to interactively set the state of the circular economy ecosystem by choosing certain values of the model parameters.

## 3 Theory/Calculation

The fundamental difference among available models of the circular economy [8–12] is primarily the selection of several functional (and later process) components, namely: extraction, production, consumption, service, reuse, processing, and disposal. This set of circular economy functions allows modeling at the process, design, and simulation levels. In addition to the proposed components, the study authors see it mandatory to consider the 9R principles. Therefore, Fig. 1 depicts the concept of circular economy ecosystem functioning.

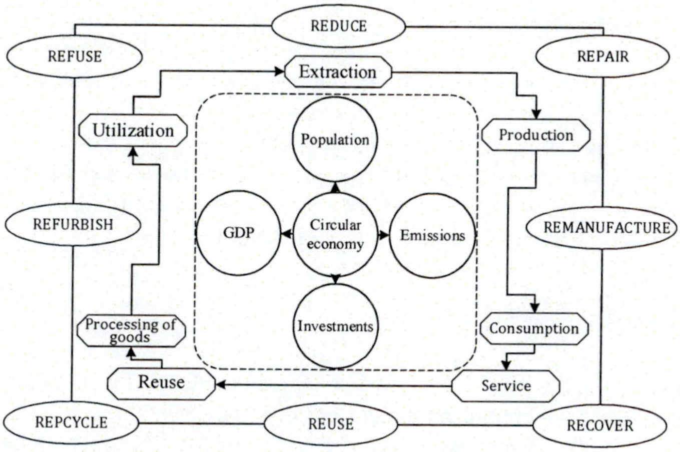


**Fig. 1.** Ecosystem model of the circular economy.

According to the IPAM [13, 14] logic, the circular economy's basic elements are the population, which acts as a labor resource, and, on the other hand, consumers of the products produced by them. Clearly, the population is an essential resource for preserving material resources during production, improving labor efficiency in all functional areas, and carefully consuming and using goods.

The second main element of the circular economy ecosystem is an investment, which is present in every economic model. The difference in investments lies primarily in use, structural redistribution, and sources of formation. Thus, based on the 9R principles, it becomes clear that investments should primarily be allocated to energy-saving technologies that sparingly use resources and minimize the amount of waste. The formation of investments into the circular economy and the owner of these investment funds are not so clearly reflected. A significant role can be assigned to the state, offering at least two scenarios of their formation. The first includes a significant amount of taxation of business entities that do not use the principles of the 9R circular economy to the possible extent, and those funds should be redistributed among enterprises by state procurement of environmental technologies. In turn, enterprises that use ecological technologies have a lower tax burden, allowing them to form their investment portfolio for independent investment in their enterprise development. The main challenge in this process will be the legislative framework [15] and the professionalism of state structures. Also, it should be noted that state investment should not be considered point by point, but a systemic vision of the circular economy development in the country should be used.

The system's next element is emissions into the atmosphere, land, ocean, etc. As noted in paper [16], the more developed the country's economy, the greater the ecological footprint it leaves. Furthermore, this contradicts the existing principles of circular economy development. A systemic vision of all functional links of the circular economy ecosystem and balanced state support should break the direct dependence between the country's development and its larger ecological footprint.

As a result of any country's activity, the GDP is formed; the classic economic indicator [17, 18] received a new indicative meaning when it was compared to the amount of environmental pollution, which allowed comparing the GDP per capita and the amount of emission, this indicator can be used as a baseline at certain stages of development ecosystems of the country's circular economy.

Therefore, after considering the proposed ecosystem of the circular economy, the authors suggest the 9R principles applied to the functional links of the system and as indicators of the development of the circular economy to calculate the indicators: population, GDP, the volume of emissions and investments.

## 4 Results and Discussion

In order to understand how business structures get changed in the circular economy, opening up new opportunities and eliminating some threats, a cognitive model was developed, which was transformed into a system-dynamic model of the circular economy ecosystem functioning.

The main concept of the developed conceptual model is the active contour of the circular economy ecosystem model; therefore, the model consists of 7 blocks corresponding to the functions mentioned above: extraction, production, consumption, service, reuse, processing of goods, and disposal presented by Fig. 2. The given model requires an **explanation and disclosure of the specifics of each component of the circular economy** ecosystem. That provides an opportunity to understand better and formulate the study's model basis.

For every country, an essential question arises as to where and how to get raw materials to ensure the production and consumption of raw materials and resources. **The first component of the cognitive model of the circular economy ecosystem is the** "Extraction" block. The "Extraction" component allows considering the possibilities of obtaining fossil raw materials (Fr), carbon emissions (Ce), primary resources for extraction (Rate_4), production energy generation (Prg), and alternative energy generation (Ag), their structure in the country's energy generation balance. The country's generated energy consumption and the realization of its excess energy reserves (Er) or

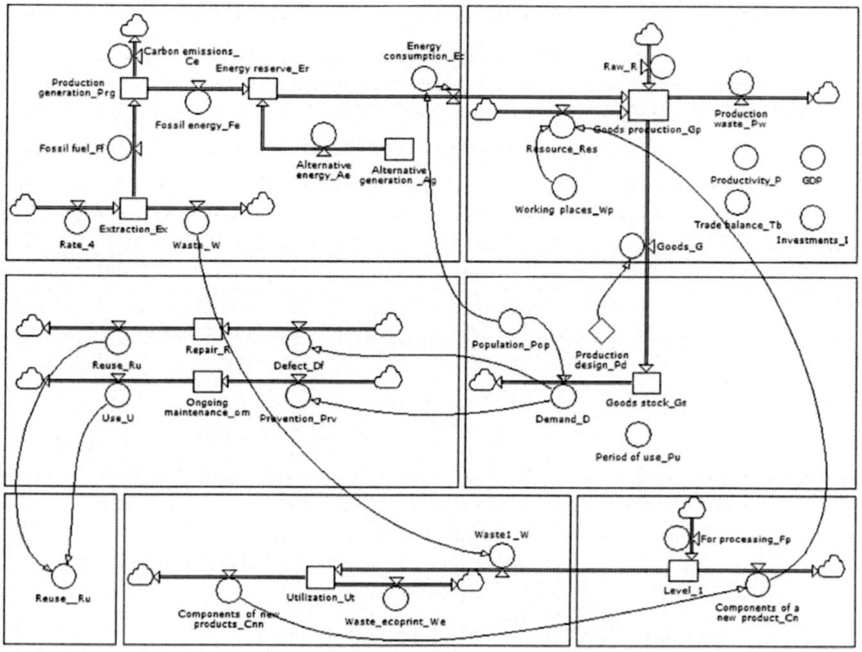


**Fig. 2.** Conceptual system-dynamic model of circular economy functioning.

additional energy purchases from partner countries are also considered. In this model block, considering the 9R principles in the model allows getting two alternatives. The **first consists of obtaining a reduced share of fossil energy (Fe) to alternative energy** (Ae) and a possible decrease in total energy reserve generation (Er) due to its economic energy consumption (Ec) in the following blocks; this will lead to a decrease in mining and generating capacities, which will allow releasing workplaces (Wp), investment (I) and other resources (Res). The second alternative also includes reducing the fossil energy in the structure of the alternative one, but not reducing the total generation, which will allow the realization of the remaining energy on international markets or to support domestic energy projects. The main goal of this link of the circular economy ecosystem is to reduce waste (W) during energy extraction and generation and increase labor productivity (P). The graphical view of the model of circular economy ecosystem functioning gave an analytical view of the key equations of the model. According to the system dynamics principles, the models' elements are represented in the form of differential equations. Let us bring it:

$$\begin{cases} Prg = (Fr - Ce - Fe)dt \\ Ex = (Rate4 - W)dt \\ Er = (Fe + Ae - Ec)dt \\ Ag = -Aedt \end{cases} \tag{1}$$

The second component of the "Production of goods" model is the most complex of the functional blocks, which is proved by the variety of technologies used in the goods' production (Gp). Almost all 9R principles can be used at this stage. The manufacturer needs to consider the efficiency and economy of input resources (Res), energy, material, investment and labor, and raw materials (R). In addition, at each link of the product production, the specifics of future processing, production waste (Pw), maintenance, repair, and extension of the period of use (Pu) of the finished goods (G) should be considered. Those thrifty principles enter the modern confrontation paradigm of "production for the sake of production," which marketers significantly support today. Thanks to production design (Pd), there is an opportunity to consider further participation of the product in the following stages after consumption, extend the product's life path, and postpone its disposal. Analytically, according to the principles of system dynamics, the proposed level in the block can be depicted as follows:

$$Pp = (Ec + R + Res - Pw - G)dt \tag{2}$$

From this stage, new opportunities for entrepreneurs appear in new forms and specifics of work. For instance, an organizational form for the development of industrial design, when there is a powerful collaboration with the manufacturer within the limits of the technical specifications and the consumer within the limits of one's desires (D) and trends. The standardization direction and product certification according to the requirements of the circular economy also become relevant, allowing the state to establish the overall production efficiency. New business structures at this stage are designed to extend the product's life and not lose any economic opportunity to use the product as a resource.

The third model block is the implemented function "Consumption" (D). At this stage, the circular economy challenges the consumer for the sake of economic and careful use of the product; the main value is the extension of the period of use (Pu) without loss of consumer properties. That will lead to an increase in the stock of goods (Gs) at the first stage, but in subsequent periods there will be a tendency to decrease production. For this, it is necessary to change the consumer's mentality, to provide one with an understanding of how to extend the product's period of use (this is usually indicated in the instructions for the product) and how the current maintenance, repair, and disposal of the product will be carried out. With a deep understanding of the product's life cycle and possibly a demonstration to the consumer of a certain amount of money that one saves when using the product sparingly, the consumer should change the paradigm of using the product. For example, countries with developing economies are the most unthrifty, while developed economies often have thrifty consumers, so they often share knowledge, cases, and logic of the principles of a thrifty use of goods. Analytically, the key dependence considering the system's elements is presented by Eq. (3).

$$Gs = (G - D)dt \tag{3}$$

New entrepreneurial formations are possible at this stage. Thanks to the economical quality product, it is clear that the demand (D) for the new product will decrease, which leads to a decrease in the production volume (Pv), and at this stage, new opportunities are formed specifically for the workplaces (Wp), who are no longer needed in these quantities at the enterprise. According to the extension of the product period of use (Pu), several specialists can immigrate to the next stage of the product's life, namely service and repair. At present, there are experience in the Bentley company (which sells aircraft engines not as a product but as an hourly service of use), several medical equipment companies (which sell equipment at a low cost, and operations on it can be done under some purchased licenses-permits for a certain type. The given examples prompt the entrepreneur to be ready to rethink their business models and, first of all, to find their place in the new reality of the circular economy.

The fourth model block is the implementation of the current maintenance (Cm) and repair (R) of the defective product (Df). Analytically, the key elements of the block are represented by the system (4). At the system entrance, the product is used by the consumer after a certain period of use (U), and at the exit, the consumer receives it in a condition close to new, if there was ongoing maintenance (Om) or in reuse (Ru), if repair and restoration work has been performed. The main result of this function is an increase in the period of use (Pu).

$$\begin{cases} R = (Df - Ru)dt \\ Cm = (Om - U)dt \end{cases} \tag{4}$$

This activity forms a significant reserve in restorative technologies and provides workplaces (Wp) released from production in a new form of business structure—service. Several world economies at a certain period of development used the approach of service and repair restoration of the consumer properties of the product, but this stage was 10–30 years. Within the next stage, this concept of post-sales service retreated due to the appearance of available raw materials and resources for producing an increasing number of goods. In addition, there was no connection between service activities and post-sales service; only warranty workshops were a continuation of the circular economy chain; however, they were not a profitable business entity at all. Today in the world, some well-known companies sell a service, not a product, transferring the right to use it under certain conditions of current maintenance (Cm) and repair (R). Such structures have significant organizational features and a complex business model. According to the authors, this function in the circular economy ecosystem is the connecting link between the producer and the recycler, ensuring the product life cycle extension. However, the service business model will be effective only in combination with all other functions of the circular economy.

The next model block is "Processing," which is the stage of the product's life cycle when the molecular structure of the product, which is no longer suitable for its intended use, does not change. During for processing stage (Fp), an attempt is made to restore several components of new products (Cn) and transfer them to the manufacturer, it is clear that the manufacturer also controls the quality of the delivered elements at the entrance to its business process, but waste generation (W) also occurs. Such processes can occur thanks to "industrial design" when, at the stage of new product development, such product components are calculated and created by engineers that would be easy to dismantle, repair, or prepare for reuse.

$$Level1 = (Fp - Cn - W)dt \tag{5}$$

For entrepreneurship, there is an opportunity at this stage to combine not only the organizational principles of processing activities but also to include the previous "Service" stage thanks to a single material and technical base; it allows to ensure the performance of exactly two functions and reduce logistics costs for the creation of the separate structures.

The last block in the circular economy ecosystem model chain is "Utilization"; the main function and purpose are clear. However, in this process, diversification of technological approaches to utilization (Ut) is required, considering the specifics of the goods elements being disposed of. Accordingly, the model block considers waste (W), which comes as a raw material resource to the manufacturer of a new product, emissions into the environment, and waste ecoprint (We); at this stage, the formation is completed. Analytically, the key dependence is given by Eq. (6).

$$Ut = (W - We - Cn)dt \tag{6}$$

Organizational business structures at this stage are quite simple in their composition and typical among themselves, so establishing one's business will be quite formulaic and clear. Of course, first, it is necessary to establish a supplier of the recycling resource, to forecast the volumes and loading of equipment, and what is no less significant, to find a consumer of the product of this activity. That is, the main function of this stage is to minimize emissions and maximize the volume of useful disposal products.

## 5 Conclusions

Limitations in use and features of the given model development will be presented in our university. First, it should be understood that the system-dynamic model is a continuous formation of indicator values during simulation, considering the period and step of modeling. Therefore, when setting up a simulation experiment, considering the available dynamics of indicators, which are presented in mathematical dependencies or data arrays considering forecasting, is necessary.

Currently, the use of the scheme of the circular economy ecosystem and its functioning model is conceptual and represents the foundation for further research. The authors consider the presence of the above functions (model blocks) of the circular economy ecosystem to be a necessary and sufficient condition; however, in the middle of each block, the number of elements (variables) may increase, which will be done by the author team in the next stages of the research. With an increased number of elements of the circular economy functioning model, a corresponding change in the analytical form of the system of differential equations will occur.

## References

1. Webster, K.: A circular economy is about the economy. Circ. Econ. Sustain. **1**(1), 115–126 (2021). https://doi.org/10.1007/s43615-021-00034-z
2. The circularity gap report 2022. https://www.circularity-gap.world/2022. Accessed 15 Aug 2022

3. Kannikar, K., Prateep, P., Lampong, K., Chainarong, K.: The 9Rs strategies for the circular economy 3.0. Psychol. Educ. J. **58**(1), 1440–1446 (2021). https://doi.org/10.17762/pae.v58i1.926
4. Palm, J., Bocken, N.: Achieving the circular economy: exploring the role of local governments, business and citizens in an urban context. Energies **14**(4), 875 (2021). https://doi.org/10.3390/en14040875
5. Charter, M. (ed.): Designing for the Circular Economy, 1st edn. Routledge, London (2018). https://doi.org/10.4324/9781315113067
6. Mayer, A., Haas, W., Wiedenhofer, D., Krausmann, F., Nuss, P., Blengini, G.: Measuring progress towards a circular economy: a monitoring framework for economy-wide material loop closing in the EU28. J. Ind. Ecol. **23**(1), 62–76 (2018). https://doi.org/10.1111/jiec.12809
7. Saeed, K.: System dynamics: a disruptive science a conversation with Jay W. Forrester, founder of the field. SSRN Electr. J. (2013). https://doi.org/10.2139/ssrn.3504274
8. Modak, P.: Practicing Circular Economy, 1st edn. CRC Press, Boca Raton (2021). https://doi.org/10.1201/9781003107248
9. Tuladhar, A., Iatridis, K., Dimov, D.: History and evolution of the circular economy and circular economy business models. In: Stefanakis, A., Nikolaou, I. (eds.) Circular Economy and Sustainability, pp. 87–106. Elsevier, Amsterdam (2022). https://doi.org/10.1016/b978-0-12-819817-9.00031-4
10. De Angelis, R.: Business Models in the Circular Economy: Concepts, Examples and Theory, 1st edn. Palgrave Pivot, Cham (2018). https://doi.org/10.1007/978-3-319-75127-6
11. Jabłoński, A., Jabłoński, M.: Creating sustainability business models in the digital and network economy. In: Social Business Models in the Digital Economy, pp. 139–147. Palgrave Macmillan, Cham (2020). https://doi.org/10.1007/978-3-030-29732-9_6
12. Jabłoński, M.: The use of sustainable business model archetypes in the design of circular business models in the digital economy. In: Nogalski, B., Szpitter, A., Jabłoński, A., Jabłoński, M. (eds.) Networked Business Models in the Circular Economy, pp. 1–18. IGI Global, Hershey (2020). https://doi.org/10.4018/978-1-5225-7850-5.ch001
13. Holdren, J.: Brief history of IPAT. J. Populat. Sustain. **2**(2), 66–74 (2018). https://doi.org/10.3197/jps.2018.2.2.66
14. Smith, E.A., Wishnie, M.: Conservation and subsistence in small-scale societies. Ann. Rev. Anthropol. **29**(1), 493–524 (2000). https://doi.org/10.1146/annurev.anthro.29.1.493
15. Pavlova, T., Zarutska, E., Pavlov, R., Kolomoichenko, O.: Ethics and law in Kant's views: the principle of complementarity. Int. J. Ethics Syst. **35**(4), 651–664 (2019). https://doi.org/10.1108/IJOES-04-2019-0080
16. Wijkman, A.; Skånberg, K.: The Circular Economy and Benefits for Society: Jobs and Climate Clear Winners in an Economy Based on Renewable Energy and Resource Efficiency. Club Rome: Winterthur, Switzerland (2015). https://clubofrome.org/wp-content/uploads/2020/03/The-Circular-Economy-and-Benefits-for-Society.pdf
17. Pavlov, R., Pavlova, T., Lemberg, A., Levkovich, O., Kurinna, I.: Influence of non-monetary information signals of the USA on the Ukrainian stock market volatility. Invest. Manag. Financ. Innov. **16**(1), 319–333 (2019). https://doi.org/10.21511/imfi.16(1).2019.25
18. Pavlov, R., Grynko, T., Pavlova, T., Levkovich, O., Pawliszczy, D.: Influence of monetary information signals of the USA on the Ukrainian stock market. Invest. Manag. Financ. Innov. **17**(4), 327–340 (2020). https://doi.org/10.21511/imfi.17(4).2020.28

# Energy Consumption at Home: Insights for Sustainable Smart Home Marketing

Şirin Gizem Köse[1](✉), Ece Özer Çizer[2], and İpek Kazançoğlu[3]

[1] MEF University, 34396 Istanbul, Turkey
siringizemkose@gmail.com
[2] Yıldız Technical University, 34200 Istanbul, Turkey
[3] Ege University, 35040 Izmir, Turkey

**Abstract.** Energy consumption has been a vital subject for both energy producers and consumers. The intersection of energy consumption and home words are of increasing importance in both literature and practice. Households try to utilize energy in the most efficient and sustainable way. On the other hand, smart home technologies which let the households control their houses are on the rise. Those technologies also help balance the energy consumption and live in a more sustainable way. This study aims to underline the importance of smart home technologies to increase energy efficiency and pave the way for a more sustainable energy management. In line with this purpose, a bibliometric study has been conducted to enlighten the literature development in energy consumption and home subjects. The results are expected to be helpful for both literature and practice as well as energy providers and consumers.

**Keywords:** Energy consumption · Sustainable energy management · Smart home

## 1 Introduction

Home energy management (HEM) refers to a group of technologies like smart thermostats, sensors, and feedback devices, that aim to control domestic energy consumption patterns and so lower peak electricity demand as well as consumer electricity costs [1]. Home energy systems are helpful to conserve energy while at the same time reducing the cost of energy and increasing comfort [2, 3]. Households can monitor and regulate their electricity consumption with the system because it is possible to visualize real-time price information on electricity [4].

As new home energy management systems continue to appear along with technological development, the growing adoption of home automation technologies cannot be overlooked. Smart home marketing is inevitably growing. Increasing internet usage, the rising need to monitor home and accelerating requests for low-carbon emission and energy-saving-oriented products have paved the way for market size development of


V. Koval et al. (Eds.): ISCMEE 2022, Lect. Notes in Managem., Industrial Engineering, pp. 104–114, 2023.
https://doi.org/10.1007/978-3-031-23463-7_7

smart homes. The market is projected to grow from USD 99.89 billion in 2021 to USD 380.52 billion in 2028 [5]. A smart home platform includes groups of software programs that enable communication between different tools [6]. The ability of smart products to promote energy savings and demand-side control is one of their significant advantages [7], but smart technology's assistance is not limited to these. Smart homes not only help in energy consumption and management, but are also beneficial for assisted living, security and remote monitoring [8].

Therefore, marketing sustainable home energy systems is crucial more than ever. In the light of these developments, this study aims to shed light on the literature side of the energy consumption and home concepts and provide understanding for sustainable smart home marketing. The place of those concepts in the literature is explored via bibliometric analysis.

## 2 Background

Almost all activities we do at home have a potential to consume energy. Households do have some habits to decrease energy consumption, for instance, Pierce et al. [9] developed a vocabulary to present energy-conserving activities that include the words "cutting, trimming, switching, upgrading, shifting." Both consumers, researchers and industry have devoted many hours to study on how to decrease energy consumption. There has been a concentration on numerous methods to address needless energy consumption and guarantee a healthy living environment for green smart cities. In order to successfully monitor energy usage, one of these strategies is Smart Home Energy Management Systems (SHEMs), which make electric home appliances and sensor nodes autonomous tools [10]. Automation, multi-functionality, adaptability, interactivity and efficiency are the five essential characteristics that describe smart homes [11]. It can be inferred that smart home technology provides the benefit of saving time and cost, which links those systems to sustainable energy management. However, assessing the sustainability advantages of smart home technology is often neglected [12].

There has been a significant alteration in the energy industry in recent years. As a result of the emergence of a decarbonized economy, solar households that both produce and consume energy have become popular [13]. The term to describe people who both produce and consume energy is "prosumer", whose increase in the marketplace leads novel approaches to almost every industry and energy management is no exception. The electric utility industry of today is projected to experience considerable changes over the next few decades as the number of prosumers increases, bringing opportunities for system greening [14]. Prosumers are defined as more proactive [15]. Parag and Sovacool [14] suggested peer-to-peer, prosumer-to-grid and organized prosumer groups as prosumer market models in the energy industry.

Improving energy efficiency is useful for decreasing energy consumption that helps households to cut down energy-related expenses [16]. Building and home automation have the potential to have an impact on the energy efficiency of existing and future buildings by informing users about their current consumption patterns, encouraging more energy-efficient behaviors, and proactive changing user actions for reducing the associated energy wastes [17]. As the product of digital and connected consumerism, smart home promises new ways of convenience and also reduced and more efficient energy consumption is an accompanying benefit [18].

In fact, consumer decisions regarding whether to buy or not, which products consumers prefer to buy and how they use those products will determine how much energy is saved and, eventually, how much greenhouse gas emissions are produced [19]. Thus, the appropriateness for smart home technology for the households are being evaluated by households. For instance, Korneeva et al. [20] presented that the most significant aspect to increase satisfaction of smart home is to ensure the reliability of the smart home systems. Balta-Ozkan et al. [21] worked on the social obstructions to the acceptance of smart homes. They found that suitability to lifestyle, administration, interoperability, reliability, privacy and cost are the hindering factors. On the other hand, Paetz et al. [22] investigated consumers' responses to an energy administration system which is capable of optimizing electricity consumption via utilizing several information and communication technologies. The consumers generally had positive attitudes toward smart home atmosphere and found the technology helpful, particularly saving money. But the participants have privacy issues in their mind like the other studies in this area. Findings of Sanguinette et al. [23] point that the sense of protecting the household accelerates the adoption of smart homes whereas lack of technological knowledge and cost associated with those products decrease the likelihood to adopt smart home. In addition, Jensen et al. [12] formed three smart home personas in their study; which are he helper, optimizer and hedonist.

## 3 Research Approach

In this study, a bibliometric analysis was conducted on the studies in the literature from the past to the present on the concepts of "smart home" and "energy consumption". For analysis, studies between 1980 and 2022 containing the keywords "smart home" and "energy consumption" were downloaded from the Web of Science database. A total of 3516 documents were reached. Of these documents, 1976 are articles, 1431 are proceedings papers and 109 are other documents. Bibliometric analysis was applied to the articles with the R Studio analysis program.

In Fig. 1, the 50 most repetitive words in the data set were examined. Among these words, it is seen that words such as "smart home", "energy efficiency", "smart grid", "internet of the things" and "demand response" are common words in articles.

In the word tree shown in Fig. 2, the distribution of the frequencies of the words is given in order. In addition to the leading words in the word cloud "energy consumption", "energy management", "energy", "energy saving" and "home appliances" are frequently discussed in the literature.

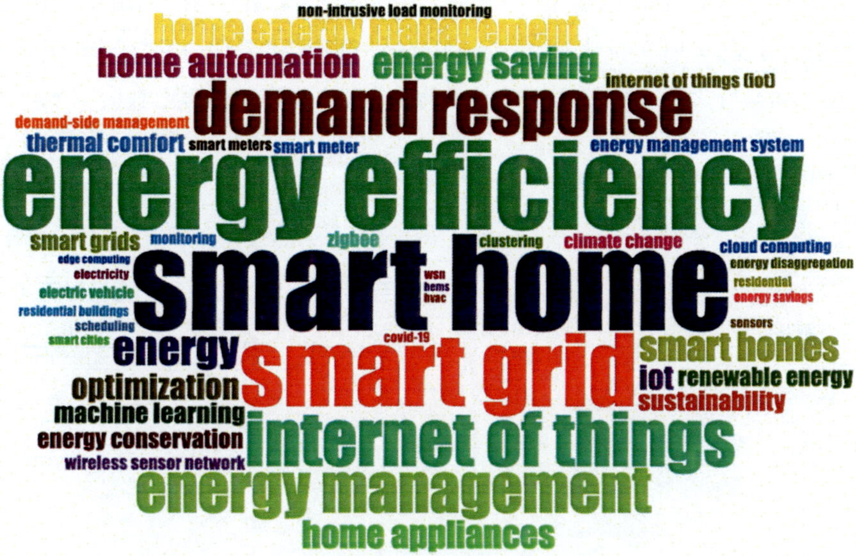


**Fig. 1.** Word cloud.

**Fig. 2.** Word tree.

Since 2012, studies on energy consumption—smart home have been carried out in the literature. In Fig. 3, the distribution of the trending topics in the last 10 years is given. During the first six-year period, "zigbee", "wireless sensor networks" and "home automation" concepts were discussed. Between 2014–2018 it is seen that the concepts of "energy saving" and "energy management" have come to the fore. There

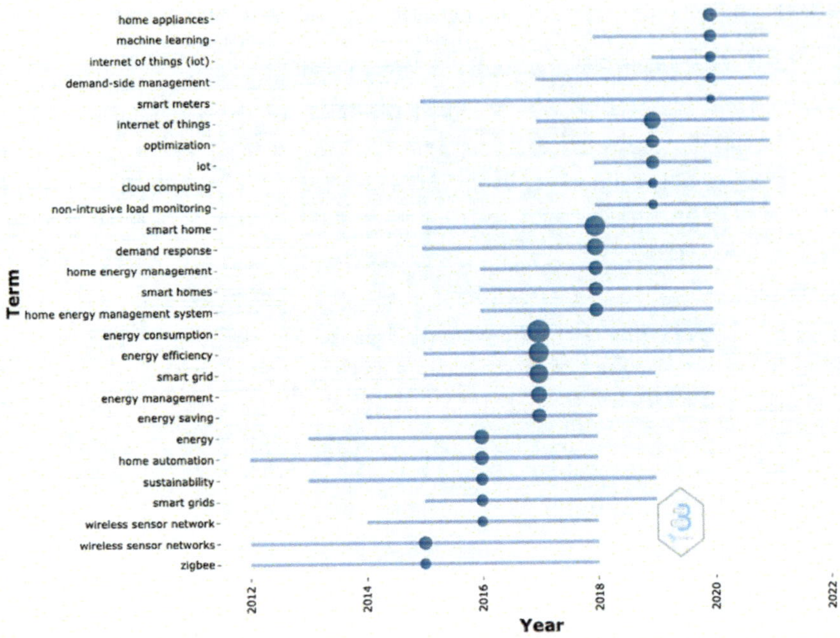


**Fig. 3.** Trend topics.

has been an increase in the number of studies on "sustainability" in 2016. Between 2016–2020, publications on "energy consumption", "energy efficiency", "smart grid", "energy management" and "energy saving" topics were produced. In 2018, there was an increase in studies on "smart home". Between 2018 and 2020, it is seen that "internet of things" studies accelerated. After the 2020 "home appliances", "machine learning", "demand-side management", and "smart meters" topics gain popularity.

In the study, a thematic map divided into four topological regions based on density and centrality was created and is shown in Fig. 4. The upper right quadrant shows "motor" or "driving" themes indicated with high intensity and centrality; These topics, which include "energy consumption", "home appliances" and "smart meter", need to be further developed given their importance for future research. In the upper left quadrant, there are "energy efficiency", "energy" and "sustainability" topics, which are high density but low centrality. These topics illustrate specific and underrepresented issues that are areas of rapid development. Themes in the lower left quadrant are shown with low centrality and density. This area, which includes "smart grid", "demand response" and "home energy management" topics, now includes topics with a downward trend. Finally, the lower right quadrant contains the core themes shown with high centrality but low intensity. This field, which includes the subjects of "smart home", "internet of things" and "energy management", includes subjects of general importance for the studies to be carried out.

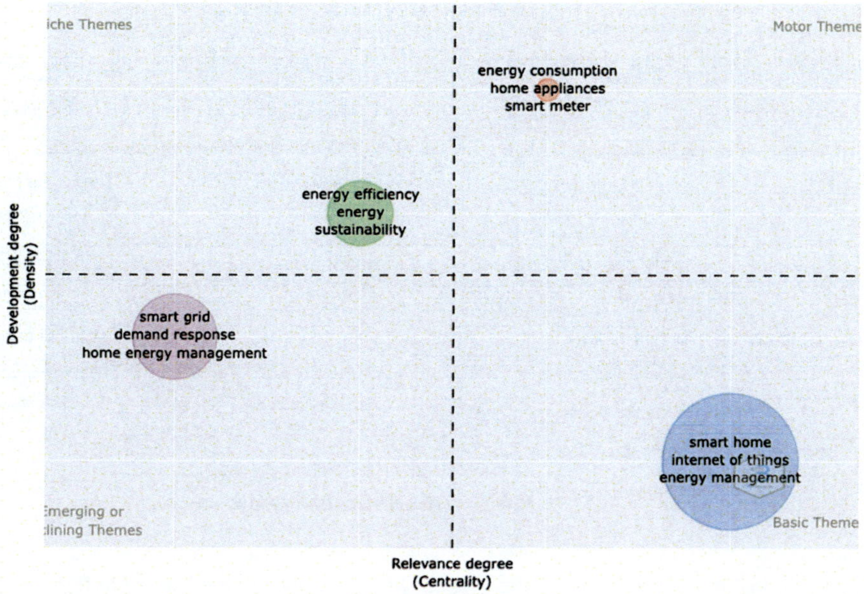


**Fig. 4.** Thematic map.

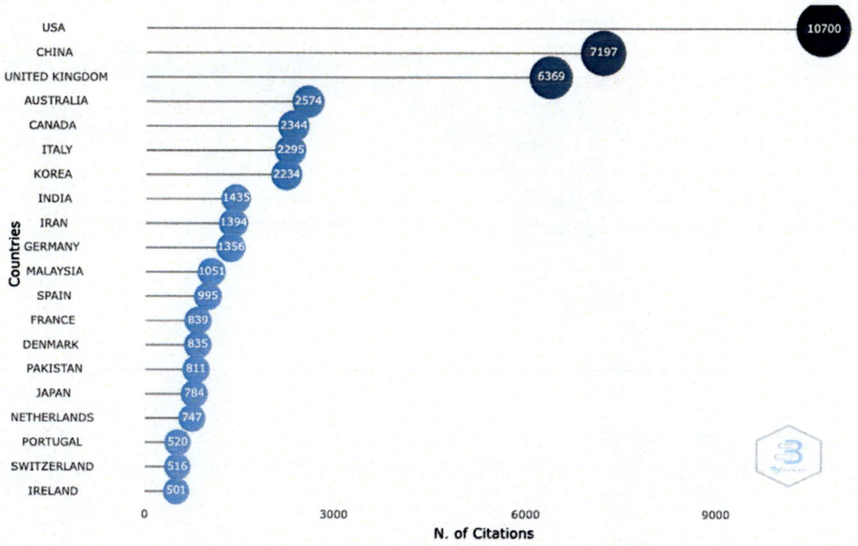


**Fig. 5.** Most cited countries.

According to Fig. 5, it is seen that the most cited publications in studies on "smart home" and "energy consumption" are presented by authors from USA, China, United Kingdom and Australia, respectively.

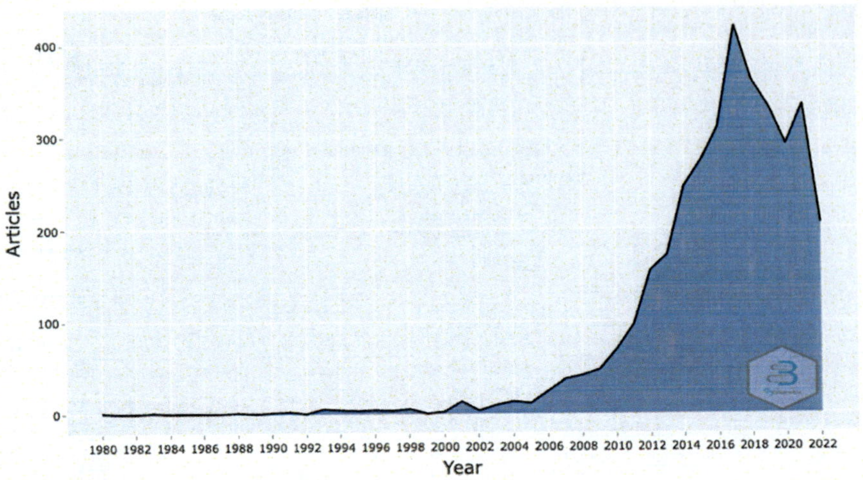


**Fig. 6.** Annual scientific production.

In Fig. 6, it is seen that the first studies on "smart home" and "energy consumption" were put forward in 1993. In addition, the number of studies increased rapidly starting from 2010 and reaching the highest level in 2016.

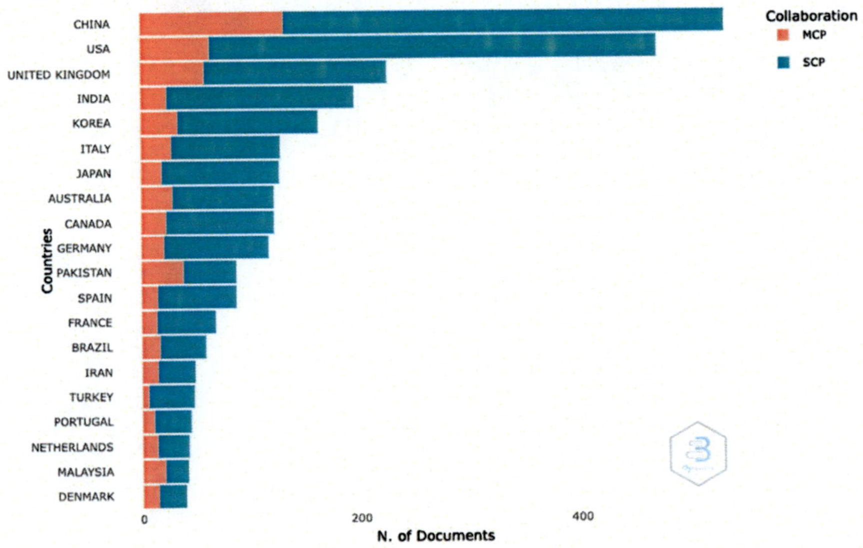


**Fig. 7.** Corresponding author's country.

The countries of the responsible authors for the publications were examined and shown in Fig. 7. According to Fig. 7, it is seen that the country with the most publications is the China. China is followed by United States of America, United Kingdom, and India, respectively. While the green colors in the table show that the broadcasts are made in a single country, the orange-colored parts indicate the number of broadcasts from multiple countries.

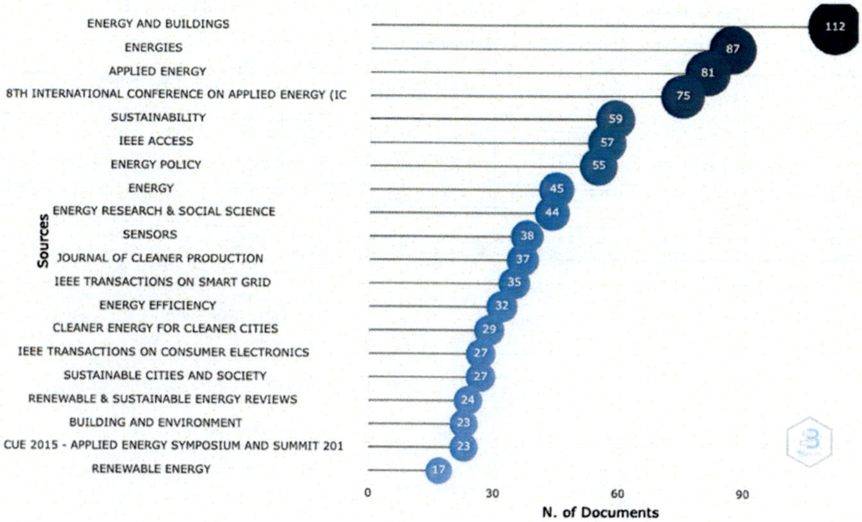


**Fig. 8.** Most relevant sources.

The most preferred journals were searched to publish studies on "smart home" and "energy consumption". According to the most relevant sources in Fig. 8, "Energy and Buildings", "Energies", and "Applied Energy" are the most popular scientific journals, respectively.

Figure 9 shows the closeness of the keywords used in the studies in the literature with each other. Keywords are divided into four separate clusters. Each subset is colored with a different color. The red cluster shows the most frequently used word network and the words in this network are concentrated under the keyword "energy consumption", while in the blue cluster it is observed that the keywords are distributed around "smart home". The words in the purple cluster are concentrated around the keyword "smart grid" and the words in the green cluster are concentrated around the keyword "energy efficiency".

## 4 Conclusion

The increase in home energy prices has led to an increased desire of households to reduce costs by using less energy. In addition, the correct management of energy is also important for environmental sustainability. At this point, smart homes that contribute to both reducing costs and environmental sustainability come to the fore.

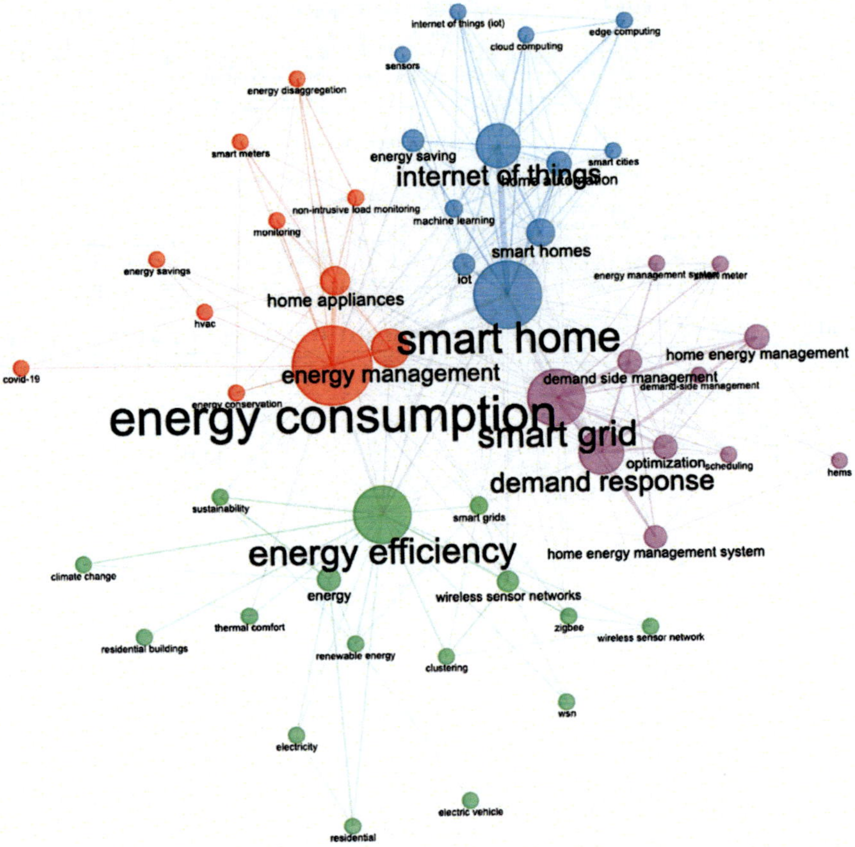


**Fig. 9.** Co-occurrence network.

Technologies used in homes have facilitated many tasks in the life of households and have also saved time. However, the desire for convenience often faced problems in advancing at the common point with energy saving. Nowadays, smart home technologies bring convenience to households and the advantage of saving energy. For this reason, sustainable smart home technologies continue to increase their importance both in practice and in the literature. In this regard, this research aims to shed light on the literature of energy consumption at home through the lens of suitable smart home marketing via bibliometric analysis.

**The findings of the bibliometric analysis show that the most repetitive words are "smart home", "energy efficiency", "smart grid", "internet of the things" and "demand** response". Those words demonstrate that when it comes to smart home and energy consumption, smart home is considered as an answer in the literature. The words "the internet of things" also underlines the effect of technology in sustainable energy consumption. In addition, the word tree presents that energy consumption", "energy management", "energy", "energy saving" and "home appliances" are frequently explored in the literature.

Trends topics also provide a comprehensive picture for the literature development in the area. It can be seen that "energy saving" and "energy management" were the popular subjects during the years between 2014–2018 whereas between 2018 and 2020, "internet of things" studies came to front. The topics of home appliances", "machine learning", "demand-side management", and "smart meters" are gaining ground after 2020. Furthermore, according to the thematic map, includes the subjects of "smart home", "internet of things" and "energy management" have general importance while smart grid", "demand response" and "home energy management" areas present decreasing trend. On the other hand, "energy efficiency", "energy" and "sustainability" areas are promising fields that are developing expeditiously. Also, "energy consumption", "home appliances" and "smart meter" subjects need to be improved because they are still significant.

At the same time, authors from USA, China, United Kingdom and Australia have the most cited papers in smart home and energy consumption area of study. Moreover, China has the most publications worldwide. United States of America, United Kingdom, and India are in order of precedence after China. Beginning in 2010, the quantity of research gradually climbed until it peaked in 2016. It is seen that 2016 was the golden year for smart home and energy consumption studies. "Energy and Buildings", "Energies", and "Applied Energy" are the most chosen journals to publish studies in the "smart home" and "energy consumption" area. Those journals include a lot of studies focused on smart home and energy consumption.

## References


1. LaMarche, J., Cheney, K., Christian, S., Roth, K.: Home Energy Management: Products & Trends, pp. 165–175. University of California (2021)
2. Wei, Y.M., Liu, L.C., Fan, Y., Wu, G.: The impact of lifestyle on energy use and $CO_2$ emission: an empirical analysis of China's residents. Energy Policy **35**(1), 247–25 (2007)
3. Guibourdenche, J., Salembier, P., Poizat, G., Haradji, Y., Galbat, M.: A contextual approach to home energy management systems automation in daily practices. In: Proceedings of the European Conference on Cognitive Ergonomics (2015)
4. Park, E.S., Hwang, B., Ko, K., Kim, D: Consumer acceptance analysis of the home energy management system. Sustainability **9**(12), 2351 (2017)
5. Smart Home Market Size, Share & COVID-19 Impact Analysis, By Product (Home Monitoring/Security, Smart Lighting, Entertainment, Smart Appliances and Others) and Regional Forecast, 2021–2028, Fortune Business Insight, February 2022
6. Ford, R., Karlin, B., Sanguinetti, A., Nersesyan, A., Pritoni M.: Assessing Players, Products, and Perceptions of Home Energy Management, Pacific Gas and Electric, San Francisco, CA (2016)
7. Ford, R., Pritoni, M., Sanguinetti, A., Karlin, B.: Categories and functionality of smart home technology for energy management. Build Environ. **123**, 543–554 (2017)
8. Balta-Ozkan, N., Davidson, R., Bicket, M., Whitmarsh, L: The development of smart homes market in the UK. Energy **60**, 361–372 (2013)
9. Pierce, J., Schiano, D.J., Paulos, E: Home, habits, and energy: examining domestic interactions and energy consumption. In Proceedings of the SIGCHI conference on human factors in computing systems, 1985–1994 (2010)
10. Aliero, M.S., Qureshi, K.N., Pasha, M.F., Jeon, G.: Smart home energy management systems in Internet of Things networks for green cities demands and services. Environ. Technol. Innov. **22**, 101443 (2021)

11. Lê, Q., Nguyen, H.B., Barnett, T.: Smart homes for older people: positive aging in a digital world. Fut Internet **4**(2), 607–617 (2012)
12. Jensen, R.H., Strengers, Y., Kjeldskov, J., Nicholls, L., Skov, M.B.: Designing the desirable smart home: a study of household experiences and energy consumption impacts. In: Proceedings of the 2018 CHI Conference on Human Factors in Computing Systems, 1–14 (2018)
13. Miller, W., Senadeera, M.: Social transition from energy consumers to prosumers: rethinking the purpose and functionality of eco-feedback technologies. Sustain. Cities Soc. **35**, 615–625 (2017)
14. Parag, Y., Sovacool, B.K.: Electricity market design for the prosumer era. Nat. Energy **1**(4), 1–6 (2016)
15. Sousa, T., Soares, T., Pinson, P., Moret, F., Baroche, T., Sorin, E.: Peer-to-peer and community-based markets: a comprehensive review. Renew. Sustain. Energy Rev. **104**, 367–378 (2019)
16. Santos, C.N., Cetin, K.S., Salehi, H.: Energy-efficient technology retrofit investment behaviors of Midwest households in lower and higher income regions. Sustain. Cities Soc. 104141 (2022)
17. Bonino, D., Corno, F., De Russis, L.: Home energy consumption feedback: a user survey. Energy Build. **47**, 383–393 (2012)
18. Strengers, Y., Nicholls, L: Convenience and energy consumption in the smart home of the future: industry visions from Australia and beyond. Energy Res. Soc. Sci. **32**, 86–93 (2017)
19. Yu, B., Zhang, J., Fujiwara, A.: Representing in-home and out-of-home energy consumption behavior in Beijing. Energy Policy **39**(7), 4168–4177 (2011)
20. Korneeva, E., Olinder, N., Strielkowski, W.: Consumer attitudes to the smart home technologies and the Internet of Things (IoT). Energies **14**(23), 7913 (2021)
21. Balta-Ozkan, N., Davidson, R., Bicket, M., Whitmarsh, L.: Social barriers to the adoption of smart homes. Energy Policy **63**, 363–374 (2013)
22. Paetz, A.G., Dütschke, E., Fichtner, W.: Smart homes as a means to sustainable energy consumption: a study of consumer perceptions. J. Consum. Policy **35**(1), 23–41. (2012)
23. Sanguinetti, A., Karlin, B., Ford, R.: Smart home consumers: comparing self-reported and observed attitudes (2017)

# Development of Sustainable Transport in Ukraine: Evolution of the Concept, Actions and Indicators

Olena Nykyforuk(✉), Olha Stasiuk, Larysa Chmyrova, and Natalia Fediai

State Organization "Institute of the Economy and Forecasting of the National Academy of Sciences of Ukraine", Panasa Myrnoho 26, Kyiv 01011, Ukraine
chaicynan@ukr.net, elena.nikiforuk@gmail.com

**Abstract.** The article reviews and singles out four stages for the concept of sustainable transport in Ukraine to be established as follows: the first is to shape the sustainable development key principles; the second is to draft the millennia objectives by 2015; the third is to identify the sustainable development key objectives by 2030; and the fourth is to focus the Ukraine's transport policy on the Green Deal. Each of the stages evaluates the key strategic documents, laying down the institutional grounds for the development of sustainable transport. A system of indicators is produced to assess the sustainable development of road transport, thus enabling the monitoring of transformational developments and, following its findings, adopting decisions facilitating the national economy's sustainable development. Key objectives are determined for Ukraine to ensure the development of its sustainable transport by observing the effort towards integration into Europe in the light of the EU Agreement and Green Deal, which constitute the underlying concept for the changes to make the EU a climate-neutral entity by reducing by 2015 the $CO_2$ emissions, including those originating from road transport activities, which account for a major proportion of such emissions..

**Keywords:** Development of sustainable transport · Institutional background · Goals sustainable development · Greenhouse gases · System of indicators

## 1 Introduction

The development of modern civilization has led to challenges related to the exhaustion of opportunities for development oriented to the growth of material consumption. The model of the economy based on long-term and unlimited economic growth, led to a rapid reduction in the reserves of many natural and energy resources, critical poverty in some countries, and increased economic and social inequality between countries. Such a model led to critical economic and social consequences in the world and exacerbated environmental problems manifested in climate change leading to global warming and environmental degradation. The aggravation of economic, social and environmental problems prompted scientists to develop new concepts, and world leaders to begin


V. Koval et al. (Eds.): ISCMEE 2022, Lect. Notes in Managem., Industrial Engineering, pp. 115–134, 2023.
https://doi.org/10.1007/978-3-031-23463-7_8

cooperation aimed at coordinating environmental policy on a global scale. This began the elaboration of the concept of sustainable development in the world, including in Ukraine.

Environmental changes are caused by, among other things, the use of fossil fuels, inefficient energy consumption due to mass consumption of non-renewable energy resources, and increased motorization causing the production of greenhouse gases (GHG), which intensify the greenhouse effect with an extremely negative impact on the environment, threatening people's health and welfare.

Ensuring the sustainable development of road transport is one of the strategic tasks of Ukraine's economy, as road transport is one of the most important sectors providing fast and convenient transportation, and meeting economic and socio-cultural needs of society. However, it is the main consumer of oil (50–60% of total consumption), and is often a source of local air pollution and environmental acidification in most developing countries, while further expansion of automobile markets due to population growth and urbanization will significantly increase atmospheric air pollution, especially in cities. Permanent modernization of road transport, in particular, that of the worn-out infrastructure of roads and vehicles, must be carried out with due regard to the global trends of sustainable development aimed at progress in greening economy. The primary basis for this is the formation of an appropriate institutional basis at the national level.

The purpose of this work is to identify a gap in the current system of strategic documents, to provide recommendations on the development of sustainable transport of Ukraine based on the study of peculiarities in the formation of an appropriate institutional basis, with due consideration of Ukraine's Eurointegration course, and to monitor the achievement of the Sustainable Development Goals. The paper is organized as follows. First, we made a literature review of scientific works revealing the theoretical and practical results of research on the development of sustainable transport (Sect. 2). Then we described our research methodology (Sect. 3). In the research approach (Sect. 4), we presented analytical data for Ukraine and the EU on greenhouse gas emissions, including those produced by transport, revealed the peculiarities in the formation of the institutional basis (according to the identified stages), and provided the results of assessment of the economic, social and environmental components of ensuring sustainable development of Ukraine's road transport based of the proposed extended system of monitoring indicators. Then, we drew various conclusions based on the conducted research and formulated various tasks identified for Ukraine as to the development of sustainable transport at the current stage (Sect. 5).

## 2 Literature Review

The methodology of provision and conceptual foundations of sustainable development are quite widely represented in the scientific works of researchers from many countries in the context of the scale of climatic change, which actualizes these issues. Most authors, when defining sustainable development, emphasize a qualitative change in the development characteristics as regards the combination of social, economic, and environmental components.

Thus, Buryk [1] defined conceptual principles for the implementation of a strategic approach to the state regulation of sustainable development of Ukraine, as a result of which the need is emphasized to develop a system of strategic documents, which should include an industry strategy for the sustainable development of road transport.

Soergel et al. [2] quantified the results of measures to achieve the Sustainable Development Goals, by showing that they significantly speeded up the progress in many aspects, using for evaluation an integrated modeling framework that covers 56 indicators for all 17 the Sustainable Development Goals At the same time, the authors concluded that there is an insufficient number of identified indicators to achieve the goals.

The research results by Ruggerio [3] show the potential of the concept of sustainability as a still-developing framework for scientific research and environmental management.

Reyers et al. [4] review the past decade's widespread application of resilience science in sustainable development practice and examine whether and how resilience is reshaping this practice to better engage in complex contexts.

Based on the methods of quantitative scientometric analysis and qualitative discussion, Zhao et al. [5] analyzed current research topics in the field of sustainable transport, and identified four promising areas of research related to the social sustainability of transport, one of which is the application of information and communication technologies.

Scientists cover a wide range of issues related to sustainable development in European countries and Ukraine, including: reduction of greenhouse gas emissions via identifying the potential of industry policies and finding a compromise between them [6]; Ukraine's potential opportunities thanks to its participation in the new global climate agreement (Paris Agreement), which can be used to modernize this country's economy and raise the welfare of its population, and will contribute to the implementation of a low-carbon development strategy [7, 8]; the impact of different carbon pricing mechanisms on poverty and distribution, which corresponds to the achievement of the goals of the Paris Agreement [9]; economic assessment of Ukraine's low-emission development scenarios [10], etc.

A number of scientists deal with general issues of sustainable development of transport, some investigate the problems of managing the sustainable development of railway transport companies [11], others develop a conceptual approach to the implementation of the paradigm of sustainable development of waterway transport [12], study theoretical approaches to defining the concept of "development of sustainable transport" [13], investigate the issue of transport development [14] and, in particular, the sustainable development of road transport [15]. There are also scientific works that cover the theoretical foundations of institutional support for sustainable development [16].

## 3 Research Methodology

To study the formation of institutional background for the development of sustainable transport of Ukraine, a methodological approach was developed, which made it possible to achieve the established goal. The essence of the developed methodological approach is as follows:

1. analytical substantiation of the importance of researching greenhouse gas emissions in the energy sectors of Ukraine and the EU, which include transport;
2. periodization in the formation of institutional foundation for the development of sustainable transport of Ukraine and the use of elements of content-analysis method for the adoption of strategic documents related to the development of sustainable transport for the evaluation of the identified stages. Evaluation is carried out in terms of certain aspects that characterize each stage, in particular: intensity of the adoption of strategic documents; content and compliance with the Sustainable Development Goals; availability of regulatory provisions regarding sustainable development, including the development of sustainable transport; and systematic extension of adopted strategic documents on sustainable development, including on the development of sustainable transport;
3. extended systems of indicators for assessing the sustainable development of road transport, within the nine goals of the sustainable development of road transport identified in [15], which allows monitoring transformational changes, assessing the sustainable development of road transport in dynamics and, on this basis, making decisions about success in ensuring the sustainable development of road transport;
4. assessing the achievement of the goals of sustainable development of road transport based on a methodology that focuses on overtime change and considers the direction of the indicator's progress (whether the indicator approaches the goal of sustainable development or moves away from it), and the speed of this progress based on the formula of "compound annual growth rate" (CAGR);
5. determining the guidelines of ensuring the sustainable development of road transport of Ukraine, considering the course of Eurointegration.

## 4 Research Approach

The main source of GHG emissions in Ukraine is the energy sectors (which include transport) [17, p. 8], whose share in 1990 was 45.6%, and in 2018 increased to 51.8%. Transport accounted for 18.7% and 20%, respectively, in the overall structure of energy-related sectors. In Ukraine, GHG emissions from all types of transport in 2018 amounted to 34.96 million tons of $CO_2$-eq, which is by 68.7% less than in 1990. The largest share of GHG emissions by the transport sector is accounted for by road transport, which in 2018 amounted to 70.7% (24.72 million tons of $CO_2$-eq), while in 1990 it was 54.9%. Starting from 1990 and until 2000, in absolute terms, GHG emission by road transport sharply decreased (from 61.37 million tons of $CO_2$-eq in 1990 to 15.78 million tons of $CO_2$-eq in 2000), while the general trend towards increased GHG emission by road transport was further observed.

It should be noted that today, GHG emissions from road transport have decreased by 59.7% compared to 1990, which corresponds to the goals established in the Paris Agreement [17]. As of today, in the EU, transport accounts for about 25% of greenhouse gas emissions, and the share of greenhouse gas emissions from road transport in total structure of emissions by transport mode is about 72%, while in Ukraine these shares are respectively 10%, (however, in large cities, this share is about 85%) and 71%. At the same time, total GHG emission in the EU is decreasing in almost all industries, while emission from transport in the overall structure of emissions is on the contrary increasing, which is related to increased number of purchased cars.

Due to the growing GHG emissions, in many developed countries, which constitute 80% of the global automobile market, there is a strengthening of environmental regulations for motor vehicles. In this regard, most countries and car manufacturers have committed to abandon the sale and production of cars with gasoline and diesel engines by 2040 and to switch to the use of cars operating on renewable energy, which produce no atmospheric emissions, being this a major advantage in view of global warming. In the context of strengthening environmental standards, the EU introduced Euro-6 standards in September 2015, while Ukraine switched to the Euro-4 standard in 2014, and to the Euro-5 standard in 2016 (established in the EU in 2009). So, in Ukraine, Euro eco-standards are implemented with an 8 to 14-year delay, but compared to the EU, at a much faster pace. The first reason for such a situation is the fact that when eco-standards began to be introduced in Europe, Ukraine was establishing a new independent state, so the introduction of the above standards was not among top priorities. Ukraine only began to introduce European standards starting with Euro-2 in 2006; however, it should be noted that, while, in the EU, the transition from Euro-3 to Euro-5 took 11 years, Ukraine had to overcome this path in 4 years. Ukraine's significant lag in the introduction of eco-standards is due to the fact that even now adjusting production to new requirements remains a difficult task demanding new technological processes.

According to Annex B to the Doha Amendment to the Kyoto Protocol, Ukraine had a permitted volume of greenhouse gas emissions in 2020 at 76% of the 1990 amount. The environmental component played an important role in achieving the goals of the Paris Agreement. Ukraine initially undertook not to exceed, in 2030, 60% of the amount of greenhouse gas emissions of 1990. However, later, in July 2021, Ukraine's Updated Nationally Determined Contribution to the Paris Agreement was approved, which foresees the need to reduce, by 2030, this country's greenhouse gas emissions to 35% compared to 1990 via decarbonization and development of renewable energy [18]. In Ukraine's greenhouse gas emissions, the largest share (about 70%) is accounted for by carbon dioxide ($CO_2$).

For a better understanding of the process of establishing the institutional foundation for the development of Ukraine's sustainable transport, depending on the intensity and efficiency of decision-making, the authors have distinguished four stages (Fig. 1). During the first stage (1987–2000) Ukraine's efforts, like those of the world, were aimed at ensuring sustainable development via overcoming negative environmental phenomena and included the following guidelines: (1) introduction of an environmental tax for water and air pollution; (2) inclusion in the state budget in 1994 of a special expenditure item of "Environmental protection", accounting for about 0.12% of GNP.

Also, Ukraine received grant funds to ensure sustainable development. In 1999, with the aim of ensuring a way out of the crisis and creating proper conditions for the sustainable development of settlements, the Concept of Sustainable Development of Settlements was adopted, which provided for the improvement of transport infrastructure, in particular the development of all types of public passenger transport, provision of urban and intercity transport, as well as transport for connecting rural settlements with urban ones and with each other.

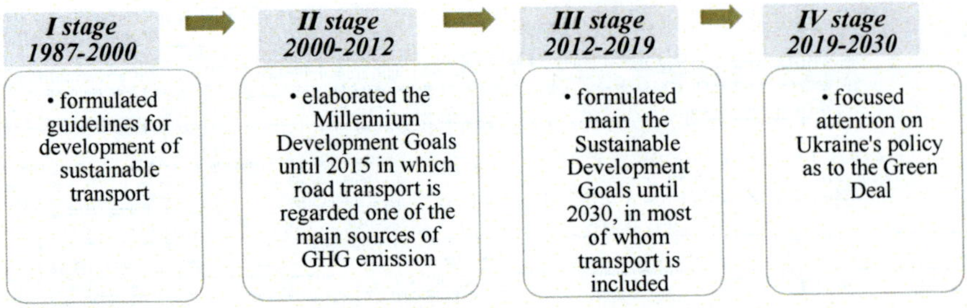


**Fig. 1.** Stages in development of the institutional foundation of sustainable transport of Ukraine.

At the same time, an important innovation in the world was the holding of conferences on transport and the environment, at which it was emphasized that the main polluter in large cities is road transport, which accounted for 75–90% of total pollution.

During the second stage (2000–2012) (see Fig. 1), Ukraine joined the UN Millennium Declaration, which defined eight Millennium Development Goals for 2010–2015 aimed at eliminating the main obstacles on the way to a dignified life for every person in society. Within the global goals, Ukraine selected seven Millennium Development Goals for Ukraine, with regard to the peculiarities of this country's economy. At this stage, the development of sustainable transport was not considered in the Millennium Development Goals, either in the world or in Ukraine. Only the task of stabilizing greenhouse gas emissions by 2020 at a level by 20% lower than the 1990 level was defined, and an indicator of the amount of harmful atmospheric emissions from mobile pollution sources was singled out for its monitoring (see Table 1) [19].

**Table 1.** Amount of harmful atmospheric emissions from mobile pollution sources, million tons per year*.

| 2000 | 2001 | 2002 | 2003 | 2004 | 2005 | 2006 | 2007 | 2008 | 2009 | 2010 | 2011 | 2012 | 2013 | 2014 | 2015 |
|---|---|---|---|---|---|---|---|---|---|---|---|---|---|---|---|
| … | 1.99 | 2.03 | 2.10 | 2.17 | 2.15 | 2.20 | 2.56 | 2.68 | 2.51 | 2.54 | 2.50 | 2.48 | 2.40 | 2.00 | 3.20 |

* For 2001–2002, data on road transport are displayed; since 2003 - for road, railway, air, and waterway transport; and since 2007 - on road, railway, air, and waterway transport and production equipment. Data from 2015 is still incomplete.
*Source* [17]

Despite individual measures aimed at improving the environmental condition, atmospheric air pollution by road transport remained one of the main challenges. This situation was primarily due to the lack of strategic documents on sustainable development, such as the Strategy and Concept. At the same time, sustainable development was a basis for shaping the policies of the EU countries and a number of post-Soviet countries, such as Belarus, Kyrgyzstan, Uzbekistan, Kazakhstan, etc., in which Sustainable Development Strategies were adopted.

Starting with 2001, several attempts were made in Ukraine to formulate and approve the Concept of Sustainable Development at the legislative level, which at that stage was not implemented. Instead, the Comprehensive Program for the Implementation of Decisions for the Period 2003–2015 adopted at the World Summit on Sustainable Development was approved at the national level. The implementation of this program was one of the priority areas in the activities of Ukraine's central and local executive authorities. However, in 2011 the program was closed [20].

In order to study the progress in the implementation of the national policy of sustainable development, to assess the progress in the creation of economic, social, environmental and other prerequisites for the transition to sustainable development of Ukraine, and to prepare the Government's proposals for institutional support, at the Cabinet of Ministers of Ukraine, the National Council for Sustainable Development of Ukraine was set up as a consultative advisory body, which was however dissolved in 2013. To solve the issues related to the increase in $CO_2$ emissions that have a negative impact on climate change, Ukraine ratified the Kyoto Protocol in 2005, according to which this country was classified as one making a transition to a market economy. In this regard, Ukraine's obligation, at the first stage of the Kyoto Protocol (2008–2012), was not to exceed the 1990 amount of greenhouse gas emissions.

During the third stage (2012–2019) (Fig. 2), a number of strategic documents were adopted, which contained guidelines to ensure both the sustainable development of Ukraine and the sustainable development of road transport. The Association Agreement between Ukraine and the EU was ratified on September 16, 2014. The main areas of cooperation in the transport sector are defined as follows [21]:

- development of a sustainable national transport policy covering all modes of transport;
- development of sector strategies in light of the national transport policy;
- development of the multimodal transport network connected to the Trans European Transport Network (TEN-T) and improvement of infrastructure policy;
- development of funding strategies focusing on maintenance, capacity constraints and missing link infrastructure as well as activating and promoting the participation of the private sector in transport projects;
- accession to relevant international transport organisations and agreements;
- scientific and technical cooperation and exchange of information for the development and improvement of technologies in the field of transport;
- scientific and technical cooperation and exchange of information for the development and improvement of technologies, such as intelligent transport systems;
- promotion of the use of intelligent transport systems and information technology in managing and operating all modes of transport.

After the Rio + 20 Conference, a national vision of the key goals and objectives in the development of Ukraine after 2015 was formed in Ukraine, and the Sustainable Development Strategy "Ukraine – 2020" [19] was adopted, which identified four vectors of movement towards sustainable development, namely: development vector; security vector; responsibility vector; and pride vector, within which the implementation of 62 reforms is provided. Transport was included in the development vector, which included the reform of transport infrastructure. Along with this, the priority reforms defined in the Strategy-2020, which influence the development of sustainable transport, include

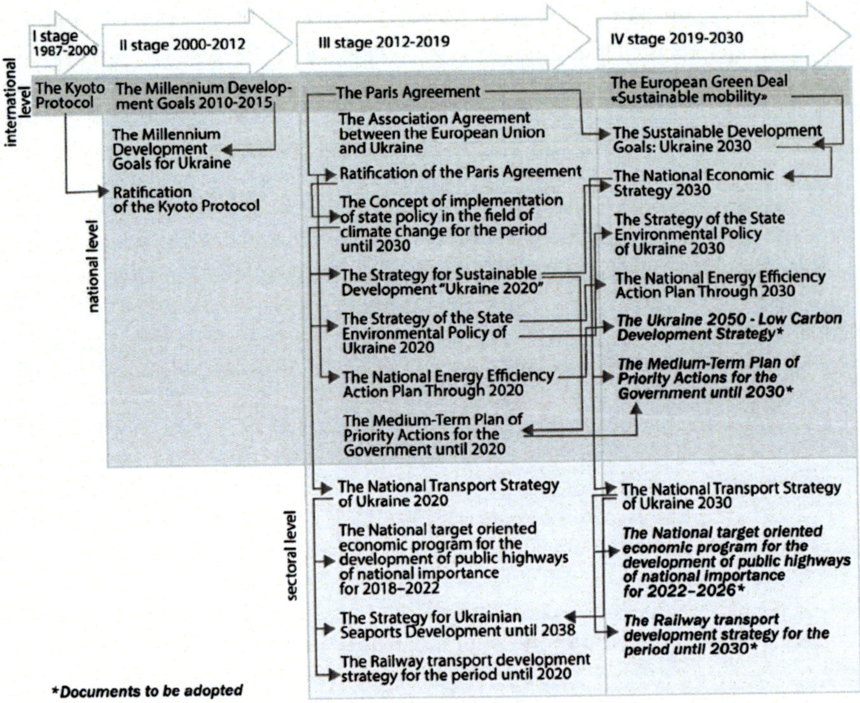


**Fig. 2.** System of strategic documents of development of sustainable transport.

decentralization and reform of public administration with the use of the latest information and communication technologies, as well as the energy independence program.

To overcome negative environmental impacts, the Paris Agreement was adopted at the supranational level, which Ukraine ratified in 2016. The Agreement is aimed at developing the economy via decarbonization policy, including in the field of transport by gradually phasing out oil products in favor of electricity, hydrogen, biofuel, etc. [22], according to which countries undertake to adopt climate plans providing for certain steps towards drastic reduction of harmful atmospheric emissions.

The continuation of this policy was the adoption in 2016 of the "Concept for the implementation of national policy in the field of climate change for the period until 2030" to ensure a gradual transition to low-carbon development subject to economic, energy and environmental security and increased welfare, and in 2017 the Action Plan for the implementation of the mentioned Concept was approved, which emphasized the need to consider the climate change factor during the development and implementation of the National transport strategy.

Ukraine's vision of achieving the Sustainable Development Goals by 2030 was presented in September 2017 in the National Report on "Sustainable Development Goals: Ukraine" (hereinafter – SDGU) [23] presenting 17 sustainable development goals adapted for Ukraine, which include 86 tasks of national development and 172 indicators for their monitoring.

The medium-term plan of the Government's priority actions until 2020 [24], approved by the Cabinet of Ministers of Ukraine in April 2017, according to the authors, contained a series of important measures aimed at achieving medium-term goals of the development of sustainable transport, which correspond to SDGU: improving national policy in the field of climate change to achieve sustainable development of the state, creating legal and institutional prerequisites to ensure a gradual transition to low-carbon development subject to economic, energy and environmental security and improved well-being; reforming the system of public administration; harmonizing national legislation and standards with the EU; defining transport development priorities; creating proper conditions for attracting private investment and public-private partnership; improving the security system in accordance with EU standards; implementing innovations and further transition to digital economy.

In order to define strategic directions for the transition of Ukraine's economy to the path of low-carbon growth based on sustainable development, in accordance with national priorities, in 2018, the Low-Carbon Development Strategy of Ukraine until 2050 was developed [25], one of whose tasks being to encourage the use of alternative petroleum products for motor fuels, including for cargo and passenger transportation thanks to more environmentally friendly transport modes. However, at present, the mentioned Strategy is not fixed by any legislative act.

The next important strategic step for the development of sustainable transport in Ukraine was the adoption of the "National Transport Strategy of Ukraine for the period until 2030" [26] in 2018, in which all tasks are grouped into four areas: a competitive and efficient transport system; innovative development of the transport sector and global investment projects; safe for society, ecologically clean and energy-efficient transport; unhindered mobility and interregional integration.

A single strategic document for road transport in Ukraine is the National Target Economic Program for the Development of Public Roads of National Importance for 2018–2022, approved by the Government with the aim to define clear strategic guidelines for the development of public roads based on the socio-economic priorities of the development of regions and the state in general.

During the fourth stage (2019–2030) (see Fig. 2), attention is focused on rethinking Ukraine's policy in the field of clean energy supply in the economy, industry, production and consumption, transport, agriculture, construction, taxation and social assistance (Green Deal).

To continue the EU transport policy in ensuring sustainable mobility, in December 2019, The European Green Deal ("Sustainable mobility", The European Green Deal) [27] was adopted with the main goal to reduce greenhouse gas emissions by 90% by 2050.

With the aim of improving the quality of atmospheric air, strengthening the response to the consequences of climate change and achieving the goals of sustainable low-carbon development of all sectors of Ukraine's economy, ensuring the implementation of ratified international documents on combating climate change and improving the quality of atmospheric air, the Law of Ukraine "On Basic Principles (Strategy) of Ukraine's national environmental policy for the period until 2030" was adopted in early 2019. This Law determines that motor vehicles are one of the main polluters of atmospheric air and a source of greenhouse gas emissions in Ukraine, and outlines the tasks, expected results and target values of indicators for the evaluation of implementation of this country's national environmental policy for medium term (until 2025) and long term (until 2030) periods between which is however a certain essential inconsistency. Also, the indicators for the transport industry are defined in a quite general and ambitious way for Ukraine at the current stage.

In late 2019, the Decree of the President of Ukraine "On the Sustainable Development Goals of Ukraine for the period until 2030" [28] was approved, which: a) legislated the Sustainable Development Goals of Ukraine until 2030; b) introduced a system for monitoring the implementation of the Sustainable Development Goals of Ukraine for the period until 2030.

In 2020, a strategic environmental assessment was introduced to define, describe and assess the consequences of the implementation of national planning documents for the environment, which include strategies, plans, schemes, urban planning documentation, nationwide programs, state target programs and other programs and policy documents regarding the transport sector regulated by The Law of Ukraine "On Strategic Environmental Assessment".

In 2021, the Sustainable Development Strategy "Ukraine - 2020" was replaced by the National Economic Strategy for the period until 2030 (hereinafter NES), which states that it is based on the principles of sustainable development and considers the SDGU and the need to achieve climate neutrality by no later than 2060. However, it has a number of disadvantages, namely: a weak connection with the SDGU; the goal defined for the development of sustainable transport, "decarbonization of the sector in order to achieve the goals of climate neutrality by no later than 2060" is not properly revealed in the Strategy (in terms of problem analysis, tasks, indicators, and expected results); the indicators selected to monitor realization of the defined goals do not correlate with the global indicators of development of sustainable transport; there are no forecast indicators for the period until 2030.

In order to introduce European practice of medium-term planning of state policy in the field of energy efficiency, in 2021 the Order "On the National Energy Efficiency Action Plan for the period until 2030" was approved, which was aimed at achieving the national energy efficiency goal - primary and final energy consumption in Ukraine in 2030 should not exceed respectively 91,468,000 and 50,446,000 tons of oil equivalent, and the Action Plan for its implementation for 2021–2023 should be approved. At the same time, the National Energy Efficiency Action Plan for the period until 2030 itself has not yet been adopted and remains only a draft.

The institutional integrity and compliance of Ukraine's strategic legislation with European legislation is assessed with the help of such a tool as monitoring, which allows assessing the current state of implementation of the Sustainable Development Goals until 2030, which requires a clearly defined system of indicators.

As of today, there is no single systematic approach to assessing the sustainable development of transport, and the choice of approach is largely determined by the set goals. We proposed to monitor the progress in achieving the goals of sustainable development of transport at the international level based on the "Global Tracking Framework for

Transport" (GTF), which consists of 29 main indicators and 70 secondary desired indicators for all modes of transport and provides the possibility of comparison between all countries based on the same indicators. According to this assessment system, the Index of sustainable development of transport for Ukraine is 59.2 out of 100, and the rating of sustainable development of transport is 40 out of 183 countries, which is a good result. At the country level, the basis for monitoring the sustainable development of road transport are the indicators for the assessment of the Sustainable Development Goals until 2030, defined in the National Report on "Goals of Sustainable Development: Ukraine 2017".

For the industry level, the indicators specified in the National Report are not sufficient, therefore, we proposed to improve the system of indicators by expanding them and classifying them in accordance with those defined by SDGU. The system should consider the structure and dynamics of passenger and cargo transportation by transport mode, the condition of transport infrastructure, which primarily includes roads and vehicles, the use of digital technologies based on the implementation of IT and ICT, etc.

To evaluate the current state of the implementation of the goals of sustainable development of Ukraine's road transport, the EU methodology was used, which focuses on overtime changes, considers the direction of the indicator's progress (whether the indicator has approached a goal of sustainable development or moved away from it), and the speed of the indicator's progress and is calculated according to the formula of "compound annual growth rate" (CAGR) (1).

$$CARG = \left(\frac{y_t}{y_{t0}}\right)^{\frac{1}{t-t0}} - 1 \tag{1}$$

where: yt is the indicator's value for previous year; yt0–indicator's value for base year; t–previous year; and t0 – base year.

The assessment of the indicator's dynamics is visualized in the form of arrows (see Table 2) [29], whose direction shows trends towards more sustainable development or in the opposite direction; so that up arrows indicate positive progress, and down arrows - a negative trend.

**Table 2.** Direction and speed of progress in achieving the goals of sustainable development.

| | | | |
|---|---|---|---|
|  | Significant progress in achieving the Sustainable Development Goals |  | Moving away from achieving the Sustainable Development Goals |
|  | Moderate progress in achieving the Sustainable Development Goals |  | Trend assessment is not possible (for example, the time series is too short) |
|  | Insufficient progress in achieving the Sustainable Development Goals | | |

In the assessment process, it is important to compare the rate of the indicator's progress with the limit values (see Table 3). A progress of 1% per year or more is considered "significant": (1) if the progress occurred in the desired direction, it means that "progress in achieving the goal of sustainable development is significant"; (2) if in the opposite direction - "significant deviation from achieving the goal of sustainable development". A progress in the desired direction of less than 1% (including 0%) per year is considered "moderate progress towards achieving the sustainable development goal", and a progress in the opposite direction of less than 1% per year is considered "moderate deviation from the achievement of the sustainable development goal".

**Table 3.** Limit values of speed of progress in achieving the goals of sustainable development.

| Limit values | Symbols | Limit values | Symbols |
|---|---|---|---|
| > = 1% |  | <0% and > = −1% |  |
| <1% and > = 0% |  | <−1% |  |

Indicators are divided into two categories: (1) S is stimulator (indicator, whose increase means improved sustainable development of road transport); and (2) D is destimulator (indicator whose increase means worsened sustainable development of road transport).

Indicators for assessing the progress in sustainable development of road transport are identified based on the nine goals of sustainable development of road transport defined in scientific study [15], which are divided into economic, social and environmental components. The identified goals cover the following recognized in the world and the EU guidelines in ensuring sustainable development of road transport: increasing the volume of transport operation, development of high-quality, reliable and accessible infrastructure, introduction of environmentally friendly vehicles; use of ecologically clean energy supplied for charging electric cars; finding ways to dispose of batteries from electric cars; negative impact on the environment from the operation of road transport; and reducing the level of serious injuries and deaths due to road accidents, which depends on the quality of road surface, condition of vehicles and the efficiency of managing traffic flows within cities; ensuring proper working conditions for drivers in compliance with regulations, which requires the implementation of Directives in accordance with the Association Agreement; increasing the efficiency of public authorities and local self-governments; mobilizing additional financial resources based on encouraging foreign and domestic investments; developing partnership relations between authorities and businesses via the implementation of public-private partnership mechanisms; acceleration and support of digital transformations in transport [30–32], which contributes to the re-planning of flows towards greater use of e-logistics; the introduction of intelligent transport systems; the use of information and communication technologies, and IoT; creation and development of new specialties and reorientation of employees to new types of activities; and women's economic empowerment via increased incomes.

Following are the results of calculations grouped by the main components of sustainable development, namely: economic, social and environmental ones (Tables 4, 5, 6).

The assessment of the economic component of the implementation of sustainable development of road transport in Ukraine, carried out based on the system of indicators proposed by the authors, shows that on the economic component, there is a positive trend towards the sustainable development of road transport, with the exception of indicators that show a reorientation between modes of transport in favor of automobiles.

The assessment of the social component of the implementation of sustainable development of road transport in Ukraine, carried out based on the system of indicators proposed by the authors, shows that on the social component, according to most indicators, negative trends are observed, which are manifested in decreased passenger transportation, and in the low availability of transport services, with the exception of indicators that show increased safety.

The assessment of the environmental component of the implementation of sustainable development of road transport in Ukraine, carried out based on the system of indicators proposed by the authors, shows that: on the environmental component, according to all indicators, a positive trend is observed towards sustainable development of road transport. The improvement observed in the assessment of indicators of sustainable development of Ukraine's road transport has been achieved, among other things, thanks to the institutional basis created in Ukraine, which however needs further improvement.

The proposed system of indicators is considered as a tool for monitoring the achievement of the goals of sustainable development of road transport in Ukraine's economy, which will contribute to balanced management decision-making.

## 5 Discussion

The main contribution of the presented research consists in the analysis of the prerequisites for the establishment of the concept of sustainable development in the world, on which basis the authors identify four stages in the elaboration of the concept of the development of sustainable transport in Ukraine, in each of which they reveal gaps in the current system of strategic documents. The authors provide recommendations on further development and introduction of the principles of sustainable transport in Ukraine based on a relevant institutional base, taking into account the trends of European integration, and develop and propose a system of indicators for assessing the sustainable development of road transport at the state level and to monitor the achievement of the goals of sustainable development.

However disputable aspects of the study include the possibility of expanding the system of indicators for assessing the sustainable development of road transport, (which can be expanded in necessity); and a difficult access to information produced by a number of institutions due to hostilities in the country. The future research suggestions include a deepee assessment of institutional support for the sustainable development of transport in Ukraine in the context of the European Green Deal and various aspects of the implementation of digital technologies, including intelligent transport systems.

**Table 4.** Estimation of the economic component in ensuring the sustainable development of road transport.

| Indicators | Estimate | | S/D |
|---|---|---|---|
| Freight volume (road transport), million tons | ↑ | 2.962 | S |
| Cargo turnover (road transport), billion t-km | ↑ | 5.086 | S |
| Share of freight (road transport) in total inland freight transport, % | ↘ | 1.325 | D |
| Share of road cargo turnover in total inland cargo turnover transport, % | ↓ | 5.484 | D |
| Average distance of transported of one ton of goods by road transport, km | ↓ | 2.322 | D |
| Ukraine's ranking of in the Global Competitiveness Index by sub-index "Quality of roads" | ↑ | −3.599 | D |
| Amount of the State Road Fund, UAH billion | ↑ | 46.445 | S |
| Share of State Road Fund funding on public roads of national importance, % | ↑ | 10.554 | S |
| Share of State Road Fund funding on public roads of local importance, % | ↓ | −4.382 | S |
| Share of State Road Fund funding on repaying for loans raised by the government under state guarantees for the development of public roads network, % | ↑ | −27.239 | D |
| **Restored roads of national significance, km** | ↑ | 89.073 | S |
| Share of innovative items in total industrial sales, % | ↓ | −1.836 | S |
| Output production of motor vehicles, trailers and semi-trailers, units | ↑ | 5.103 | S |
| Output production of other vehicles, units | ↓ | −3.477 | S |

*(continued)*

**Table 4.** (*continued*)

| Indicators | Estimate | | S/D |
|---|---|---|---|
| Internet subscribers, per 100 people | ↑ | 63.406 | S |
| Internet subscribers, per 100 people, urban areas | ↑ | 25.770 | S |
| Internet subscribers, per 100 people, rural areas | ↑ | 29.099 | S |

*Source* developed by the authors

## 6 Conclusions

Despite the fact that the concept of sustainable development combines the economic, social and environmental components, at different stages of its evolution, different components prevailed, which was reflected in the relevant strategic documents. At the first stage, the main principles of sustainable development were formulated, and the decisive role was given to environmental components, so road transport was considered as one of the main polluters of the environment; at the second - the Millennium Development Goals for the period until 2015 focused on overcoming social problems, and transport began to be considered in the context of ensuring access to basic services, and only at the third stage were the Sustainable Development Goals until 2030 formulated, which take into account all three components. At this stage, a sub-concept of sustainable development of transport emerges within the general concept of sustainable development.

This trend is institutionally consolidated at the fourth stage, in which world's transport policy focused on energy conservation and reduction of greenhouse gas emissions from mobile pollution sources, which largely corresponds to the principles of the Paris Agreement and the concept of sustainable development.

During the first stage (1990–2000) after gaining independence in Ukraine, the formation of the institutional base began, in which the first steps were taken by joining international environmental agreements, namely the Kyoto Protocol, adopting the Concept of Sustainable Development of Settlements, etc. However, the economic recession, the decrease in GDP and reduced population in Ukraine did not contribute to the real modernization of the transport system, and the performance of road transport sharply worsened.

During the second stage (after 2000) Ukraine set a goal of stabilizing greenhouse gas emissions at a level by 20% lower than the 1990 level by 2020. Despite certain measures aimed at improving the environmental condition, atmospheric air pollution by road transport remained one of the main challenges. One of the reasons for this was the actual absence in Ukraine of legally approved strategic documents and institutional basis for promoting sustainable development, while the guidelines of Ukraine's sustainable

**Table 5.** Estimation of the social component in ensuring the sustainable development of road transport.

| Indicators | Estimate | | S/D |
|---|---|---|---|
| Passengers transported by road transport, million people | ↓ | −5.363 | S |
| Passenger turnover (road transport), billion t-km | ↘ | −0.509 | S |
| Share of passengers transported by road transport in total transport overall structure of transportation by transport mode, % | ↗ | −0.586 | D |
| **Share of road (bus) transportation in intercity traffic, %** | ↑ | −1.94 | D |
| **Share of car (bus) transportation in suburban traffic,%** | ↑ | 6.88 | S |
| **Share of road (bus) transportation in intra-city traffic,%** | ↑ | −1.54 | D |
| Share of road transport in total passenger turnover, % | ↑ | −2.902 | D |
| Share of rural households who suffered from deprivation due to the lack of regular daily transport connections with another settlement with developed infrastructure, % | ↓ | 1.822 | D |
| **Number of deaths due to traffic accidents, per 100,000 persons** | ↑ | −5.839 | D |
| **Number of injured persons as a result of traffic accidents per** 100,000 persons in road transport | ↗ | −0.487 | D |
| The ratio of average wages of women and men engaged in "transport, warehousing, postal and courier activities", % | ↘ | −0.433 | S |

*Source* developed by the authors

development were determined by the Millennium Development Goals for Ukraine. No long-term strategic documents were adopted in the area of road transport that would identify the vectors of its development.

During the third stage (2012–2019), the main goals of sustainable development **until 2030 were defined, in which the economic, social and ecological components** were represented in equal measure, and transport was recognized as one of the main means of achieving the goals of sustainable development. Ukraine joined the concept of sustainable development by adopting a series of normative legal acts to improve the condition of the environment at the national level. Along with this, there is a lack of long-term strategic documents on the sustainable development of road transport, which has the

**Table 6.** Estimation of the environmental component in ensuring the sustainable development of road transport.

| Indicators | Estimate | | S/D |
|---|---|---|---|
| Depreciation rate of fixed assets in economic activity "Transport, warehousing, postal and courier activities", % | ↑ | 1.141 | D |
| Depreciation rate of fixed assets in land and pipeline transport, % | ↑ | 2.041 | D |
| The share of energy produced from renewable sources in total final energy consumption, % | ↑ | 12.491 | S |
| The share of electric transport in domestic traffic, % | ↑ | 3.762 | S |
| The share of electric cars in total number of cars, % | ↗ | 0 | S |
| The share of electric cars in total car sales, % | ↑ | 31.607 | S |
| Number of charging stations for electric vehicles, units | ↑ | 139.239 | S |
| Greenhouse gas emissions by road transport, % to 1990 amount level | ↑ | −52.109 | D |
| Share of greenhouse gas emissions by road transport in total transport emissions, % | ↑ | −1.300 | D |
| Share of greenhouse gas emissions by road transport in total energy industry emissions, % | ↘ | 0.371 | D |

greatest negative environmental impact on sustainable development, while playing an important role for the national economy. At the same time, individual decisions are taken and legal acts are adopted, which are aimed at improving the environmental, social and economic components, but they are not united into a single comprehensive system. The assessment of strategic documents on the implementation of sustainable development in Ukraine demonstrates institutional weakness and fragmentation, both at the national and sectoral level, especially regarding the solution of issues of more environment friendly functioning of road transport.

During the fourth stage, half of the strategic documents on implementing sustainable development and environmental policy expired in 2020, which poses the problem of their prolongation. Starting from this stage, Ukraine's transport policy is based on the **Sustainable Development Goals until 2030. The National Energy Efficiency Action Plan** for the period until 2030, which considers the tasks of encouraging the introduction of the latest technologies, developing the infrastructure of charging stations for electric vehicles, updating private car fleet to increase its electric mobility, optimization of the structure of passenger and freight transport in cities and renewal of urban transport with a predominant transition to electric transport.

Obtaining the status of a candidate country for joining the EU on June 23, 2022 sets new goals and tasks for Ukraine as to the harmonization of transport policies in the context of the Green Deal, namely: involvement of society in solving issues of sustainable development of road transport and the use of political participation; the use of artificial intelligence, the Internet of Things and the implementation of full digitization of transport; implementation of a carbon-neutral economy; focusing on a sustainable society, as well as on corporate social responsibility and responsible business practices; ensuring responsible and sustainable mining and finding new deposits; introduction of sustainable financing, use of public-private partnership and crowdfunding. Based on this, the tasks for Ukraine in the development of sustainable transport at this stage are to strengthen the monitoring of the main indicators (ecological, social and ecological components) of the development of transport in Ukraine, to achieve the desired results with the help of a developed and tested system of indicators.

## References


1. Buryk, Z.: State regulation of sustainable development of Ukraine: theory and practice. Prostir-M, Lviv. 383 (2017)
2. Soergel, B., Kriegler, E., Weindl, I., et al.: A sustainable development pathway for climate action within the UN 2030 Agenda. Nat. Clim. Chang **11**, 656–664 (2021). https://doi.org/10.1038/s41558-021-01098-3
3. Ruggerio, C.: Sustainability and sustainable development: a review of principles and definitions. Sci. Total Environ. 786 (2021). https://doi.org/10.1016/j.scitotenv.2021.147481
4. Reyers, B., Moore, M.L., Haider, L.J., et al.: The contributions of resilience to reshaping sustainable development. Nat. Sustain. **5**, 657–664 (2022). https://doi.org/10.1038/s41893-022-00889-6
5. Zhao, X., Ke, Y., Zuo, J., Xiong, W., Wu, P.: Evaluation of sustainable transport research in 2000–2019. J. Clean. Product. **256** (2020). https://doi.org/10.1016/j.jclepro.2020.120404
6. Deetman, S., Hof, A.F., Pfluger, B., van Vuuren, D.P., Girod, B., van Ruijven, B.J.: Deep greenhouse gas emission reductions in Europe. Energy Policy **55**, 152–164 (2013). https://doi.org/10.1016/j.enpol.2012.11.047
7. Diachuk, O.: Ukraine's contribution to the new Global agreement on climate change. Econ. Forecast. **1**, 129–141 (2016). http://nbuv.gov.ua/UJRN/econprog_2016_1_11
8. Petrovi´c, S., Diachuk, O., Podolets, R., Semeniuk, A., Bühler, F., Grandal, R., Boucenna, M., Balyk, O.: Exploring the long-term development of the Ukrainian energy system. Energies **14**, 7731 (2021). https://doi.org/10.3390/en14227731
9. Chepeliev, M., Osorio-Rodarte, I., van der Mensbrugghe, D.: Distributional impacts of carbon pricing policies under the Paris Agreement: inter and intra-regional perspectives. Energy Econ. **102** (2021). https://doi.org/10.1016/j.eneco.2021.105530
10. Chepeliev, M., Diachuk, O., Podolets, R.: Economic assessment of low-emission development scenarios for Ukraine. In: Giannakidis, G., Karlsson, K., Labriet, M., Ó Gallachóir, B. (eds.) Limiting Global Warming to Well Below 2 °C: Energy System Modelling and Policy Development. LNE, vol. 64, pp. 277–295. Springer, Cham (2018). https://doi.org/10.1007/978-3-319-74424-7_17

11. Dvulit, Z.: Economic and environmental management of sustainable development of railway transport enterprises: theory, methodology, practice. DETUT, Kyiv (2016)
12. Ivanova, N.: The potential of water transport of Ukraine in the context of sustainable development paradigm. Electron. Sci. Pract. J. Develop. Product. Forces Region. Econ. **17**, 278–285 (2018)
13. **Dorofieieva, K.: Transportation as a Catalyst for Sustainable Development of Society and the Economy. Scientific Bulletin of Kherson State University 20, vol. 1, pp. 24–26 (2016).** http://www.ej.kherson.ua/journal/economic_20/1/7.pdf
14. Stasiuk, O., Chmyrova, L., Fediai, N.: The markets of freight and passenger transport in Ukraine: problems and trends. In: Efektyvna Ekonomika, vol. 9 (2020). http://www.economy.nayka.com.ua/?op=1&z=8199. https://doi.org/10.32702/2307-2105-2020.9.54
15. Chmireva, L.: Sustainable Development of Road Transport in the National Economy. Kyiv (2021)
16. Gerasymchuk, Z., Polishchuk, V.: Theoretical bases of institutional support promoting sustainable development of the region. Region. Econ. **4**, 30–47 (2011)
17. Ukraine 2020. National Inventory Report (NIR). Ukraine's greenhouse gas inventory 1990–2018. United Nations Climate Change. https://unfccc.int/documents/228016 (2020). Last Accessed 19 Aug 2022
18. Nationally determined contribution of Ukraine. Ministry of Ecology and Natural Resources of Ukraine. https://mepr.gov.ua/news/33080.html (2021). Last Accessed 19 Aug 2022
19. The Strategy for Sustainable Development "Ukraine 2020". Verkhovna Rada of Ukraine. https://zakon.rada.gov.ua/laws/show/5/2015#n10 (2015). Last Accessed 19 Aug 2022
20. Martiusheva, O.: Projects of the concept of sustainable development of Ukraine: the possibility of their improvement and application: Analytical note. The National Institute for Strategic Studies. https://niss.gov.ua/doslidzhennya/nacionalna-bezpeka/proekti-koncepcii-stalogo-rozvitku-ukraini-mozhlivist-ikh (2014). Last Accessed 19 Aug 2022
21. Association Agreement between the European Union and the European Atomic Energy Community and their member states, of the one part, and Ukraine, of the other part. Verkhovna Rada of Ukraine. https://www.kmu.gov.ua/storage/app/media/ugoda-pro-asociaciyu/TITLE_V.pdf (2014). Last Accessed 19 Aug 2022
22. Chepeliev, M., Diachuk, O., Podolets, R. Trypolska, H. et al.: Transition of Ukraine to renewable energy by 2050. ART BOOK LLC, Kyiv. https://ecoaction.org.ua/wp-content/uploads/2017/10/perehid_ukraini_na_vidnovlyuvanu_energetiku_do_2050_roku.pdf (2017). Last Accessed 19 Aug 2022
23. App, J., Scheurer, J.P.: Ukraine. In: Ternès, A., Towers, I. (eds.) Interkulturelle Kommunikation, pp. 259–273. Springer, Wiesbaden (2017). https://doi.org/10.1007/978-3-658-10237-1_23
24. Medium-Term Plan of Priority Actions for the Government until 2020. Verkhovna Rada of Ukraine. https://zakon.rada.gov.ua/laws/show/275-2017-p#Text (2017). Last Accessed 19 Aug 2022
25. Ukraine 2050 - Low Carbon Development Strategy. Ministry of Ecology and Natural Resources of Ukraine. https://mepr.gov.ua/files/docs/Proekt/LEDS_ua_last.pdf (2018). Last Accessed 19 Aug 2022
26. National Transport Strategy of Ukraine 2030. Verkhovna Rada of Ukraine. https://zakon.rada.gov.ua/laws/show/430-2018-p#Text (2018). Last Accessed 19 Aug 2022
27. Sustainable mobility. European Commission. https://ec.europa.eu/commission/presscorner/detail/en/fs_19_6726 (2019). Last Accessed 19 Aug 2022
28. About the Sustainable Development Goals of Ukraine until 2030. Verkhovna Rada of Ukraine. https://zakon.rada.gov.ua/laws/show/722/2019 (2019). Last Accessed 19 Aug 2022

29. Sustainable development in the European Union. Overview of progress towards the SDGs in an EU context. 2020 edition. Eurostat supports the SDG s. https://ec.europa.eu/eurostat/documents/4031688/11010788/KS-01-20-192-EN-N.pdf/ae63aff0-a6f3-1d47-da83-c6886b9daaab?t=1592486268000 (2020). Last Accessed 19 Aug 2022
30. Nykyforuk, O., Stasyuk, O., Chmyrova, L., Fediaj, N.: Digitization in the transport sector: development trends and indicators. Part 1. Stat. Ukraine **86**(3), 70–81 (2019). https://doi.org/10.31767/su.3(86)2019.03.08
31. Nykyforuk, O., Stasiuk, O., Chmyrova, L., Fediai, N.: Digitization in the transport sector: development trends and indicators. Part 2. Stat. Ukraine **87**(4), 48–64 (2019). https://doi.org/10.31767/su.4(87)2019.04.06
32. Nykyforuk, O., Stasiuk, O., Chmyrova, L., Fediai, N.: System of digital transformation indicators in transport sector. Eur. J. Intell. Transp. Syst. **1**, 3–12 (2019). https://doi.org/10.31435/rsglobal_ejits/31072019/6576

# Sustainable Households Financial Behaviour in EU and Implications on Developing Financial Literacy

Lyubomir Todorov(✉), Aleksandrina Aleksandrova, and Marin Marinov

D. A. Tsenov Academy of Economics, 2 Em. Chakarov Str, Svishtov, Bulgaria
lubomir.ivanov@uni-svishtov.bg

**Abstract.** Avoiding financial problems and good management of one's income and assets is connected to people's well-being, accumulation of capital in the economy and is essential element for stability and sustainable growth. The paper focuses on the relations between financial literacy of the population and the key indicators used by Eurostat to assess the financial status and sustainable behaviour of the households in European Union, linked to savings, investment, indebtedness, and wealth. Partial correlation coefficients and multiple regression models were used to control for the influence of countries' geographical position and economic development. The results showed that financial literacy score was related to the Gross household saving rate and Household net financial assets ratio, while its connection with Gross investment rate of households, Households investment to Gross Domestic Product ratio and Gross debt-to-income ratio of households remained insignificant. In two models with significant financial literacy effect the division of the countries on the West-East axis was also crucial factor. Both further analysis in this area and policy making at European Union level focused on sustainable development, would benefit a lot if Eurostat starts collecting panel data about member states' financial education status.

**Keywords:** Financial literacy · Sustainability · Households financial behaviour · Multiple regression · Partial correlation

## 1 Introduction

In the last two decades, financial literacy of the population is an issue that has been of increasing interest, not only in academic circles, but has also become an invariable part of policy makers' agenda. According to Agarwal and Setiawan, this is largely because financial literacy plays a crucial role for individuals who consume financial products and the national and global economy, especially in the era of digitalization [1]. The question of financial literacy rate becomes even more relevant today, in a situation of global economic crisis, caused by the Covid-19 pandemic. Good financial literacy would support individuals to avoid financial problems, well-managed income, and depression conditions, especially during pandemics [2].


V. Koval et al. (Eds.): ISCMEE 2022, Lect. Notes in Managem., Industrial Engineering, pp. 135–154, 2023.
https://doi.org/10.1007/978-3-031-23463-7_9

During economic growth and as well as financial difficulties and challenges, financial problems can be avoided through financial literacy [3]. Improving the population's financial knowledge is crucial for all governments when many financial services are delivered [4]. Financial literacy includes financial skills and capability, financial knowledge and awareness, financial risk management and management of individuals' finances. These are also fundamental elements of stability and economic and financial growth in the economy. Financial literacy has been recognized worldwide as a significant element of stability and economic and financial growth [5].

## 2 Literature Review

### 2.1 Financial Literacy and Sustainability

The Organization for Economic Cooperation and Development (OECD) [6], defines financial literacy as a combination of awareness, knowledge, skill, attitude and behaviour necessary to make sound financial decisions and attain individual financial well-being. The European Commission defines financial literacy as "the knowledge and skills needed to make important financial decisions" [7]. Nowadays, low financial literacy is one of the main factors that create barriers to the full and informed participation of people in the economic life of the state. It is often associated with the individual's low income and belonging to a lower social class, which are a prerequisite for social exclusion. Poor financial literacy implies insufficiently adequate management of personal wealth, manifested in negative individual and general economic consequences such as gaps in individual economic development, social exclusion, excessive indebtedness, sub-optimal market structures and growth and additional burden on the social welfare system (Fig. 1).

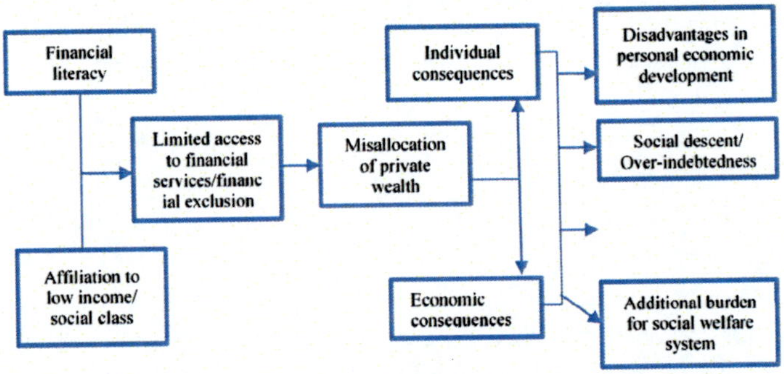


**Fig. 1** Risks deriving from financial illiteracy.

To a significant extent, financial literacy defines the financial inclusion of individuals. The financial inclusion of the population plays a crucial role in the economic development of the state. Financial inclusion has the potential to eradicate poverty which is why a cornerstone of eight of the seventeen Sustainable development goals 2030. Financial inclusion promises to enhance poor families' ability to minimize financial shocks, undertake human capital investments in health and education and/or engage in modest asset accumulation in order to take advantage of promising investment opportunities in their economies [9]. There is a significant similarity between this definition and the definition of financial literacy. The relationship between financial inclusion and development plays a crucial role in achieving the Sustainable development goals 2030 as financial literacy, and more specifically, improving its' rates is among the factors that will hence SDG achievement. Financial literacy affects Goal 1. End poverty in all its forms everywhere, Goal 4 Quality education and Goal 8. Promote sustained, inclusive and sustainable economic growth, full and productive employment and decent work for all [10]. It is directly related to the well-being of the individuals since knowledge and skills in the field of managing personal finances are extremely important for ensuring a good standard of living. Due to a lack of financial literacy, individuals are unable to optimize their welfare [11]. Financial literacy affects all financial affairs, including borrowing, saving, investment, and money management [12]. Financial literacy as a human capital investment educates people on how to handle finances during an economic crisis and on how to process economic information to make reasonable decisions on financial planning, wealth accumulations, borrowings, and pensions [13]. A higher financial literacy of individuals contributes to their individual financial development, which means that the economic development of the country is connected to the level of financial literacy of country's population.

### 2.2 Financial Literacy and Its' Correlation to Debt, Income and Savings

Often, the literature that examines financial literacy associates it with income. Disney & Gathergood finds that those with superior financial literacy have higher household incomes and savings [14]. A recent study on Financial Capability in the United States shows that younger respondents, those with lower incomes, those with lower education levels, and Black/African American and Hispanic/Latino respondents show higher levels of vulnerability across multiple measures of financial capability. The data collected and analysed in the five editions of the National Financial Capability Study prove the hypothesis that income is positively correlated with financial capability [15]. According to the research results, respondents with higher financial literacy are better prepared to meet their short-term financial needs. People with lower financial literacy and education report spending more than income. On the other hand, respondents with higher financial literacy report spending less than their income and setting aside three months' worth of emergencies.

Financial difficulties are not just a function of income only (low income) [3]. Many authors suggest that there is a positive correlation between poor financial literacy and debt problems. A lack of financial literacy is correlated with higher debt burdens, incurring greater fees, loan defaults and loan delinquency [16]. Poor money management skills have been found to contribute to higher levels of indebtedness. Bucks & Pence; Campbell; Disney & Gathergood; Duca & Kumar and Gerardi have considered how financial literacy affects debt. Poor financial literacy caused by failing to understand interest rate calculations is correlated with higher debt burdens, incurring greater fees, and defaults and delinquency [17–21]. Poor financial literacy might cause unsecured personal loans [22] which is a prerequisite for deterioration of the personal financial situation, and in the longer term for possible bankruptcy [23], which these individuals own, they are not inclined to borrow funds at higher interest rates. In addition to managing their debt in an informed and correct manner, they have the opportunity to identify potential financial risks and minimise the chances of indebtedness.

At the same time, higher financial literacy assumes a good knowledge of financial instruments, the opportunities for saving and investing and increasing wealth. Many authors such as Christelis, Van Rooij and Cardak and Wilkins examine the relationship between financial literacy and stock market participation [24], and risky asset holdings [25], and portfolio choice [26]. Measures of financial literacy correlate with multiple outcomes indicative of lifetime financial well-being [27]. Lusardi and Mitchell claim that financially literate individuals are more likely to plan for retirement and accumulate wealth (savings) [28]. Anderson, Baker and Robinson prove that financial literacy correlates with retirement planning and precautionary savings behaviour [29], Wagner and Walstad - with spending versus saving for the future [30]. De Bassa Scheresberg finds the tendency that financially literate people are more likely to set aside savings for emergencies [31] and to invest in different financial assets e.g. stocks [32], saving deposits, investments in equity, shares, bonds etc.

### 2.3 Financial Literacy in the EU

Having in mind the correlations between financial literacy, the well-being of individuals, indebtedness, income and economic development, savings and investments, the authors focus this research on financial literacy in the European Union (EU). Financial literacy is a problem that is increasingly placed on the agenda of EU policymakers. This is largely caused by the fact that the world's best performers (Sweden, Denmark), as well as those that score below the global average (Romania, Portugal) in financial literacy rankings, are located in the EU. Financial literacy rates vary widely across the European Union. [33]. According to Standard & Poor's Global Financial Literacy Survey EU [34] countries that score the highest financial literacy rates are located in northern Europe and the countries that score the lowest are in eastern Europe [35]. This is disturbing for many reasons, most essential of which is that low financial literacy is a prerequisite for the deterioration of well-being, deepening of inequality, regional differences, and hence divergence within the Community.

From September 2020, the European Commission and the OECD are jointly developing financial literacy framework. The idea of introducing this framework is to foster the national public authorities and stakeholders to develop policies, programs and educational materials for financial literacy. In 2022 the European Commission introduced the "Financial competence framework for adults in the European Union" jointly published by the European Union and the OECD. The framework contains the skills and knowledge that individuals need to possess in the context of managing and administering their finances with an emphasis on technological change and achieving sustainability. The main objective of the framework is to foster financial education and training, thereby increasing the financial literacy of the population of the EU.

# 3 Data and Methods

## 3.1 Data

The key indicators, that describe the financial status, behaviour and competence of the households, were computed from non-financial Annual Sector Accounts (ASA) data and partially – by combining data from non-financial with financial accounts. The households, according to the European System of Accounts 2010 were defined as an institutional sector, that combined institutional units with broadly similar characteristics and behaviour [36]. The households sector comprised not only households, but also included unincorporated household enterprises, defined as sole proprietorships and partnerships that didn't have legal status independent from their owners. Non-profit institutions serving households (NPISHs) were part of the sector too, but it was noteworthy to mention that their economic weight was very limited.

For the study the following key indicators were selected from the list, being important when performing the econometric analysis for the financial education and its relation to the households' behaviour:

- Gross household saving rate (noted thereafter by $y_1$) – represented the abilities and desire of the households to save, relative to their income.
- Gross investment rate of households (noted thereafter by $y_2$) and Households investment to GDP ratio (noted thereafter by $y_3$) – both represented the investment behaviour of the households relative to their income and as contributor to the country's economic growth (via investments).
- Gross debt-to-income ratio of households (noted thereafter by $y_4$) – represented the predisposition, propensity and eagerness of the households to take loans (both for consumption and for household enterprises.
- Household net financial assets ratio (noted thereafter by $y_5$) – represented the wealth of the households relative to their income. It was important to divide this indicator by 100 to get the better interpretation similar to that was used by Piketty – how many years of income the households had accumulated [37].

All key indicators were computed in percentage as follows:

$$\text{Gross household saving rate} = \frac{\text{GS}}{(\text{GDI} + \text{ACPE})} \times 100, \tag{1}$$

$$\text{Gross investment rate of households} = \frac{\text{GFCF}}{(\text{GDI} + \text{ACPE})} \times 100, \tag{2}$$

$$\text{Households investment to GDP ratio} = \frac{\text{GFCF}}{\text{GDP}} \times 100, \tag{3}$$

$$\text{Gross debt to income ratio of households} = \frac{\text{L}}{(\text{GDI} + \text{ACPE})} \times 100, \tag{4}$$

$$\text{Household net financial assets ratio} = \frac{\text{FNW}}{(\text{GDI} + \text{ACPE})} \times 100, \tag{5}$$

where: GS was gross saving of the households;
GDI – gross disposable income of the households;
ACPE – the adjustment for the change in pension entitlements (receivable/payable);
GFCF – gross fixed capital formation of the households;
GDP – gross domestic product;
L – loans of the households;
FNW – financial net worth.

The data used in the study were for the year 2014, in accordance with the timing of the study of financial literacy rating of S&P [34], that was the primary source of information for the countries in the European Union. The data for the key indicators were available in Eurostat for 27 members except for Bulgaria, Romania and Malta. Thus, remaining 24 countries were included in the study [38].

### 3.2 Methods

The main methodology of empirical study comprised the tools of regression and correlation analysis. The key indicators were regarded as dependent variables ($y$) and the financial literacy score was used as independent or explanatory variable ($x$). The simple linear regression was formalized as:

$$y = \alpha + \beta x + \varepsilon, \tag{6}$$

where: $\alpha$ was the constant;
$\beta$ – regression parameter, representing the influence of financial literacy score over the studied key indicator – that is how many percentage points change in the indicator was worth 1-point change of financial literacy score;
$\varepsilon$ – disturbance term that captures influence of all other factors than the independent variable, including errors.

The strength of the relation was measured by the correlation coefficient ($r_{xy}$), computed as Pearson product-moment correlation coefficient:

$$r_{xy} = \frac{\sum (x_i - \overline{x})(y_i - \overline{y})}{\sqrt{\sum (x_i - \overline{x})^2 \sum (y_i - \overline{y})^2}}, \tag{7}$$

that takes values between –1 and + 1.

It was important to note that correct application of regression analysis required several assumptions to hold: (a) linearity; (b) constant variance; (c) no multicollinearity (d) independence of errors. The latter twos were very difficult to achieve especially in economics when relations and inter-relationships were diverse, complex and complicated. Many economic variables were related to each other, and when used together would lower the efficiency of the estimates. On the other hand, when important variables were omitted from the regression, their influence was redistributed either to the other variables (omission bias) or into the disturbance term (making it not random).

To prevent such problems and achieve better estimates, several control variables were used in the regression analysis:

GDP per capita – as important measure of economic development, linked both to financial education (ability to study) and financial behaviour of the households;

North-South and West-East division – as members of the European Union were different in traditions and consumer/spending behaviour.

In order to control for the influence of GDP per capita, the partial correlation coefficients were used:

$$r_{xy.z} = \frac{r_{xy} - r_{xz}r_{zy}}{\sqrt{\left(1 - r_{xz}^2\right)\left(1 - r_{zy}^2\right)}}, \tag{8}$$

where: $z$ was the controlling factor (GDP per capita).

## 4 Findings

The scatter-diagrams revealed (Fig. 2) that the points were somewhat dispersed and the relation is any would not be too strong. Still there were no indication of either outliers or non-linearity in the variables' behaviour.

The correlation coefficients between FINLIT and key indicators were presented in Table 1. The number were low and only two of the coefficients appeared to be statistically significant at 5% level: between FINLIT and *Gross household saving rate*, and between FINLIT and *Household net financial assets ratio*. The observed correlation was positive for both indicators suggesting at first that the relation between financial literacy and those two indicators is also positive – higher financial literacy was associated with bigger saving rate and larger net wealth of the households.

The picture changed drastically when accounted for the control variables. Table 2 showed that the correlation between FINLIT score and GDP per capita was **0.490**, that was significant at 5%. On one hand that result supported the importance of the financial literacy as it was related to the GDP, but on the other there was no exact conclusion where financial literacy influenced GDP per capita or higher GDP per capita allowed receiving better education (including financial). Francis and Sunday claim that countries with higher per capita gross national income record higher investments in education [39]. According to Goczek; Witkowska and Witkowski the better the quality of education is, the higher the expected economic growth, because there exist confirmed returns - economies with better education quality grow faster [40]. The relationship between economic development, proxied by per capita GDP, and financial literacy is stronger for developed economies than developing economies [33]. This achieving sustainable growth could be easier to developed countries while more efforts would be required in developing ones, including special attention to financial education of the population.

Additionally, there was significant relation between FINLIT and geographical division among member countries (**0.645**) along the North-South axis. Still such dependence complicated the analysis of relation between key indicators and financial literacy, as those control variables were related to the key indicators too (Table 3). GDP per capita was primary control variable and showed significant correlation to 4 of the 5 key indicators – excluding *Households investment to GDP ratio*. North-South division had significant correlation only with one key indicator – *Gross investment rate of households*, while

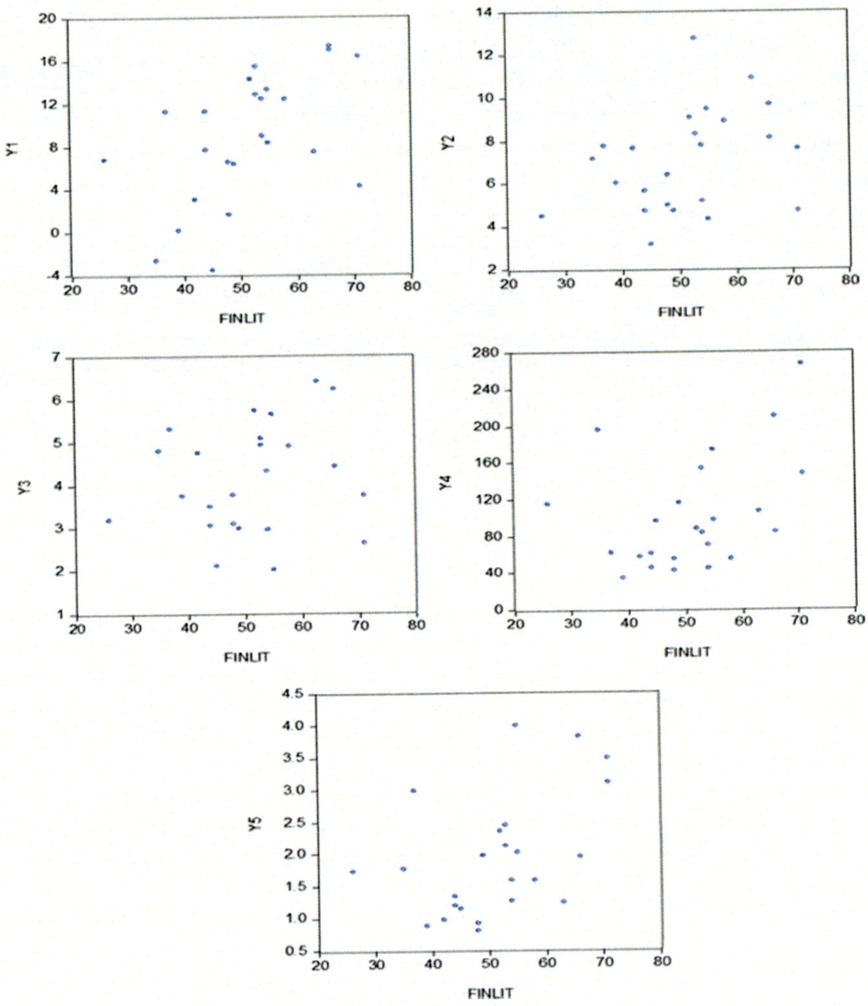


**Fig. 2.** Scatter diagrams representing the linear relationships between FINLIT (Financial Literacy Score) and selected key indicators

**Table 1.** The correlation coefficients between FINLIT and key indicators.

| Key Indicators | $y_1$ | $y_2$ | $y_3$ | $y_4$ | $y_5$ |
|---|---|---|---|---|---|
| Correlation coefficient* | **0.527** | 0.339 | 0.171 | 0.395 | **0.464** |

*Correlation coefficients that were significant at 5% level were bolded

West-East division proved to be correlated significantly with 3 of the key indicators – *Gross household saving rate, Gross debt-to-income ratio of households* and *Household*

**Table 2.** The correlation coefficients between FINLIN and control variables.

| Variables | FINLIT | GDP per capita | N-S | W-E |
|---|---|---|---|---|
| FINLIT | 1.000 | **0.490** | **0.645** | 0.152 |
| GDP per capita | **0.490** | 1.000 | 0.308 | **0.596** |
| N-S | **0.645** | 0.308 | 1.000 | 0.038 |
| W-E | 0.152 | **0.596** | 0.038 | 1.000 |

*Correlation coefficients that were significant at 5% level were bolded

**Table 3.** The correlation coefficients between key indicators and control variables.

| Key Indicators | Control Variables | | |
|---|---|---|---|
| | GDP per capita | N-S | W-E |
| $y_1$ | **0.491** | 0.371 | **0.446** |
| $y_2$ | **0.585** | **0.465** | 0.311 |
| $y_3$ | 0.266 | 0.299 | 0.256 |
| $y_4$ | **0.608** | 0.038 | **0.454** |
| $y_5$ | **0.608** | 0.146 | **0.647** |

*Correlation coefficients that were significant at 5% level were bolded

*net financial assets ratio*. Obviously, the EU population located in different geographical regions have different financial culture and understanding about personal finance management which implies in investing, saving, indebting.

Controlling for the influence of the aforementioned variables gave the correlation in Table 4. The differences were substantial. While correlation between FINLIT and key indicators was significant for two, when controlling for GDP per capita no partial correlation coefficient were significant. The other two control variables had only marginal effect, as they didn't change the significance of the correlation. In fact, when we used control variable for North-South division among countries, one of the nonsignificant correlation coefficients became significant.

The results gave very important conclusion – the correlation itself was not enough to establish the relationship between financial literacy score and studied key indicators. Complex nature of the relations among the variables required building of at least multiple regression model that included all control variables, solving and performing appropriate diagnostic.

The models are solved in full form as:

$$y_j = \alpha + \beta \times FINLIT + d_1 \times GDPPC + d_2 \times NS + d_3 \times WE + \varepsilon, \tag{9}$$

where: *GDPPC* was the GDP per capita;

*NS* – dummy variable for North-South division (1 – Northern country; 0 – Southern country);

**Table 4.** The partial correlation coefficients between FINLIN and key indicators.

| Key Indicators | Correlation | Partial correlation | | |
|---|---|---|---|---|
| | | Controlling for: | | |
| | | GDP per capita | North-South | West-East |
| $y_1$ | **0.527** | 0.378 | **0.406** | **0.520** |
| $y_2$ | 0.339 | 0.074 | 0.058 | 0.311 |
| $y_3$ | 0.171 | 0.048 | −0.030 | 0.138 |
| $y_4$ | 0.395 | 0.140 | **0.485** | 0.370 |
| $y_5$ | **0.464** | 0.240 | **0.489** | **0.485** |

*Correlation coefficients that were significant at 5% level were bolded

*WE* – dummy variable for West-East division (1 – Western country; 0 – Eastern country).

During the evaluation of the models some variable proved to not had significant influence over the key indicators. Those variables were removed from the equations via step-wise procedure one variable at the time. The only exception was FINLIT score. As the aim of the study was to establish the financial literacy relation with key variables of household financial behaviour, FINLIT variable was left in the equations even when its corresponding parameters were not significant.

Full information about the models solving was presented in the Appendix. The diagnostic check of the models' residuals revealed that they can be treated as having common variance – there were not heteroscedasticity problems as all White [41] test statistics were not significant at 5% level. In four of the five models the residuals could be regarded as normally distributed – the Jarque-Bera [42] test statistics were not significant at 5%. Only in one model – for the indicator *Gross debt-to-income ratio of households* the test statistics was significant at 5% but not significant at 1%. There could be some deviations from normal distribution for that model.

In four of the models the F-test proved to be significant at 5%, thus establishing presence of relationship between key indicators and explanatory variables. The R-squared statistics showed that variance explained by the models and variables in them varied between 38% (for *Gross debt-to-income ratio of households*) and 55% (for *Household net financial assets ratio*). In the model for *Households investment to GDP ratio* there was not enough evidence for such relationship.

The summary of the results was presented in Table 5. For the *Gross household saving rate* the FINLIT score showed positive significant influence – increase in score by 1 point was associated with increase in saving rate by 0.25 percentage points. The West-East division also proved to be significant as western countries experienced higher saving rate by about 4.4 percentage points. This result is significantly related with the households' income. The income rate in Western and Northern European countries traditionally is higher than the Eastern [43], which means the Western countries population have higher

**Table 5.** Summary of Models for the Key indicators.

| Variable | Coefficient | Std. Error | t-Statistic | Prob |
|---|---|---|---|---|
| Dependent Variable: *Gross household saving rate* | | | | |
| Constant | −6.011 | 4.609 | −1.304 | 0.206 |
| WE | **4.403** | 1.983 | 2.221 | 0.038 |
| FINLIT | **0.248** | 0.089 | 2.788 | 0.011 |
| Dependent Variable: *Gross investment rate of households* | | | | |
| Constant | **5.682** | 1.957 | 2.904 | 0.009 |
| NS | **2.181** | 1.100 | 1.982 | 0.061 |
| FINLIT | −0.044 | 0.049 | −0.890 | 0.384 |
| GDP | **0.000** | 0.000 | 2.932 | 0.008 |
| Dependent Variable: *Households investment to GDP ratio* | | | | |
| Constant | **3.172** | 1.219 | 2.602 | 0.016 |
| FINLIT | 0.019 | 0.023 | 0.814 | 0.425 |
| Dependent Variable: *Gross debt-to-income ratio of households* | | | | |
| Constant | 15.762 | 48.360 | 0.326 | 0.748 |
| FINLIT | 0.680 | 1.047 | 0.650 | 0.523 |
| GDP | **0.002** | 0.001 | 2.773 | 0.011 |
| Dependent Variable: *Household net financial assets ratio* | | | | |
| Constant | −0.139 | 0.630 | −0.221 | 0.827 |
| WE | **1.086** | 0.271 | 4.007 | 0.001 |
| FINLIT | **0.031** | 0.012 | 2.544 | 0.019 |

*Regression coefficients that were significant at 5% level were bolded

disposable income to manage, e.g. to save, invest etc. Saving is one of the most popular financial decisions and it does not require in and depth knowledge and understanding of personal finance management or financial literacy, which explains that there is a positive significant influence in all EU countries analysed. Even the poor financial literacy of population means that people understand the significance of savings for their well-being and they save. While the effects seemed small they were none the less evidence that good level of financial literacy is connected with savings and thus – formed sustainable financial behaviour of the households.

*Gross investment rate of households* was influenced by two factors: this time important was division North-South – northern countries had 2.2 percentage points higher gross investment rate, while every 1000 euros of GDP per capita led to increase in investment rate of households by about 0.08 percentage points. The influence of FINLIT though was not significant for this indicator. According to the results of Standard and Poor's Global financial literacy survey, Denmark, Germany, the Netherlands, and Sweden have the highest literacy rates in the European Union: at least 65 per cent of their adults are financially literate [34], which explains why northern countries point out 2.2 percentage points higher gross investment rate. Their population is financially literate which fosters their willingness to invest. They have more profound sustainable financial behaviour – knowledge of different financial and investing instruments and the skills to exploit them to increase their wealth, thus improving further their well-being.

The key indicator *Households investment to GDP ratio* was not influenced by FINLIT score neither by any of the control variables. Obviously, there were other factors that could be used to explain country's variation in household investment to GDP, because it was affected by GDP per capita, North-South or West-East divisions. Further research is needed in that area.

*Gross debt-to-income ratio of households* was influenced by GDP per capita. For every 1000 euros, gross debt ratio increased by about 2 percentage points. The financial literacy score effect was not significant. Indicator *Household net financial assets ratio* was positively by FINLIT score. In countries with 1 point, higher financial literacy score the net assets ratio was higher by about 0.03 years of income. The effect of the West-East division was bigger though, as in western counties household assets were larger than in the eastern by about 1.09 years of income. Across the EU, the financial assets of households mainly comprise currency and deposits, equity and investment fund shares, and assets held with life insurance companies and pension funds [44]. To a wide extent, the skills for managing these financial assets is related to the financial literacy rate.

Poor financial literacy means that individuals have a limited understanding of financial assets and the possibilities of the distribution of wealth. This statement is confirmed by the results – the more financially literate western European countries households' assets are larger than in the eastern. While effect of the financial education was smaller compared to the West-East division, it was important factor for the sustainable growth as contributor of wealth.

## 5 Conclusion

The financial literacy of the population defines to a big extent the economic development and sustainability of the countries. Poor financial literacy is among the factors that influence negatively not only individuals' or households' development, but also the economic development of the state. Our research examined the key indicators, that describe the financial status, behaviour and competence of the households. The conclusion from the implemented models' summary pointed to the fact that financial education was affecting the Gross household saving rate and Household net financial assets ratio but on both occasions, the West-East division played also a significant role and was more important than the financial literacy score. On the one hand, the western European countries are better financially literate than the eastern ones, on the other – most western countries have a better understanding of the importance of financial education and implement different policies aiming to foster people of all ages to be financially literate. The better understanding of financial instruments and the preconditions their appropriate usage creates for increasing personal wealth, the bigger the willingness to invest and distribute your income to gain extra value.

The major limiting factor in the study was connected to the available information, as the last worldwide survey about the financial literacy of the population had been performed in 2014 and no new data could be obtained for the European Union member states while the research was performed. The potential future research could be focused on the rising problem of financial literacy of graduate students. Given the high rate of unemployment of young people in European Union, low financial literacy is big obstacle not just for their realization in the labour market, but also when starting their own business, generating income and savings, and managing their financial flows. This research area will help with the investigation of the relation between youth's financial literacy and the transitional stage from education to career development. Having in mind the differences between eastern and western European countries in their economic development and financial literacy rates, and the results of this study, we confirm the necessity of designing and implementing a financial education curriculum in each educational stage starting from the early education, going through all formal educational phases and going to the education of adults. Financial literacy's influence will continue to grow in all economic sectors which means its' observation and analysis will become of more importance to the countries. In this regard, we assume that the EU policymakers and institutions should perform more adequate actions regarding collecting and analysing data about the financial literacy rate among the EU member states.

## Appendix

Model results (EViews 7.0).

| | | | | |
|---|---|---|---|---|
| Dependent variable: Y1 | | | | |
| Method: Least squares | | | | |
| Date: 08/02/22 Time: 11:54 | | | | |
| Sample: 1 24 | | | | |
| Included observations: 24 | | | | |
| Variable | Coefficient | Std. Error | t-Statistic | Prob |
| C | −6.011133 | 4.609493 | −1.304077 | 0.2063 |
| WE | 4.403195 | 1.982852 | 2.220638 | 0.0375 |
| FINLIT | 0.248023 | 0.088972 | 2.787636 | 0.011 |
| R-squared | 0.415476 | Mean dependent var | | 8.6975 |
| Adjusted R-squared | 0.359807 | S.D. dependent var | | 5.979022 |
| S.E. of regression | 4.783937 | Akaike info criterion | | 6.084874 |
| Sum squared resid | 480.6072 | Schwarz criterion | | 6.23213 |
| Log likelihood | −70.01848 | Hannan-Quinn criter | | 6.123941 |
| F-statistic | 7.463344 | Durbin-Watson stat | | 1.791544 |

(*continued*)

(*continued*)

| Dependent variable: Y1 | | | |
|---|---|---|---|
| Prob(F-statistic) | 0.00356 | | |
| Heteroskedasticity Test: White | | | |
| F-statistic | 1.984927 | Prob. F(4,19) | 0.1378 |
| Obs*R-squared | 7.073312 | Prob. Chi-Square(4) | 0.1321 |
| Scaled explained SS | 5.849963 | Prob. Chi-Square(4) | 0.2106 |

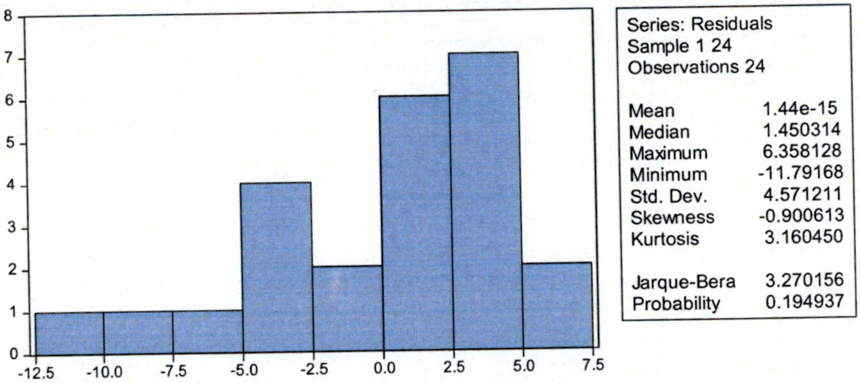


| Dependent variable: Y2 | | | | |
|---|---|---|---|---|
| Method: Least squares | | | | |
| Date: 08/02/22 Time: 11:54 | | | | |
| Sample: 1 24 | | | | |
| Included observations: 24 | | | | |
| Variable | Coefficient | Std. Error | t-Statistic | Prob |
| C | 5.68191 | 1.95679 | 2.90369 | 0.0088 |
| NS | 2.180811 | 1.100141 | 1.982301 | 0.0614 |
| FINLIT | −0.04374 | 0.04914 | −0.89003 | 0.384 |
| GDP | 7.88E-05 | 2.69E-05 | 2.931995 | 0.0082 |
| R-squared | 0.453466 | Mean dependent var | | 7.060417 |
| Adjusted R-squared | 0.371485 | S.D. dependent var | | 2.360391 |
| S.E. of regression | 1.871293 | Akaike info criterion | | 4.242148 |
| Sum squared resid | 70.03471 | Schwarz criterion | | 4.43849 |
| Log likelihood | −46.9058 | Hannan-Quinn criter | | 4.294237 |
| F-statistic | 5.531408 | Durbin-Watson stat | | 2.026911 |

(*continued*)

(*continued*)

| Dependent variable: Y2 | | | |
|---|---|---|---|
| Prob(F-statistic) | 0.006235 | | |
| Heteroskedasticity Test: White | | | |
| F-statistic | 0.501425 | Prob. F(8,15) | 0.8369 |
| Obs*R-squared | 5.063996 | Prob. Chi-Square(8) | 0.7507 |
| Scaled explained SS | 3.701978 | Prob. Chi-Square(8) | 0.883 |

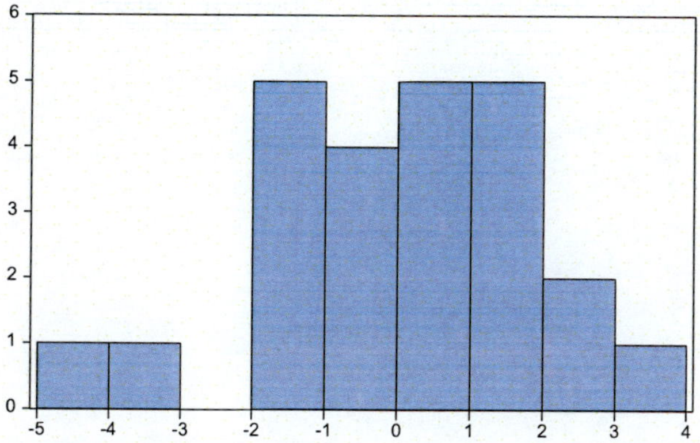


| Series: Residuals | |
|---|---|
| Sample 1 24 | |
| Observations 24 | |
| Mean | -7.40e-17 |
| Median | 0.112139 |
| Maximum | 3.062665 |
| Minimum | -4.280095 |
| Std. Dev. | 1.744989 |
| Skewness | -0.562433 |
| Kurtosis | 3.105392 |
| Jarque-Bera | 1.276430 |
| Probability | 0.528234 |

| Dependent Variable: Y3 | | | | |
|---|---|---|---|---|
| Method: Least Squares | | | | |
| Date: 08/02/22 Time: 12:51 | | | | |
| Sample: 1 24 | | | | |
| Included observations: 24 | | | | |
| Variable | Coefficient | Std. Error | t-Statistic | Prob |
| C | 3.171617 | 1.218746 | 2.602362 | 0.0163 |
| FINLIT | 0.018942 | 0.023277 | 0.813769 | 0.4245 |
| R-squared | 0.029221 | Mean dependent var | | 4.140833 |
| Adjusted R-squared | -0.01491 | S.D. dependent var | | 1.256946 |
| S.E. of regression | 1.266278 | Akaike info criterion | | 3.389697 |
| Sum squared resid | 35.27614 | Schwarz criterion | | 3.487868 |
| Log likelihood | −38.6764 | Hannan-Quinn criter | | 3.415742 |
| F-statistic | 0.66222 | Durbin-Watson stat | | 2.0414 |
| Prob(F-statistic) | 0.424498 | | | |

(*continued*)

(*continued*)

| | | | |
|---|---|---|---|
| Dependent Variable: Y3 | | | |
| Heteroskedasticity Test: White | | | |
| F-statistic | 1.302553 | Prob. F(2,21) | 0.2929 |
| Obs*R-squared | 2.648687 | Prob. Chi-Square(2) | 0.266 |
| Scaled explained SS | 1.047821 | Prob. Chi-Square(2) | 0.5922 |

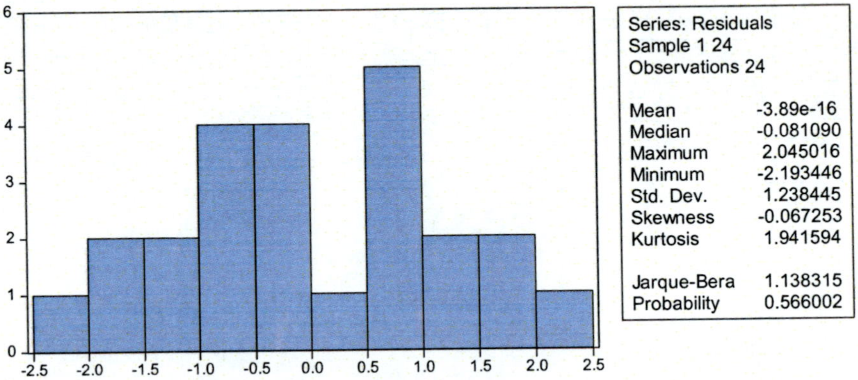


| | | | | |
|---|---|---|---|---|
| Dependent variable: Y4 | | | | |
| Method: Least squares | | | | |
| Date: 08/02/22 Time: 11:54 | | | | |
| Sample: 1 24 | | | | |
| Included observations: 24 | | | | |
| Variable | Coefficient | Std. Error | t-Statistic | Prob |
| C | 15.7621 | 48.35953 | 0.325936 | 0.7477 |
| FINLIT | 0.680185 | 1.046664 | 0.64986 | 0.5228 |
| GDP | 0.001977 | 0.000713 | 2.772665 | 0.0114 |
| R-squared | 0.38223 | Mean dependent var | | 102.5275 |
| Adjusted R-squared | 0.323395 | S.D. dependent var | | 60.34667 |
| S.E. of regression | 49.63877 | Akaike info criterion | | 10.76389 |
| Sum squared resid | 51744.15 | Schwarz criterion | | 10.91115 |
| Log likelihood | −126.167 | Hannan-Quinn criter | | 10.80296 |
| F-statistic | 6.496614 | Durbin-Watson stat | | 2.559884 |

(*continued*)

(*continued*)

| Dependent variable: Y4 | | | |
|---|---|---|---|
| Prob(F-statistic) | 0.006363 | | |
| Heteroskedasticity test: White | | | |
| F-statistic | 1.360773 | Prob. F(5,18) | 0.2849 |
| Obs*R-squared | 6.583361 | Prob. Chi-Square(5) | 0.2535 |
| Scaled explained SS | 6.878252 | Prob. Chi-Square(5) | 0.2299 |

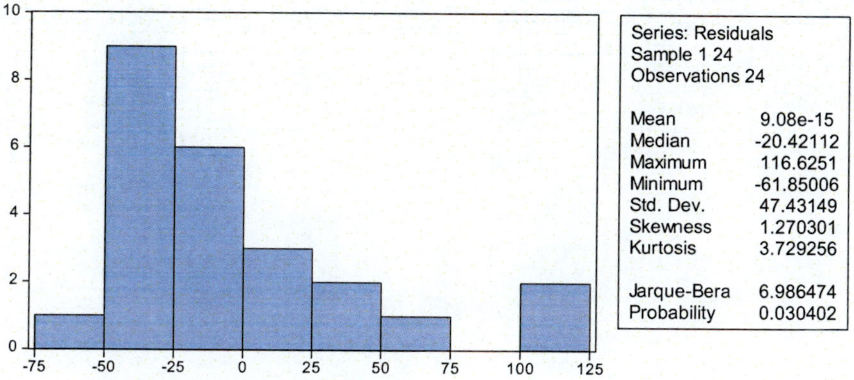


| Dependent variable: Y5 | | | | |
|---|---|---|---|---|
| Method: Least squares | | | | |
| Date: 08/02/22 Time: 11:55 | | | | |
| Sample: 1 24 | | | | |
| Included observations: 24 | | | | |
| Variable | Coefficient | Std. Error | t-Statistic | Prob |
| C | −0.13912 | 0.630016 | −0.22081 | 0.8274 |
| WE | 1.085972 | 0.271012 | 4.007102 | 0.0006 |
| FINLIT | 0.030933 | 0.012161 | 2.543692 | 0.0189 |
| R-squared | 0.55535 | Mean dependent var | | 1.941346 |
| Adjusted R-squared | 0.513003 | S.D. dependent var | | 0.936958 |
| S.E. of regression | 0.653858 | Akaike info criterion | | 2.104616 |
| Sum squared resid | 8.978144 | Schwarz criterion | | 2.251873 |
| Log likelihood | −22.2554 | Hannan-Quinn criter | | 2.143684 |
| F-statistic | 13.11409 | Durbin-Watson stat | | 1.544486 |
| Prob(F-statistic) | 0.000201 | | | |

(*continued*)

(*continued*)

| | | | |
|---|---|---|---|
| Dependent variable: Y5 | | | |
| Heteroskedasticity test: white | | | |
| F-statistic | 1.476518 | Prob. F(4,19) | 0.2486 |
| Obs*R-squared | 5.69121 | Prob. Chi-Square(4) | 0.2234 |
| Scaled explained SS | 4.651064 | Prob. Chi-Square(4) | 0.325 |

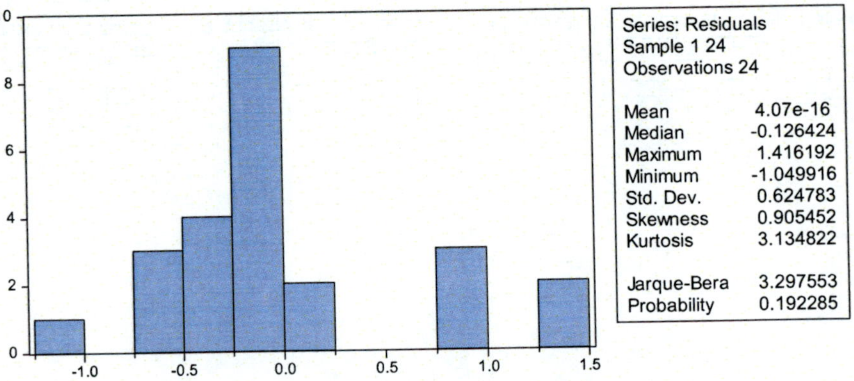


## References


1. Agarwal, S., Driscoll, J.C., Gabaix, X., Laibson, D.: The age of reason: financial decisions over the life cycle and implications for regulation. Brookings Pap. Econ. Activ. **2**, 51–117 (2009)
2. Yuesti, A., Ni, W.R., Suryandari, N.N.A.: Financial literacy in the COVID-19 pandemic: pressure conditions in Indonesia. Entrepreneur. Sustain. Issues. **8**(1), 884–898 (2020)
3. Damayanti, S.M., Murtaqi, I., Pradana, H.A.: The importance of financial literacy in global economic era. Busin. Manag. Rev. **9**(3), 435–441 (2018)
4. Van Nguyen, H., et al.: Understanding financial literacy and associated factors among adult population in a low-middle income country. Heliyon **8**(6), e09638 (2022). https://doi.org/10.1016/j.heliyon.2022.e09638
5. Potrich, A.C., Vieira, K.M., Kirch, G.: Determinants of financial literacy: analysis of the influence of socioeconomic and demographic variables. Account. Financ. Rev. **26**(69), 362–377 (2015)
6. OECD/INFE: OECD/INFE Toolkit to Measure Financial Literacy and Financial Inclusion: Guidance, Core Questionnaire and Supplementary Questions (2013)
7. European Commission: Financial competence framework for adults in the European Union. https://ec.europa.eu/info/sites/default/files/business_economy_euro/banking_and_finance/documents/220111-financial-competence-framework-adults_en.pdf (2022) Accessed 29 July 2022
8. Habschick, M., Seidl, B., Evers, J.: Survey of Financial Literacy Schemes in the EU27. VT Markt/2006/26H - Final Report (2007). https://edufinanciera.com/wp-content/uploads/2014/02/report_survey_en.pdf

9. Kuada, J.: Financial inclusion and the sustainable development goals. In: Makina, D. (ed.) Extending financial inclusion in Africa, pp. 259–277. Elsevier, Academic Press (2019)
10. United Nations, Economic and Social Council: Progress towards the Sustainable Development Goals Report of the Secretary-General (2022). https://sustainabledevelopment.un.org/content/documents/29858SG_SDG_Progress_Report_2022.pdf
11. Hastings, J.S., Madrian, B.C., Skimmyhorn, W.L.: Financial literacy, financial education, and economic outcomes. Ann. Rev. Econ. **5**, 347–373 (2013)
12. Widdowson, D., Hailwood, K.: Financial literacy and its role in promoting a sound financial system. Reserve Bank New Zealand Bull. **70**(2), 37–47 (2007)
13. Lusardi, A., Mitchell, O.: The economic importance of financial literacy: theory and evidence. J. Econ. Literat. **52**(1), 5–44 (2014). https://doi.org/10.1257/jel.52.1.5
14. Disney, R., Gathergood, J.: Financial literacy and consumer credit portfolios. University of Nottingham, Mimeo (2012)
15. Lin, J. T., Bumcrot, C., Mottola, G., Valdes, O., Ganem, R., Kieffer, C., Lusardi, A., Walsh, G.: Financial Capability in the United States: Highlights from the FINRA Foundation National Financial Capability Study (5th Edition). FINRA Investor Education Foundation (2022). www.FINRAFoundation.org/NFCSReport2021
16. French, D., McKillop, D.: Financial literacy and over-indebtedness in low-income households. Int. Rev. Financ. Anal. **48**, 1–11 (2016)
17. Bucks, B., Pence, K.: Do borrowers know their mortgage terms? J. Urban Econ. **64**, 218–233 (2008)
18. Campbell, J.: Household finance. J. Financ. **61**, 1553–1604 (2006)
19. Disney, R., Gathergood, J.: Financial literacy and consumer credit portfolios. J. Bank. Financ. **37**, 2246–2254 (2013)
20. Duca, J.V., Kumar, A.: Financial literacy and mortgage equity withdrawals. J. Urban Econ. **80**, 62–75 (2014)
21. Gerardi, K., Goette, L., Meier, S.: Financial literacy and subprime mortgage delinquency: evidence from a survey matched to administrative data. Federal Reserve Bank of Atlanta Working Paper 2010–10 (2010)
22. Wang, H., Kou, G., Peng, Y.: Multi-class misclassification cost matrix for credit ratings in peer-to-peer lending. J. Oper. Res. Soc. **72**(4), 923–934 (2021)
23. Bourova, E., Anderson, M.E., Ramsay, I., Ali, P.: Impacts of financial literacy and confidence on the severity of financial hardship in Australia. Austr. Account. Bus. Fin. J. **12**(4), 4–23 (2018)
24. Van Rooij, M., Lusardi, A., Alessie, R.: Financial literacy and stock market participation. J. Financ. Econ. **101**(2), 449–472 (2011)
25. Cardak, B.A., Wilkins, R.: The determinants of household risky asset holdings: Australian evidence on background risk and other factors. J. Bank. Financ. **33**(5), 850–860 (2009). https://doi.org/10.1016/j.jbankfin.2008.09.021
26. Christelis, D., Jappelli, T., Padula, M.: Cognitive abilities and portfolio choice. Eur. Econ. Rev. **54**(1), 18–38 (2009). https://doi.org/10.1016/j.euroecorev.2009.04.001
27. Lin, C.-A., Bates, T.C.: Smart people know how the economy works: cognitive ability, economic knowledge and financial literacy. Intelligence **93**, 101667 (2022). https://doi.org/10.1016/j.intell.2022.101667
28. Lusardi, A., Mitchell, O.S.: 2011 Financial literacy around the world: an overview. J. Pens. Econ. Financ. **10**(4), 497–508 (2011). https://doi.org/10.1017/S1474747211000448
29. Anderson, A., Baker, F., Robinson, D.T.: Precautionary savings, retirement planning and misperceptions of financial literacy. J. Financ. Econ. **126**(2), 383–398 (2017)
30. Wagner, J., Walstad, W.B.: The effects of financial education on short-term and long-term financial behaviors. J Consum Aff. **53**, 234–259 (2019). https://doi.org/10.1111/joca.12210

31. de Bassa Scheresberg, C.: Financial Literacy and Financial Behavior among Young Adults: Evidence and Implications. Numeracy **6**(2), Article 5 (2013). http://dx.doi.org/https://doi.org/10.5038/1936-4660.6.2.5
32. Yoong, J.: Financial illiteracy and stock market participation: evidence from the RAND American life panel. Financ. Literacy: Implicat. Retirem. Secur. Financ. Marketplace **76**, 39 (2011)
33. Batsaikhan, U., Demertzis, M.: Financial literacy and inclusive growth in the European union. Policy Contrib. **8**, 25536 (2018)
34. Standard & Poor's: Financial Literacy Around the World: Standard & Poor's Ratings Services Global Financial Literacy Survey (2014). https://gflec.org/wp-content/uploads/2015/11/3313-Finlit_Report_FINAL-5.11.16.pdf?x70470. Accessed 29 July 2022
35. Klapper, L., Lusardi, A., Van Oudheusden, P.: Financial literacy around the World: insights from the S&P Global FINIT Survey. W. Bank (2015). https://responsiblefinanceforum.org/wp-content/uploads/2015/12/2015-Finlit_paper_17_F3_SINGLES.pdf
36. Consolidated text: Regulation (EU) No 549/2013 of the European Parliament and of the Council of 21 May 2013 on the European system of national and regional ac-counts in the European Union (Text with EEA relevance), ELI: http://data.europa.eu/eli/reg/2013/549/2015-08-24 (2013). Accessed 29 July 2022
37. Piketty, T.: Capital in the Twenty-First Century. Harvard Business School (2014)
38. Eurostat: Economy and Finance. National Accounts (ESA 2010). Annual sector accounts (ESA 2010). Key indicators - annual data (2022). https://ec.europa.eu/eurostat/databrowser/view/nasa_10_ki/default/table?lang=en (2022). Accessed 29 July 2022
39. Francis, B., Sunday, I.: Education and development in the Caribbean: a cointegration and causality approach. Econ. Bull. **15**(2), 1–13 (2006). https://econpapers.repec.org/article/eblecbull/eb-05o10022.htm. Accessed 29 July 2022
40. Goczek, Ł, Witkowska, E., Witkowski, B.: How does education quality affect economic growth? Sustainability. **13**, 6437 (2021). https://doi.org/10.3390/su13116437
41. White, H.: A heteroskedasticity-consistent covariance matrix estimator and a direct test for heteroskedasticity. Econometrica **48**(4), 817–838 (1980). https://doi.org/10.2307/1912934
42. Jarque, C.M., Bera, A.K.: Efficient tests for normality, homoscedasticity and serial independence of regression residuals. Econ. Lett. **6**(3), 255–259 (1980). https://doi.org/10.1016/0165-1765(80)90024-5
43. Eurostat: Mean and median income by household type - EU-SILC and ECHP surveys (2022). https://appsso.eurostat.ec.europa.eu/nui/show.do?dataset=ilc_di04. Accessed 30 July 2022
44. Eurostat: Households - statistics on financial assets and liabilities (2022). https://ec.europa.eu/eurostat/statistics-explained/index.php?title=Households_-_statistics_on_financial_assets_and_liabilities. Accessed 30 July 2022

# Evaluation of the Business Process Sustainable Value Chain Based on Enterprise Cost Management

Olena Sova[1(✉)], Nataliia Bieliaieva[2], Victoriia Khmurova[2], and Valentyna Khrapkina[3]

[1] Ptoukha Institute for Demography and Social Studies of the, National Academy of Sciences of Ukraine, 60 Tarasa Shevchenka Blvd, Kyiv 01032, Ukraine
sovy@ukr.net

[2] State University of Trade and Economics, 19 Kyoto St, Kyiv 02156, Ukraine

[3] National University of "Kyiv-Mohyla Academy", 2 Skovorody St, Kyiv 04070, Ukraine

**Abstract.** The article is about investigating of evaluation's peculiarities of the business process value chain based on cost management at the example of industrial enterprise in a circular economy. The economy should urgently be redesigned so that resources aren't wasted. Intentions to provide maximum level of energy efficiency of production, green occupations and sustainability, reduce the level of resources spending are promising areas of development each industry and the primary tasks of advanced countries and companies. It is investigated the production costs formation in the production sector and analysed factors influencing profit of industrial enterprises for sustainability. The results of this research indicate that the application of ABC can improve the accuracy of cost per unit calculation. It is proved that cost management of industrial enterprises should be considered as an independent module in the sytem of enterprise economic management, especially in terms of sustainability, the basis of which is the decisions making process grounded on organizational and methodological system of formation, cost allocation and cost industrial of products in accordance with long-term enterprise's goals. On this basis, activity-based management tools can contribute to a more accurate calculation of environmental costs, in line with a sustainable product value chain. The ways of reducing the cost of products based on the cost formation centers have been substantiated. It is obtained that general economic costs and their distribution by types should combined with the management theory and the idea of sustainability to view business as contributing to society's sustainable development.

**Keywords:** Cost managment · Business process value chain · Activity-based costing · Economic modelling · Sustainable development and green economy


V. Koval et al. (Eds.): ISCMEE 2022, Lect. Notes in Managem., Industrial Engineering, pp. 155–169, 2023.
https://doi.org/10.1007/978-3-031-23463-7_10

## 1 Introduction

The cost of production is the most important indicator of the impact on production efficiency and economic activity of the industrial enterprises. The formation of an environmentally oriented and lowcarbon economy in modern conditions in many countries is a priority and is associated with the advanced develoment of new types of employment in the labour market and the creation of new jobs. Opportunities to create millions of green jobs are an answer and a result of transition to a decarbonised economy, which is environmentally friendly, and the circular economy, which involves reusing, repairing or recycling, increasing sustainable manufacturing and consumption. However, implementation of circular economy model isn't so easy due to a policy, technology and public involvement issues.

For instance, long-run growth of agricultural activity in the context of market competition stems from the introduction of new strategic cost management methods focused on the efficient use of all available resources. For instance, the high technical complexity of the product leads to the fact that up to 90% of production costs are determined precisely at the R&D stage. Anyway, cost management is the part of the management of the enterprise as a whole. Thus, it is important to choose the optimal method of cost accounting and cost management that will help to be stable for the enterprise in all operating conditions and make stable its business process value chain.

In order to grow in a sustainable environment, businesses must take a close look at all their business processes, approval mechanism and cost analysis, supply chain systems, automation levels, purchasing policies, and employee performance. Advanced weakness analysis will open up opportunities for more sustainable cost management in the enterprise.

Especially it's important for industrial enterprises as, for example, agricultural sector, due to its specificity, which can include dependence on natural conditions, the presence of constant risks, high labor costs and dependence on world prices for certain types of agricultural products, always struggle not only from non-efficient management system, but also from these and others external circumstances. Stable business process value chain based on effective cost management is that tool of the management system, which can provide industrial enterprise with a high economic result for Sustainability and Circularity.

It's also important to note, that in times of sustainable development and cost reducing, it's worth not to forget about influence of green economy. Greening the small farm sector through the promotion and dissemination of sustainable agriculture can be the most effective way to increase the availability of food available to the poor people, carbon storage and the availability of growing international markets for green products. The introduction of green jobs in small enterprises and especially in the field of agriculture should be influenced by the state as a guarantor of the implementation of the support system for agriculture, especially in certain agricultural areas, in particular, on the legal regulation of this process.

The purpose of the article is to investigate peculiarities of evaluation of the sustainable value chain based on cost management at the example of industrial enterprise's business process.

## 2 Literature Review

DIfferent scientists give different results, so let's look deeper and stop at the most relevant for this research. Scientific works that made a great impact in area of business process value chain based on cost managing optimization and sustainable development, are listed further.

There are numerous scientific researchings connected with industrial enterprises' activity, and both agrarian, but mostly they are about peculiarities of activity, but not about cost management [1–7]. More grounded are works of such scientists as I. Pushkar [8] and A. Fitó-Bertran, J. Llobet & N. Cugueró-Escofet [9] talked about activity-based costing model as progressive method of cost management, and it's a good base to use it for industrial enterprises. Further R. Sodoma, H. Skhidnytska, A. Shvorak, T. Shmatkovska & I. Zhurakovska [10] obtained the necessity of industrial receipts introducing as a new efficient instrument. S. Mustafiz, A. Nakayasu & M. Itabashi [11] pointed also, and it's important for this research, that a lack of market information, poor institutions and arrangements, poor marketing infrastructures, transportation system, and high and unfair profit margin distribution among the value chain actors influence a lot at the activity of industrial enterprises. However, M. Stępień [12] made a conclusion that there are no evidential solutions in the scope of uniform identification of costs incurred for solutions implementing the sustainable development strategy in the company. D. Malone [13] has offered two cost management tools that can assist the firm in seeking to be environmentally sustainable while still operating cost effectively. And he also highlighted another critical component of environmental management – decisions related to managing an sustainability initiative are evaluated in the same way as all other decisions across the organization. For most managers, that common language is dollars. That is important issue, when talking about sustainable value chain based on cost management.

However, that there are many scientific works in the field of cost management in the enterprise, but the impact of the cost management mechanism on the stability of the value chain of the process in the context of the concept of sustainability in a business remains insufficiently developed.

## 3 Research Methodology

Research methods are a set of theoretical and methodological, economic, mathematical and applied issues of definition, development and implementation of effective cost management in business process value chain decisions at an industrial enterprise with consideration of sustainable development. It was analysed scientific articles connected with managing an organization on the basis of its business processes, cost management, features of industrial enterprises activity, sustainability.

Carrying out functional-cost analysis of peculiar subjects implies gradual transition from use the method for solution of separate specific tasks and problems to it applying as the main lever for control of expenses and products' quality at all stages of their business process.

Economic and mathematical modelling methods used to asses that the number of equations in system depends on the degree of detail of the production process and the number of links in the value chain – the value of $R_1$ is proportional to $C_1$, as is the cost of the business process $x_1$.

It is important to understand that a reduction a company turnover affects not only the well-being of the owners, but also the well-being of the whole country and the whole world. Since any company is the primary element of economy. The larger the volume of the company, the more taxes it pays to the state, and the more the nation will prosper. The main thing, that should be taken into account after conducted analysis (matrix analysis, economic and mathematical analysis, etc.), that the first step for the responsible person (manager) should be figuring out why there is such results, what is their dynamics, – and only after that – making a decision about the best options that need to be taken to change (if it is) a negative situation (especially if the indicators have deteriorated over a long period of time). For instance, the ABC methodology, combined with value chain analysis, allows an enterprise not only to cut costs step by step, but to identify excess resource consumption and redistribute them in order to increase productivity. In a case when enterprise wants to plan or optimize its sustainable value chain, it's necessary to investigate resource intensity in accordance with its indirect physical environmental costs. In such a case an enterprise gets more actual and wide information (especially from the point of view of environmental sustainability), that allows to plan strategic initiatives and react in time on environmental demands).

Value formation (final product) for the consumer and cost analysis for its creation A. Thompson & A. Strickland [14] proposes to determine on the basis of the cost chain, which in their opinion consists of: the cost of suppliers, the cost of production and economic activities of the enterprise, the cost of sales channels and marketing costs, the cost of product consumption. The cost chain shows that the competitive development strategy of the enterprise depends not only on internal costs but also on external ones (for example, supplier costs). Also, it should be taken into account than even with abundant renewable resources and well performing technologies measured using levelized cost of electricity and expected internal rate of return, investment on renewable electricity generation is quite low, so in some countries managements don't accept, for example, importance of "green" direction. To save energy, it is necessary not only to use energy-saving lamps, but also to properly plan windows in buildings, etc., and not all enterprises can afford it to themselfes. Anyway, with the help of Activity Based Costing (ABC methodology) it is possible to structurally display the process of cost and cost formation at the enterprise and to form an adequate information base for various components (divisions, processes, etc.) – to build a map of cost formation. Thus, this method allows to accurately determine production costs, control the amount and causes of costs, effectively manage costs and profits at the level of technological operations, increase the objectivity of the assessment of product profitability and business process efficiency [8].

The basis of ABC analysis in cost management is the concept of the so-called "cost center formation" (CCF), a separate functionally organizational process (business process) or phenomenon (technological operation, agreement, etc.), which is associated with the formation of a homogeneous the total cost of the enterprise. CCF can be considered on the scale of the entire enterprise, structural unit or other object of analysis. A typical set of cost centers of the enterprise, as well as the sustainable value chain of the enterprise, may include maintenance of equipment, supply of raw materials, delivery of products to customers, research and development, innovation, etc.

R. Kaplan in his work [15] gives examples of application of the Activity based costing approach for several enterprises. This method involves having information about the value of individual resources and business processes (hereinafter BP) in the value chain required to manufacture a particular product, in particular:

n – the number of BP;

m – the number of resources;

$x_i$ – the total cost of the *i*-th BP;

$R_i$ – the total cost of the *i*-th resource;

$a_{i,k}$ – the share of the value of the *i*-th resource, which is transferred to the cost of the *k*-th business process;

$b_{i,k}$ – the share of BP of the *i*-th operation, which is transferred to the cost of the *k*-th operation;

$d_{i,k}$ – the share of the value of the *i*-th BP, which is transferred to the value of the *k*-th resource;

$C_k$ – the initial cost of the *k*-th resource.

Thus, if for the production of a particular product it is necessary to attract 5 types of resources involved in four business processes, then to apply the ABC approach you need to have information on 65 different indicators, based on which you can calculate 9 more performance indicators, namely: $x_i$ – the full cost of the *i*-th business process; $R_i$ – the total cost of the *i*-th resource after the implementation of all necessary processes.

These exponents are found by solving such a system of algebraic equations:

$$\begin{cases} x_1 = \sum_{i=1}^{\infty} R_i \times a_{i,1} + \sum_{i=1}^{n} x_i \times b_{i,1}; \\ x_n = \sum_{i=1}^{\infty} R_i \times a_{i,n} + \sum_{i=1}^{n} x_i \times b_{i,n}; \\ R_1 = C_1 + \sum_{i=1}^{n} x_i \times d_{i,1}; \\ R_\infty = C_\infty + \sum_{i=1}^{n} x_i \times d_{i,\infty}; \end{cases} \tag{1}$$

The study shows that the number of equations in system (1) depends on the degree of detail of the production process and the number of links in the sustainable value chain (for example, we proposed six links).

It should be noted that the minimum number of equations in any situation will be 2, with $m = 1$ and $n = 1$, i.e., with one resource, which is convenient to express its monetary value, and with one business process, system (1) takes the form:

$$\begin{cases} x_1 = R_1 a_1 + x_1 b_{11}; \\ R_1 = C_1 + x_1 d_{11}. \end{cases} \tag{2}$$

Rewrite system (2), moving the unknown values of costs $x_1$ and $R_1$ to the left, and the known value of the resource price $C_1$ – to the right:

$$\begin{cases} (1 - b_{11})x_1 - a_{11}R_1 = 0 \\ -d_{11}x_1 + R_1 = C_1 \end{cases} \quad (3)$$

The initial price of resource $C_1$ should obviously be a positive value:

$$C_1 > 0 \quad (4)$$

since at zero resource price $C_1 = 0$ from the system (3) we also get that the cost of production can be zero ($R_1 = 0$; $x_1 = 0$), and such a case can be interesting only in theoretical terms.

Under condition (4), the system of Eq. (3) is inhomogeneous and has a unique solution if its principal determinant $\Delta$ is nonzero:

$$\Delta = \begin{vmatrix} 1 - b_{11} & -a_{11} \\ -d_{11} & 1 \end{vmatrix} \neq 0; \quad (5)$$
$$1 - b_{11} - d_{11}a_{11} \neq 0$$

Under condition (5), system (3) has the solution:

$$x_1 = \frac{\Delta_1}{\Delta}; \quad (6)$$

$$R_1 = \frac{\Delta_2}{\Delta} \quad (7)$$

where the auxiliary determinant $\Delta_1$ is calculated by the formulas:

$$\Delta_1 = \begin{vmatrix} 0 & -a_{11} \\ C_1 & 1 \end{vmatrix} \Rightarrow \Delta_1 = C_1a_{11}, \quad (8)$$

$$\Delta_2 = \begin{vmatrix} 1 - b_{11} & 0 \\ -d_{11} & C_1 \end{vmatrix}, \quad (9)$$
$$\Delta_2 = C_1(1 - b_{11})$$

Substituting the value of the determinant (8) in the formula (6), calculate the cost to operations $x_1$:

$$x_1 = \frac{C_1a_{11}}{1 - b_{11} - d_{11}a_{11}}. \quad (10)$$

Given that the sum of costs $x_1$ must be a positive value ($x_1 > 0$), based on formula (10) we obtain a condition that is more stringent than condition (5):

$$1 - b_{11} - d_{11}a_{11} > 0 \quad (11)$$

Let's analyse the dependence of the costs to operation $x_1$ on the parameters of the model. Formula (10) can be written as:

$$x_1 = q(a_{11}, b_{11}, d_{11})C_1, \tag{12}$$

where the coefficient $q(a_{11}, b_{11}, d_{11})$ depends on the parameters $a_{11}$, $b_{11}$ and $d_{11}$ and does not depend on the parameter $C_1$:

$$q(a_{11}, b_{11}, d_{11}) = \frac{a_{11}}{1 - b_{11} - d_{11}a_{11}}. \tag{13}$$

Therefore, on the basis of formulas (12) and (13) we obtain a conclusion about the directly proportional dependence of the value of $x_1$ on the initial price of the resource $C_1$. It is clear that such a conclusion is possible only if the other parameters do not change, and this is almost not the case in practice.

From the share of the cost of the business process $b_{11}$, which is transferred to the cost of the business process itself, the value of $x_1$depends nonlinearly on the monotonically increasing law, i.e:

$$b_{11}^{'} < b_{11}^{''} \Leftrightarrow x_1(b_{11}^{'}) < x_1(b_{11}^{''}). \tag{14}$$

Herewith, the lowest cost $x_1$ is reached at zero value $b_{11}$:

$$x_{1\ \min}\Big|_{b_{11}=0} = \frac{C_1 a_{11}}{1 - d_{11}a_{11}}. \tag{15}$$

The same character has the dependence of the cost of operation $x_1$ from the share of the cost of this business process, which is transferred to the cost of the resource, ie this dependence is also nonlinear and also monotonically increasing:

$$d_{11}^{'} < d_{11}^{''} \Leftrightarrow x_1(d_{11}^{'}) < x_1(d_{11}^{''}). \tag{16}$$

The lowest value of $x_1$ is reached at zero value of $d_{11}$:

$$x_{1\ \min}\Big|_{d_{11}=0} = \frac{C_1 a_{11}}{1 - b_{11}}. \tag{17}$$

Also, nonlinear and monotonically increasing dependence of the value of $x_1$ from the share of the cost of the resource, transferred to the cost of the business process, $a_{11}$,

$$a_{11}^{'} < a_{11}^{''} \Leftrightarrow x_1(a_{11}^{'}) < x_1(a_{11}^{''}). \tag{18}$$

However, the dependence of the cost of the business process $x_1$ from the parameter $a_{11}$ has a significant feature compared to the dependences of the value of $x_1$ from $d_{11}$ and $b_{11}$, namely that at zero share $a_{11}$ the cost of BP $x_1$ also becomes zero:

$$x_1 \bigg|_{a_{11}=0} = 0. \tag{19}$$

It is clear that the zero value of the business process from the value chain is not really possible to achieve, but this conclusion shows which factor can most significantly reduce the cost of the business process, which involves operations on a particular resource, i.e. parameter $a_{11}$.

Based on formulas (5), (7) and (9), calculate the total cost of the resource $R_1$ after the operation with it in a particular business process:

$$R_1 = \frac{C_1(1 - b_{11})}{1 - b_{11} - d_{11} - a_{11}}. \tag{20}$$

Thus, the value of $R_1$ is proportional to $C_1$, as is the cost of the business process $x_1$. From other parameters of the model $b_{11}$, $d_{11}$ and $a_{11}$, the value $R_1$ depends nonlinearly and in ascending order for each of them in particular.

Therefore, the application of this model will allow stabilize a business process value chain based on cost management optimization. Managing overhead costs proactively and with a long-term perspective can effectively break the circular cycle and permanently improve a business's cost competitiveness.The cost management and control component focuses on improving weaknesses in the way costs are managed to drive continuous improvement and sustainability.

## 4 Research Results and Discussion

Successful work in the global competition direction requires not only constant updating of the nomenclature and quality of products, but also a thorough analysis of the enterprise's activities to reduce unnecessary or duplicate functions (work). Often, enterprises, in pursuit of the goal of reducing costs, adopt a policy of total cost cutting. Such a solution is the worst, since with such a policy, all work is subject to reduction, regardless of its usefulness. The overall reduction can reduce the performance of major work, leading to deterioration overall quality and the productivity of the enterprise. A drop in productivity will lead to another wave of reductions, and that will once again reduce the efficiency of the enterprise.

Some authors have investigated the main common ways to reduce costs in crisis: reducing the production capacity (including the number of purchases, storage), changing the schedule of employees working day (taking into account daylight hours), reduction (closure) of sale points (which, among other things, provokes staff reduction also) [16].

Sustainability is gaining increasing public attention and attracting greater debate. And as our population becomes more environmentally aware, a host of opportunities in the sustainability and green jobs industry are becoming available. "Greening" are central to sustainable development and respond to the global challenges of enviromental protection, economic development and social inclusion. Peculiarities of the production process of, for example, an agricultural enterprise determine that its development depends on natural factors, the main means of production is land, and plants and animals simultaneously act as means and objects of labor, complicates the process of production management. For organizations to meet customers' expectations and be truly sustainable they must ensure responsible business practices inside their own organization and across their entire value chain.

At the same time, for example, in plant growing and animal husbandry, not one, but several types of products are obtained. Of these, one type of product is the main one (grain, root crops, marketable fish), the others are associated, or by-products. Associated products (milk and offspring, flax seeds and flax straw), along with the main one, have an independent value, and the secondary one is secondary, although it has consumer value (manure, straw, tops, feathers).

It should be taken into account while sustainable value chain that usual the costs of products manufacturing in a such main business process, are formed by responsibility centers, accounting objects and objects for calculating the cost of production. In this case, more often, ways to reduce costs will be as follows:

1. Increase in yield and productivity due to:

   - selection and seed production work;
   - improving technology (using the achievements of scientific and technological progress and, on this basis, increasing the technical level of production process);
   - improving animal feeding;
   - use of modern machines, combined units.

2. Better labor motivation, improvement of labor remuneration (for example, the higher the yield – the higher level of the wages).
3. Improving the organization of production process. Reduced general and production management costs.
4. Reduction of unproductive losses (elimination of losses from rejects and elimination of production costs).
5. Rational location of production and its cooperation.

At the same time, the crisis phenomena especially strongly affected the growth of production costs, especially constant ones. But in the way of their influence on the process of costs formation, environmental factors far exceeded the effect of internal factors. Proceeding from this, an important governmental task is to improve the mechanisms of state regulation in industrial sphere that will help to stabilize business processes at such enterprise.

However, for instance, there is a difficult competitive situation in land- and labour-intensive production in the EU countries (Central and Eastern Europe) and Mediterranean region. That is because the rational utilization of the production potential is limited by the structural deficit concerning with the fragmented agrarian structure and pointed in the low level of land and capital assets assigned to labour actively involved in the production process [17].

For a more efficient production of livestock products, the following cooperation scheme can be proposed, for example, in cattle breeding: raising pedigree young animals in one group of farms, raising replacement breeding stock (heifers and first-calf heifers) in another, milk production in the third, rearing super-repair young animals in the fourth, final fattening of livestock – in the fifth group of farms. The most important thing for the most industrial enterprises is the optimal concentration and specialization of production.

If at least one variety is ordered within the basic period $T$, the similar cost will be formed by one element. Similarly, it is possible to define the multiplicities with the periods $2T$, $3T$, etc. In each multiplicity there may be many elements, but there may be none too.

Optimization task lies in the following: to choose the basic period $T$ and to divide the range of products in such a way that general expenses (GE(T)) are minimal. First, we arrange ranges of products in order according the growth of the discussed ratio $\frac{C_i}{D_i C_{hi}}$, then number them. Further on we choose the original approximate value (in shares) of the basic period *(T)*. Then we group ranges of products, choosing the previous $k_i$ for each, at which delivery expenses are the smallest. In general terms, for a products range $i$ the optimal is the smallest whole $k$.

The presented methodology is advisable to be applied in cases when the range of products is big enough. Its weak point is an assumption as to the demand regularity and its definiteness.

For example, for "Eltex" Corporation, one of fabric suppliers is the suppling firm "Monevon" which offers 10 nomenclature fabric groups. The data of storage cost and demand intensity are tabulated in Table 1. The structure of order expenses is such that the cost constant equals 6 c.u. (per truck 4800), and a variable constituent for each range of products is the same and makes up 3.2 c.u.

Let's calculate general expenses according to the presented research methodology, calculations are tabulated in Table 2. The sum of 122.2690876 c.u. is taken as 100%.

Further on, according to the formula (3) we can calculate expenses on condition of general optimization: $\sqrt{2 \cdot 10 \cdot 3,2}\sqrt{112,16} = 84,7245$. This value is much lower than the sum calculated according to the separable optimization. Let's make calculations for the divisible periods' strategy. According to the algorithm, we arrange the ranges of products as to the growth of the relator $\frac{C_i}{D_i C_{hi}}$ and number them in Table 3.

Now we choose the primary approximate at product range B, as the value $\frac{C_i}{D_i C_{hi}}$ for it is the lowest. We plug the data into the formula (5); then we group product range groups, choosing a separate $k$ for each one, on the condition (6). We take into.

consideration that product range group B has $k_1 = 1$.

**Table 1.** Input information as to storage cost and demand intensity for fabrics

| № | Name of fabrics | | Articles | Manufacturers | Storage cost ($C_{hi}$), c.u | Demand intensity ($D_i$)roll/period |
|---|---|---|---|---|---|---|
| | Code | Kind of fabrics | | | | |
| 1 | A | Heavy melton | Ital01, Ital02, Ital03, Ital04, Ital05 | Italy 100% wool | 0.48 | 42 |
| 2 | B | Cashmere | к-60, к-65 | Turkey 60% wool | 0.48 | 50 |
| 3 | C | Wool | WB-13, WB-9 | Italy 50% wool | 0.48 | 30 |
| 4 | D | Wool | WB-12, WB-11 | Turkey 50% wool | 0.48 | 30 |
| 5 | E | Boucle | W-1 | Italy | 0.36 | 22 |
| 6 | F | Boucle | BT1, SN, SN2 | Turkey | 0.36 | 18 |
| 7 | G | Corduroy velvet | PB-11, PB-24, PB-28 | Turkey | 0.24 | 34 |
| 8 | H | Polyester | BT-1, BT-2 | Turkey | 0.2 | 36 |
| 9 | I | Tweed | BT-H, BT-3 | Turkey | 0.44 | 10 |
| 10 | J | Knitted fabric | | Turkey | 0.28 | 18 |

*Source* Compiled on the information provided by the representatives of Monevon LLC

Separately we calculate the products $D_i C_{hi}$ for all product range groups, starting with the second one, and the value $\frac{2C}{T_0^2}$, as variable expenses are accepted on the level of 3.2 c.u. We calculate the value $k_I$ for other product range groups. The order sheet can be presented as a matrix (Table 4).

Consequently, the application of the divisible periods' strategy allowed reducing the expenses at 28.3% as compared to the separable optimization strategy. The presented methodology is advisable to be applied in cases when the range of products is big enough. **Its weak point is an assumption as to the demand regularity and its definiteness.**

It should be taken into account that the ABC method provides a number of advantages, in particular: this method helps to make informed decisions about reducing costs, because their real picture allows you to accurately determine the types of costs that need to be optimized. We can see this in the formulas (1)-(20). Agreeing with this opinion, we also note that on the basis of this method it is possible to adequately evaluate the inventive and innovative proposals to reduce certain types of costs and, accordingly, the implementation of a competent policy to encourage incentives for inventive and innovative activities of employees.

In addition, the ABC method contributes to the implementation of an adequate pricing policy, as the accurate allocation of costs to the objects of calculation makes it possible to

**Table 2.** Expenses calculation according to product range groups (fabrics varieties)

| Product range group | Delivery expenses, c.u./lot |
|---|---|
| 1 | 17.03830978 |
| 2 | 18.59032006 |
| 3 | 14.4 |
| 4 | 14.4 |
| 5 | 10.67932582 |
| 6 | 9.659813663 |
| 7 | 10.8399262 |
| 8 | 10.18233765 |
| 9 | 7.959899497 |
| 10 | 8.519154888 |
| Total | 122.2690876 |

*Source* Calculated on the basis of Table 1

**Table 3.** Arranged succession of product range groups

| № new | № old | Name of fabric | Criterion value |
|---|---|---|---|
| 1 | 2 | B | 0.133333333 |
| 2 | 1 | A | 0.158730159 |
| 3 | 3 | C | 0.222222222 |
| 4 | 4 | D | 0.222222222 |
| 5 | 7 | G | 0.392156863 |
| 6 | 5 | E | 0.404040404 |
| 7 | 8 | H | 0.444444444 |
| 8 | 6 | F | 0.49382716 |
| 9 | 10 | J | 0.634920635 |
| 10 | 9 | I | 0.727272727 |

*Source* Compiled on the basis of the conducted analysis, Tables 1, 2

determine the lower limit of prices, further reduction of which relative to such a limit led to unprofitable products. It should be noted here that the justification of pricing policy based on the ABC method can and should be used as a counterargument in anti-dumping prosecutions by unfair competitors. However, nowadays, the advancement of human and machine intelligence enables businesses to create a digital thread that provides visibility across the entire product and operating sustainable value chain.

**Table 4.** Arranged succession of product range groups

| № new | № old | Name of fabric | Product $D_iC_{hi}$ | Value | ki |
|---|---|---|---|---|---|
| 1 | 2 | B | 24 | 4.174 | 1 |
| 2 | 1 | A | 20.16 | 4.174 | 1 |
| 3 | 3 | C | 14.4 | 4.174 | 1 |
| 4 | 4 | D | 14.4 | 4.174 | 1 |
| 5 | 7 | G | 8.16 | 4.174 | 2 |
| 6 | 5 | E | 7.92 | 4.174 | 2 |
| 7 | 8 | H | 7.2 | 4.174 | 2 |
| 8 | 6 | F | 6.48 | 4.174 | 2 |
| 9 | 10 | J | 5.04 | 4.174 | 2 |
| 10 | 9 | I | 4.4 | 4.174 | 2 |

*Source* Compiled on the basis of the conducted analysis, Table 3

## 5 Conclusions

The proposed modification of the functional approach to cost formation allows to determine the full cost of the resource by its business processes and build a cost map along the value chain, on the basis of which a strategic cost management system can be built in accordance with sustainable principles. However, cost management should be understand as more than just decreasing costs; it's more about wise use of available resources – just continuously cutting of costs can lead to the deterioration of product quality, employee dissatisfaction, frequent equipment breakdowns, etc.

Special attention in further research needs to be taken into account the general economic costs and their distribution by types of crop production, determination of the characteristic bases of distribution; research on the creation of cost centers and responsibility centers with the allocation of the controlling department in the organizational and functional management structure of the agricultural enterprise; combining the management theory and the idea of sustainability to view business as contributing to society's sustainable development with justification. By 2030, the projected demand for energy and water will increase by 50% and 40% respectively. To meet needs new jobs will be created in areas such as manufacturing alternative energy, new technologies, new product development, recycling and use secondary resources. Millions of people employed in the field of traditional energy, and the industry as a whole will succumb rapid reorganization. The cost level indicator should become a quality indicator of the agricultural enterprise sustainable management, a tool of management technology that will meet market requirements.

## References


1. Adamisin, P., Kotulic, R., Kravcakova, I., Vozarova, I.: Legal form of agricultural entities as a factor in ensuring the sustainability of the economic performance of agriculture. Agricult. Econ. (Czech Republic) **63**(2), 80–92 (2017). https://doi.org/10.17221/208/2015-AGRICECON
2. El Bilali, H.: The multi-level perspective in research on sustainability transitions in agriculture and food systems: a systematic review. Agriculture (Switzerland) **9**(4), art. no. 74 (2019). https://doi.org/10.3390/agriculture9040074
3. Heldak, M., Kucher, A., Stacherzak, A., Kucher, L.: Structural transformations in agriculture in Poland and Ukraine: towards economic sustainability. J Environ. Manag Tourism **9**(8), 1827–1841 (2018). https://doi.org/10.14505/jemt.v9.8(32).24
4. Henke, R., Benos, T., De Filippis, F., Giua, M., Pierangeli, F., Pupo D'Andrea, M.R.: The new common agricultural policy: ηow do member states respond to flexibility? J Common Market Stud **56**(2), 403–419 (2018). https://doi.org/10.1111/jcms.12607
5. Slätmo, E., Fischer, K., Röös, E.: The framing of sustainability in sustainability assessment frameworks for agriculture. Sociol. Rural. **57**(3), 378–395 (2017). https://doi.org/10.1111/soru.12156
6. Strochenko, N., Koblianska, I., Markova, O.: Structural transformations in agriculture as necessary condition for sustainable rural development in Ukraine. J. Adv. Res. Law Econ. **8**(1), 237–249 (2017). https://doi.org/10.14505/jarle.v8.1(23).27
7. Bieliaieva, N., Sova, O., Ganushchak, T., Zhuk, O., Matusova, O.: Digitalization of the financial subsystem of industrial enterprise: points of implementation. In: E3S Web Conference. International Conference on Sustainable, Circular Management and Environmental Engineering (ISCMEE 2021), vol. 255 (2021). https://doi.org/10.1051/e3sconf/202125501045
8. Pushkar, I.V.: Activity-based costing as progressive method of cost management in accounting. Bull. Zaporizhzhia Natl. Univ. **3**(43), 61–66 (2019). https://doi.org/10.26661/2414-0287-2019-3-43-09
9. Fitó-Bertran, A., Llobet, J., Cugueró-Escofet, N.: The activity-based costing model trajectory: a path of lights and shadows. Intang. Capit. **14**(1), 146 (2018). https://doi.org/10.3926/ic.1107
10. Sodoma, R., Skhidnytska, H., Shvorak, A., Shmatkovska, T., Zhurakovska, I.: Peculiarities of agrarian receipts as a modern financial tool. Econ. Ann.-XXI **169**(1–2), 46–49 (2018). https://doi.org/10.21003/ea.V169-09
11. Mustafiz, S., Nakayasu, A., Itabashi, M.: Marketing of vegetable seeds: practice and behavioral inclinations of vegetable seed sellers and farmers in selected areas of Bangladesh. Agriculture **11**, 364 (2021). https://doi.org/10.3390/agriculture11040364
12. Stępień, M.: Sustainability as a determinant of cost management in the accounts of a manufacturing industry. Glob. J. Environ. Sci. Manag. **5**(SI), 151–159 (2019). https://doi.org/10.22034/gjesm.2019.SI.17
13. Malone, D.: Cost management tools for the environmentally sustainable firm. Cost Manag. (2015). https://www.cam-i.org/docs/cost_management_tools_for_the_environmentally_sustainable_firm.pdf
14. Thompson, A., Strickland, A.: Strategic management: concepts and cases (8th Ed) International Student Edition USA: Irwin. J. Australian New Zealand Acad. Manag. **1**(1), 59–60 (1995). https://doi.org/10.1017/S1833367200006465
15. Kaplan, R., Cooper, R.: Cost and effect. Using Integrated Cost Systems to Drive Profitability and Performance Boston. Harvard Business School Press, pp. 115–305 (1998)

16. Bieliaieva, N., Krushelnytskyi, M., Voliak, L., Usata, N., Sova, O.: From survival to business prosperity: the financial aspect of managing an organizational in a crisis. Independ. J. Manag. Product. **11**(9), 2275–2290 (2020). https://doi.org/10.14807/ijmp.v11i9.1414
17. Pawlak, K., Smutka, L., Kotyza, P.: Agricultural potential of the EU countries: how far are they from the USA? Agriculture **11**, 282 (2021). https://doi.org/10.3390/agriculture11040282

Check for updates

# Clusterization of the Countries by the Level of Achieving the Sustainable Development Goals for Economic Development

Olena Trofymenko[1](✉), Olha Ilyash[1], Nataliia Koba[2], Nataliia Kuzminska[1], and Maksym Koba[2]

[1] National Technical University of Ukraine "Igor Sikorsky Kyiv Polytechnic Institute", 37, Prosp. Peremohy, Kyiv 03056, Ukraine
o.o.trofymenko@gmail.com

[2] International University of Finance, 37, Prosp. Peremohy, Kyiv 03056, Ukraine

**Abstract.** The study is devoted to the analysis of the level of achieving the sustainable development goals by different states. The goal set by the authors was to determine strategic guidelines for ensuring sustainable development of the global economy through finding the level of achieving the economic goals of sustainable development by various countries. In addition, the authors aimed to group the countries based on the level of achieving the goals using the clustering method, as well as study the cases of leading countries in this area. The research was conducted using the hierarchical cluster analysis. The results of the research revealed six clusters of the countries by their level of achieving the chosen goals in 2020 and 2021. According to the obtained results, Sustainable Development Goal 9 was chosen as the main goal of sustainable development in ensuring economic growth. The recommendations for ensuring sustainable development of the states based on circularization, digitalization and intellectualization were made.

**Keywords:** Clusterization · Sustainable development · Circularization · Digitalization · Intellectualization

## 1 Introduction

Ensuring the states' sustainable development is the main focus of the process of socio-economic global transformations in accordance with the plan "Transforming our world: the 2030 agenda for sustainable development" adopted in 2015 in New York at the UN Summit. In this document, 17 Sustainable Development Goals (SDGs) and 169 tasks were approved. Achieving the defined SDGs will ensure the development of the socio-economic sphere in the conditions of the Sixth Technological Mode, the Post-Industrial Era, the Knowledge Economy and the Fourth Industrial Revolution (Industry 4.0). Therefore, the study into the level of achieving SDGs by different countries is an interesting scientific and practical problem.

It is worth noting that a group of goals aimed at ensuring economic development deserves special attention of economist-researchers. These include Goal 8 "Decent work


V. Koval et al. (Eds.): ISCMEE 2022, Lect. Notes in Managem., Industrial Engineering, pp. 170–181, 2023.
https://doi.org/10.1007/978-3-031-23463-7_11

and economic growth", Goal 9 "Innovation and infrastructure", Goal 12 "Responsible consumption", Goal 17 "Partnerships for the goals". The selected group of goals plays a decisive role in ensuring other goals of sustainable development, especially those, which are aimed at solving social problems.

Each of the UN member states joined the global process of ensuring sustainable development, went through the process of adapting the set goals to the national needs and formed national tasks and indicators of achieving the goals of sustainable development. Ukraine is also actively working in this direction. As a result, it has developed the national SDG system (86 tasks of national development and 172 indicators for their monitoring), which will provide a solid basis for further planning of Ukraine's development and monitoring of the state of achieving the SDGs. National tasks, the indicators for monitoring their implementation and benchmarks for achievement until 2030 are reflected in the National Report "Sustainable Development Goals: Ukraine" from 2017 [1], and also studied in the Voluntary National Report [2]. It is obvious that the Russian military invasion of Ukraine has had a significant impact on the world economy and global plans to achieve the SDGs. At the same time, this necessitates strengthening cooperation between states to ensure economic and military security, which reinforces the importance of achieving SDG 17.

Therefore, the research into the level of achieving the economic SDGs by various states, their division into groups based on the level of achievement of the goals using the clustering method, as well as the study of the cases of leading countries in this area will make it possible to determine strategic guidelines for ensuring sustainable development of the global economy.

## 2 Material and Methods

Hierarchical cluster analysis was used in the study. In such analysis, each observation first forms a separate cluster; at the beginning of the analysis, two adjacent clusters are merged into one; this process continues until only two clusters remain; the distance between clusters is the average of all distances between all possible pairs of points from both clusters.

This study was carried out in the following stages:

(1) formulation of the problem. The problem of this study is to ensure the achievement of the SDGs in the field of economy. Four SDGs were chosen for the study. They relate to or have a positive impact on the economic growth and industrial development of the country, namely Goal 8 "Decent work and economic growth", Goal 9 "Innovation and infrastructure", Goal 12 "Responsible consumption", Goal 17 "Partnerships for the goals". The following countries were selected to analyse progress towards achieving the SDGs: Finland, the USA, France, Belgium, Canada, Austria, the Netherlands, Spain, Italy, the United Kingdom, Japan, Denmark, Germany, Sweden, Norway, the Czech Republic, Poland, Slovenia, Portugal, Estonia, Ireland, Slovakia, Lithuania, Latvia, Hungary, China, Moldova, Ukraine, Belarus, Romania, Bulgaria, the Russian Federation, Turkey. These countries were selected based on the following criteria: neighbouring countries, similar resource opportunities, countries with which Ukraine cooperates in various directions, countries with

a leading economy and a high level of GDP. It is important to systematise these countries according to similar characteristics in the process of achieving specific SDGs, which will make it possible to assess the level of economic development and opportunities for the implementation of a circular economy;

(2) choosing a method of distance measurement (determining the degree of similarity of the selected countries with achieving the SDGs). For the study, the measure of similarity was chosen–squared Euclidean distance:

$$d(x, y) = \sum_{i=1}^{n} (x_i - y_i)^2 \tag{1}$$

(3) choosing a clustering method. A clustering method is a method of calculating the distances between clusters. For the analysis, Ward's method [3; 4] was chosen, which consists in the fact that initially in both clusters for all observations, it is obligatory to calculate the mean values of individual variables–the indicators of achieving the defined SDGs. Secondary data from the "Sustainable Development Report" was used to determine the values of the variables. Then the squares of the Euclidean distances from the individual observations of each cluster were calculated relative to the calculated cluster mean value. All the distances were added up. Then, those clusters that give the smallest increase in the total amount of distances were grouped into one new cluster;
(4) deciding on the number of clusters;
(5) interpretation and profiling of clusters;
(6) assessing the reliability of the clusterization.

Hence, to conduct the cluster analysis with the goal to identify clusters of the countries with a close level of achieving the goals in the economic sphere, the following indicators were determined: G8–level of achievement of SDG 8 "Decent work and economic growth"; $G_9$–level of achievement of SDG 9 "Industry, innovation and infrastructure"; $G_{12}$–level of achievement of SDG 12 "Responsible consumption and production"; $G_{17}$–level of achievement of SDG 17 "Partnerships for the goals".

The cluster analysis was carried out on the basis of these indicators in accordance with the assessment in the "Sustainable Development Report" [5] for 2020 and 2021. The analysis characterises the countries' progress in achieving the SDGs and shows areas that need faster progress. According to the methodology of the reports, each country is assigned a corresponding SDG index, which is used to evaluate the overall score of the achievement of all SDGs for each country and rank countries based on the higher value of the index. The overall score of the SDG achievement index may differ significantly from the country's score for a specific goal. That is why we used the scores in accordance with individual goals and defined clusters within the goals 8, 9, 12, 17 because they are the main goals for the economy development. The initial data for conducting the cluster analysis are presented in Table 1.

The results of the cluster analysis using Ward's hierarchical method were obtained with the help of IBM SPSS Statistics 20.0. Software package.

**Table 1.** Initial data for conducting a cluster analysis of the countries for determining the clusters with a close level of achieving the SDGs in the field of economy in 2020 and 2021

| Country | 2020 | | | | 2021 | | | |
|---|---|---|---|---|---|---|---|---|
| | $G_8$ | $G_9$ | $G_{12}$ | $G_{17}$ | $G_8$ | $G_9$ | $G_{12}$ | $G_{17}$ |
| | Goal 8 Score | Goal 9 Score | Goal 12 Score | Goal 17 Score | Goal 8 Score | Goal 9 Score | Goal 12 Score | Goal 17 Score |
| Sweden | 83,60 | 96,91 | 55,90 | 91,13 | 83,27 | 97,43 | 56,88 | 88,40 |
| Denmark | 85,28 | 96,85 | 42,62 | 83,87 | 88,85 | 96,80 | 43,41 | 82,71 |
| Finland | 85,67 | 93,88 | 54,68 | 68,48 | 88,27 | 94,05 | 55,59 | 71,24 |
| France | 78,45 | 86,23 | 57,42 | 66,27 | 82,42 | 87,40 | 57,66 | 71,20 |
| Germany | 84,96 | 92,31 | 52,43 | 79,22 | 87,35 | 92,96 | 54,88 | 77,83 |
| Norway | 82,82 | 91,29 | 44,54 | 99,63 | 87,65 | 91,49 | 44,54 | 96,69 |
| Austria | 83,29 | 93,39 | 46,71 | 63,35 | 85,49 | 95,53 | 48,21 | 68,05 |
| Czech Republic | 85,98 | 78,44 | 69,28 | 50,11 | 85,14 | 80,30 | 69,42 | 54,94 |
| Netherlands | 84,08 | 89,90 | 47,71 | 61,55 | 86,88 | 91,93 | 49,06 | 63,17 |
| Estonia | 88,01 | 74,35 | 59,38 | 43,52 | 84,19 | 75,77 | 60,32 | 52,46 |
| Belgium | 82,41 | 89,04 | 55,77 | 64,14 | 85,49 | 92,19 | 56,33 | 64,86 |
| Slovenia | 88,12 | 73,20 | 57,32 | 54,51 | 88,14 | 73,30 | 58,06 | 62,25 |
| United Kingdom | 82,27 | 89,76 | 53,30 | 51,82 | 80,48 | 89,82 | 53,30 | 58,21 |
| Ireland | 86,47 | 81,56 | 44,30 | 41,75 | 87,32 | 82,01 | 45,17 | 51,30 |
| Japan | 86,83 | 90,41 | 69,09 | 65,52 | 86,81 | 89,53 | 66,61 | 71,48 |
| Belarus | 75,39 | 43,69 | 87,67 | 74,58 | 67,80 | 45,35 | 86,43 | 73,84 |
| Canada | 84,34 | 84,57 | 51,18 | 65,63 | 81,46 | 85,55 | 51,02 | 72,55 |
| Spain | 75,20 | 85,71 | 62,29 | 52,14 | 76,21 | 86,87 | 63,05 | 60,82 |
| Poland | 85,88 | 69,28 | 75,70 | 54,60 | 84,40 | 71,76 | 75,86 | 62,38 |
| Latvia | 85,81 | 59,98 | 63,49 | 36,14 | 84,28 | 62,11 | 63,76 | 46,34 |
| Portugal | 82,54 | 73,15 | 62,53 | 57,09 | 82,15 | 75,41 | 63,05 | 62,67 |
| Slovak Republic | 82,22 | 61,26 | 67,33 | 48,56 | 80,88 | 64,20 | 66,99 | 55,47 |
| Hungary | 83,16 | 62,99 | 72,96 | 38,78 | 80,64 | 66,50 | 73,08 | 51,37 |
| Italy | 80,90 | 83,04 | 62,61 | 58,18 | 76,57 | 83,60 | 63,60 | 62,95 |
| United States | 85,20 | 93,79 | 54,72 | 67,85 | 76,60 | 93,94 | 53,58 | 68,81 |
| Lithuania | 81,97 | 59,89 | 62,08 | 44,84 | 81,45 | 62,34 | 62,86 | 53,19 |

*(continued)*

**Table 1.** (*continued*)

| Country | 2020 | | | | 2021 | | | |
|---|---|---|---|---|---|---|---|---|
| | $G_8$ | $G_9$ | $G_{12}$ | $G_{17}$ | $G_8$ | $G_9$ | $G_{12}$ | $G_{17}$ |
| | Goal 8 Score | Goal 9 Score | Goal 12 Score | Goal 17 Score | Goal 8 Score | Goal 9 Score | Goal 12 Score | Goal 17 Score |
| Romania | 82,73 | 53,25 | 76,18 | 60,64 | 82,72 | 55,02 | 76,29 | 48,53 |
| Bulgaria | 81,75 | 51,61 | 74,97 | 68,01 | 78,13 | 52,55 | 74,50 | 72,06 |
| Moldova | 73,11 | 26,67 | 81,65 | 75,25 | 69,31 | 27,36 | 80,07 | 72,64 |
| Ukraine | 72,51 | 34,45 | 82,73 | 75,42 | 71,70 | 39,00 | 81,84 | 74,25 |
| China | 87,47 | 72,15 | 88,55 | 44,36 | 71,63 | 72,95 | 87,09 | 42,07 |
| Russian Federation | 79,59 | 66,26 | 77,84 | 67,78 | 75,40 | 67,73 | 76,66 | 70,29 |
| Turkey | 69,71 | 59,34 | 81,46 | 71,90 | 59,06 | 60,30 | 79,81 | 71,96 |

*Source* compiled by the author using the data [5].

## 3 Results

The clustering results for 2020 and 2021 are presented in the form of a horizontal dendrogram of the similarity of individual countries according to the level of achieving the selected SDGs in the economic sphere (Fig. 1). The countries are displayed on the vertical axis, and the distance scale is on the horizontal axis. Based on the calculations made using the method of hierarchical cluster analysis, the countries were grouped into 6 clusters and some countries that were not included in these clusters were located separately. The states were clustered according to the chosen indicators in 2020–2021 (Fig. 1).

In 2020, the countries were grouped into the following clusters: the largest cluster No. 1, which includes Finland, the United States, France, Belgium, Canada, Austria, the Netherlands, Spain, Italy, the United Kingdom, Japan. Cluster No.2 includes countries such as Denmark, Germany, Sweden, Norway. Cluster No. 3 includes the following countries the Czech Republic, Poland, Slovenia, Portugal, Estonia. The Slovak Republic, Lithuania, Latvia, Hungary belong to cluster No.4. Moldova, Ukraine and Belarus are in cluster No.5. Cluster No.6 includes Romania, Bulgaria, Turkey, and the Russian Federation. Countries such as Ireland and China formed separate clusters.

In 2021, six clusters of the countries were also observed but there was only one country separated into a cluster – China. In 2021, Spain and Italy dropped out of the group of the countries in cluster No. 1, i.e. the cluster contains the following countries Finland, the United States, France, Belgium, Canada, Austria, the Netherlands, the United Kingdom, and Japan. Germany was also included in it; however, in 2020, it was in cluster No. 2. With the exception of Germany, in 2021 the same countries remained in cluster No. 2 – Denmark, Norway and Sweden. Instead, in 2021, Spain and Italy, which were in cluster No. 1 in 2020, joined a cluster with the countries like Slovenia, Portugal, the Czech Republic, Estonia, and Ireland, which was separated into an individual cluster in

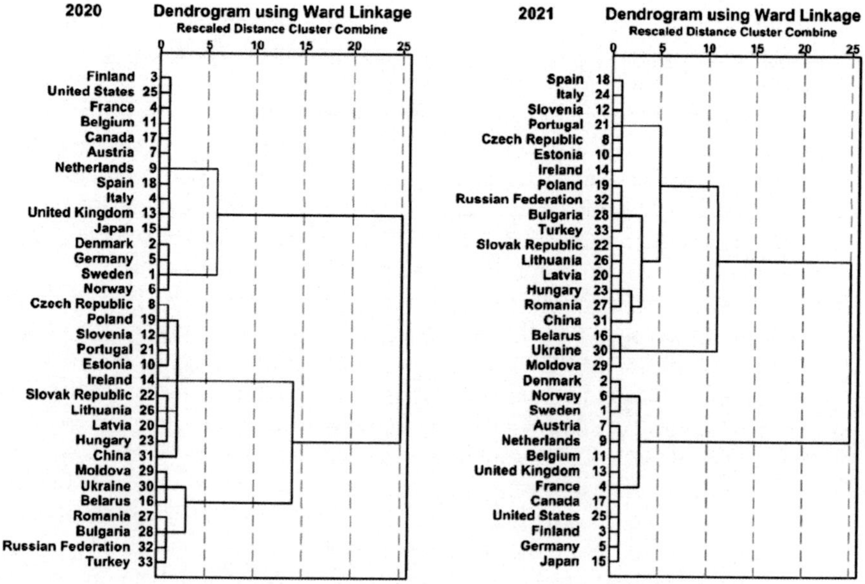


**Fig. 1.** A dendrogram showing the results of the cluster analysis of the level of achieving the SDGs in the sphere of economy in 2020 and 2021 (using Ward's method)

2020 and they together formed cluster No. 3. Such countries as the Slovak Republic, Lithuania, Latvia, and Hungary remained in cluster No. 4 and Romania joined them. Such countries as Belarus, Ukraine, and Moldova remained in cluster No. 5. Poland was added to the countries in cluster No. 6, Romania left the cluster, but the Russian Federation, Bulgaria and Turkey remained there. Of all the studied countries in 2022, China formed a separate cluster.

A group of economically and socially developed countries that make up cluster 1 and cluster 2 stands out among other groups. They have high indicators of the achievement of the selected SDGs, and the highest indicators of the achievement of Goals 8 and 9 on average among the clusters. Moreover, Cluster 1 is formed from the largest number of countries – 10, and cluster 2 consists of three countries according to the calculations for 2021. When comparing clusters 1 and 2 in 2021, on average, the highest indicator of achieving Goals 8 and 9 is observed in the association of the countries in cluster 2 –95.24 and 86.59, respectively. In cluster 1, the level of the achievement of Goal 9 in 2021 is 4% lower and amounts to 91.92, for Goal 8 it decreased by almost 3% and is 84.13. That is why it is advisable to take into account the experience of the countries of cluster 2 and cluster 1 in terms of directions and measures to achieve goals 8 and 9.

The experience of the countries from cluster 1 and 2 shows that the basis for the introduction of innovation is *digitalization and intellectualization of all spheres of the economy.* Currently, the advanced technologies of Industry 4.0 (in particular, cloud technologies, biotechnologies, Big Data collection and analysis tools, 3D printing, crowdsourcing, Blockchain technologies) are radically changing most sectors of the economy. Digital transformation contributes to the growth of the quality of industrial goods and services thus significantly reducing costs. In addition, digitalization promotes the development of innovations, transforms value creation chains and opens up new opportunities for increasing value added [6].

The experience of the countries from the identified clusters that should be mentioned is the introduction of support for innovative technologies of Industry 4.0 [7] by forming strategies and plans for the development of the economy taking into account the world programmes and global initiatives, in particular: the EU Industrial Strategy 2021–2024, European initiatives regarding ICT innovation for manufacturing SMEs (I4MC), Marshall Plan 4.0 "Priority measures for the economic redeployment of Wallonia" (Belgium), The New Industrial France programme, Produktion2030 – a strategic innovation programme supported by the Swedish Agency for Innovation Systems (Vinnova), the national platform MADE in Denmark etc. The study published within the framework of the World Economic Forum [8] states that many companies of leading countries are trying to introduce Industry 4.0 technologies into production, but few manage to do it on a scale that allows a significant financial and economic effect to be achieved. At the same time, the main directions of production transformation are determined by three global technological trends: network integration, intellectualization and flexible automation. It should be noted that those advanced enterprises, which managed to successfully use these trends (according to the study [8]), reaching a new level of efficiency, are located in the countries with high indicators of value added of production and ICT. Among the studied countries, China (39%), the Czech Republic (32%) and Norway (29%) are the leaders by the share of value added of industrial production in GDP. By the share of value added of medium and high-tech production in the total volume of industrial production, the leaders are the following countries with a share of more than 50%: Germany (61%), Denmark (58%), Hungary (54%), Sweden (52%), the Czech Republic (52%) and France (50%). At the same time, high indicators of the ICT subindex are observed in Germany, Sweden, Norway, Germany, France, and the United Kingdom.

In terms of the level of achieving the selected SDGs, all the countries in cluster No. 5 remained unchanged. In our opinion, this situation results from a very low level of achieving the same 9th SDG "Industry, innovation and infrastructure" (G9), which is 34.45 and 39.00 in Ukraine, and 26.67 and 27.36 in Moldova in 2020 and 2021, respectively. According to the indicators of 2021, Ukraine demonstrates a low position in the main indicators of the ICT infrastructure. By the ICT sub-index within the Global Innovation Index, Ukraine took 70th place (among 152 countries) with a score of 64.9 out of 100 maximum scores, which is lower than that of Romania, the Czech Republic, Belarus, Lithuania, Poland. The top 20 are South Korea, Great Britain, Japan, Denmark, the Netherlands, Estonia, the USA, etc. The Network Readiness Index ranking is also indicative – for example, in 2021, Ukraine took 53rd place with an index value of 55.7 and fell behind such countries as Brazil, Turkey, Uruguay, Malaysia, Hungary, and Poland.

The scientists [9; 10; 11] emphasize that a completely new type of industrial production is emerging; it is based on full automation of production processes, augmented reality technologies, big data sets and the Internet of Things. That is why, in the current conditions, scientists, economists-practitioners and government officials regard digitalization as an important driver of socio-economic development of both the country, the economic sector, and an individual enterprise. When digital platforms began to develop, it was expected that they would reduce transaction costs for platform participants and create conditions for business scaling. The reduction of transaction costs is ensured due to such advantages as mobility, convenience, speed, low cost of service, visualization. The issue of digitalization became the subject of discussion by world leaders at the G20 summit in Osaka (Japan), where the main priorities were defined, including support for digitalization of micro, small and medium-sized enterprises.

It is imperative to ensure the manufacture of high-tech products in the defence industry. Today, in the conditions of military aggression of the Russian Federation against Ukraine and its threat throughout the world, *the industrial innovative development* will also ensure the effective functioning of the military complex. For example, the Scandinavian countries, which, according to the results of the calculations formed cluster 2, are now developing a joint military infrastructure for defence against the Russian Federation.

Some enterprises are implementing revolutionary innovative technologies, which according to the study [8] are called "beacons", i.e. leading enterprises in the implementation of Industry 4.0 technology. Geographically, they belong to the following countries from clusters 1, 2, 3 and 7: Germany (BMW, Phoenix Contact), China (Bosch Automotive, Danfoss, Foxconn Industrial Interne, Haier, Siemens Industrial Automation Products), the Czech Republic (Procter & Gamble—Rakona), France (Schneider Electric), Italy (Rold), Sweden (Sandvik Coromant). To improve industrial production, it is advisable to form a favourable infrastructure for the development of innovative enterprises in the countries of clusters 4, 5 and 6.

Industry, as the main component of the real economy, defines one of the priority directions for achieving the 9th SDG "Industry, innovation and infrastructure". However, the experience of the countries from the clusters with a high level of achieving the 9th goal shows that the acceleration of economic growth of this particular component is largely dependent *on the processes of circularization of production and the economy*. This is because a circular economy provides value creation mechanisms that do not recognize the use of limited resources. This is a system model in which each part of the product is viewed as a valuable resource that can be reused to create value added. In a circular economy, resources circulate repeatedly.

In 2020, EU countries organized the new Incubation Forum for Circular Economy in European Defence (IF CEED), which is managed by the European Defence Agency (EDA) with a total financial contribution of €784,000 from the European Commission (EU LIFE Programme) and Luxembourg's Directorate of Defence. Austria, Belgium, Bulgaria, the Czech Republic, Estonia, Germany, Spain, Finland, France, Hungary, Italy, Lithuania, the Netherlands, Poland, Portugal, Romania, Slovenia, and Sweden are among the member states we are examining in this article. The goal of the IF CEED is the development of transnational projects in the circular economy, its application in the defence sector of the European Union. This will lessen the negative impact on the environment, as well as improve the autonomy of the EU by increasing the share of resource recycling, reducing the level of waste and, most importantly today, reducing Europe's dependence on imported resources, including in the energy sector. According to the results of our research, all the mentioned countries are in different clusters, so this proves the unity of all the countries regarding the vision of security and ensuring Europe's sustainable development in the future. Currently, two incubation clusters "Materials and Innovative Designs" and "Processes and Digitalisation" have been created with nine working areas, including critical raw materials, circular additive manufacturing, circular materials for textiles, sustainable eco-design, EU waste regulation, green public procurement, eco-management audit strategies, circular data and spare parts management. It is appropriate to note that such cooperation contributes to the achievement of SDG 17, and the circularization of economies contributes to the development of smart consumption, and, therefore, to the achievement of SDG 12.

The countries in cluster No 3 show a relatively high level of achieving Goal 12 in 2021; on average, in cluster 3, the value is 60.38, for comparison in cluster 2 it is 48.28, and in cluster 1 – 54.62. However, the indicators for the achievement of Goals 8 and 9 are lower than those in cluster No 2 and No 1. Indeed, by the Eco-Innovation Index in 2021, which demonstrates the level of implementing eco-innovations in the member countries, Slovenia, Portugal, the Czech Republic, Estonia, and Ireland are included in the category of Eco-I performers with the values of 113, 115, 111, 97 and 109 respectively. On the one hand, this testifies to the possibilities of further realizing the potential and reaching the level of the countries in the Eco-I Leader category. On the other hand, according to the indicators from the Sustainable Development Report it reflects good progress in the development.

Hence, taking into account the experience of the countries of this cluster when carrying out the tasks to achieve Goal 12 is vital for the countries located in other clusters.

With regard to Goal 17, relatively high indicators are observed in cluster No 2 and cluster No 5, which includes Ukraine. It is obvious that the chosen political vector and the Association Agreement between Ukraine and the EU, which came into force on September 1, 2017, open up opportunities for political association and economic integration with the EU, which creates opportunities for the development of new European sales markets, contributes to the liberalization and development of trade conditions with EU countries, and has a positive effect on the indicators of goals 12 and 17. For instance, it stimulates enterprises to introduce the principles of responsible production and consumption and increase the quality of products so that they meet European requirements.

The studied indicators do not refer to the period after the start of the war in Ukraine. On the one hand, the war has changed Ukraine's priorities in the direction of ensuring its military security. On the other hand, ensuring sustainable development in the field of the economy and energy will make it possible to form the necessary ecosystem of innovations for the recovery of the economy. Industry, as one of the most important components of the real economy, relies, as a rule, on exhaustive resources. Consequently, in order to accelerate the economic growth of this particular component in modern economic and security conditions and taking into account the experience of the countries we have studied, it is essential to rely on the provisions of the circular economy concept.

Therefore, in this aspect it is necessary to consider the six-month war experience and stress the importance of such a component of the real economy as infrastructure and the direction of achieving the 9th SDG. During the war, the whole world became aware of the importance of infrastructure in ensuring the development of the economy. As the practice of 2022 showed, the lack of developed infrastructure, in particular, its transport and logistics component in the west of Ukraine, caused significant problems in ensuring the sustainable development of the world economy. The existing problems raised doubts about the plans to ensure the achievement of the 2nd SDG "Zero hunger" due to a sharp drop in the volume of export of Ukrainian grain and other agricultural products. It is clear that the main cause of the threat of famine in the world is Russian aggression against Ukraine. Moreover, the military actions, the blocking of ports, the lack of transit through the territory of Belarus have significantly changed the logistics of export transportation, causing the rapid development of transport infrastructure on the western borders and a radical change in logistics routes. Neither Ukraine nor the countries of Western Europe were ready for such changes, so, in the first months of the war, the volume of transportation was reduced to a minimum. Due to the reconfiguration of logistics from the capitals of many European countries, in particular, the Baltic States, basic food products disappeared, and due to the impossibility of exporting 20 million tons of Ukrainian grain, such countries in Africa like Egypt, Yemen, Syria and Lebanon can face the threat of famine.

Consequently, the war in Ukraine has proved that the real sector of the economy i.e. industry, transport and the agricultural sector is of crucial importance. Therefore, in the current conditions of the destabilization of the entire world economy, the threat to global security and geopolitical transformations, we consider the achievement of Goal 9 to be decisive among those we have chosen to ensure sustainable development in the economic sphere.

## 4 Conclusions

In the process of conducting the cluster analysis, the 33 studied countries were divided into 6 clusters according to the achievement of the goals of sustainable development in the economic sphere, namely: Goal 8 "Decent work and economic growth", Goal 9 "Innovation and infrastructure", Goal 12 "Responsible consumption", Goal 17 "Partnerships for the goals". It was determined that the directions and measures to achieve goals 8 and 9 should take into account the experience of the countries united in clusters 1 and 2. This experience includes support for innovative technologies of Industry 4.0 by forming strategies and plans for the development of the economy, taking into consideration the programmes and global initiatives of these countries in the direction of network integration, intellectualization and flexible automation of production. It is advisable to pay attention to the experience of the countries from cluster 3 in achieving Goal 12 and the countries from clusters 2 and 5 in achieving Goal 17. Taking into account the modern needs of the economy, as well as the growing needs of ensuring military security, the achievement of Goal 9 is a high priority.

The successful experience of the studied countries suggests that business should be more oriented towards ecological cyclic production rather than linear one. This kind of production can be created if enterprises use marketing communications effectively, which will help change the culture of consumption. Emphasis should be placed on the environmental friendliness of products, reliability and cost savings in the process of using. The purpose and importance of marketing in the sphere of the circular economy is the dissemination of information and the appeal of society to use resources correctly, the possibility of increasing well-being without excessive use of natural raw materials. Thus, thanks to the introduction of ecological cyclic production into the industry, the level of achieving the 12th SDG "Responsible consumption" will increase.

It is clear that at present the development of innovation can be ensured only on the basis of digitalization and intellectualization of all spheres of the economy, which is confirmed by the successful experience of the leading countries in achieving SDG 9. Furthermore, it is extremely important for both Ukraine and Ukraine's European neighbouring countries to implement these processes specifically in the defence industry for ensuring its innovative development.

Regarding the third infrastructural direction of ensuring the achievement of SDG 9, in our opinion, the decisive way is the partnership between various market participants (farmers, carriers, ports, terminals, etc.) and various states to create unified transport corridors, unified information infrastructure for supporting business development, which will contribute to the acceleration of the European integration of the Ukrainian economy.

That is, innovation based on digitization and intellectualization of industry, the infrastructure based on partnerships and a circular economy for sustainable development of industry in the conditions of resource exhaustion should become the basis of the real economy. In addition, now they are important vectors of the development of the economic security of the EU and Ukraine.

## References


1. UNDP, The 'Sustainable Development Goals: Ukraine': 2017 National Baseline Report https://www.undp.org/ukraine/publications/sustainable-development-goals-2017-baseline-national-report. Last Accessed 18 June 2022
2. Voluntary National Report, https://ukraine.un.org/sites/default/files/2021-10/1_VNR-SDG-EN_02_20200612.pdf. Last Accessed 22 May 2022
3. Ward, J.H.: Hierarchical groupings to optimize an objective function. J. Am. Stat. Assoc. **58**, 236–244 (1963)
4. Domokos, E.-K., Bálint, C.: Definition of user groups applying Ward's method. Transp. Res. Proc. **22**, 25–34 (2017)
5. Sustainable Development Report, https://dashboards.sdgindex.org/. Last Accessed 12 Jan 2022
6. Verhoef, P.C., et al.: Digital transformation: A multidisciplinary reflection and research agenda **122**, 889–901 (2021)
7. Accelerating clean energy through Industry 4.0. https://www.unido.org. Last Accessed 12 Jan 2022
8. Industry 4.0, https://www.mckinsey.com/industries/advanced-electronics/our-insights/capturing-value-at-scale-in-discrete-manufacturing-with-industry-4-0, last accessed 2022/01/12

9. Borowski, P.F.: Digitization, digital twins, blockchain, and industry 4.0 as elements of management process in enterprises in the energy sector. Energies **14**, 1885 (2021). https://doi.org/10.3390/en14071885
10. Ilyash, O., Trofymenko, O., Dzhadan, I., Tsarova, T.: Ecological and economic effects of industrial and technological development. IOP Conf. Ser.: Earth Environ. Sci. **915**(1), 012004 (2021)
11. Coreynen, W., Matthyssens, P., Van Bockhaven, W.: Boosting servitization through digitization: Pathways and dynamic resource configurations for manufacturers. Ind. Mark. Manage. **60**, 42–53 (2017)
12. Eco-Innovation, https://ec.europa.eu/environment/ecoap/indicators/index_en. Last Accessed 17 May 2022

# Advanced Bioengineering Applications with Eco-Technology Approach for Sustainable Development

Ece Yıldız-Ozturk and Ruhan Askin-Uzel(✉)

Department of Food Processing, Yasar University, 35100 Izmir, Turkey
ruhan.uzel@yasar.edu.tr

**Abstract.** Problems such as excessive consumption habits, depletion of natural resources and global climate change keep the concepts of both eco-entrepreneurship and bioeconomy on the agenda. Within the scope of the study, agro-food-oriented bioengineering practices, in which biotechnology is at the center, were examined in the circular economy line. In the review, eco-technological applications under field-specific titles have progressed in the form of examining academic and sectoral studies in the world. Hence, it has been determined that the need for interdisciplinary improvement in biosystems has become an important issue on international scale, especially in recent years. In this way, it has been seen that significant contributions will be made in finding solutions to the productivity and efficiency problems that arise in eco-technological bioengineering applications, especially with the contribution of entrepreneurship ecosystem studies. For this reason, it is essential to carry out studies in which all actors involved in agriculture-food-biotechnology systems can come together. Therefore, existing problem areas in bioengineering applications for sustainable development should be identified and joint solution proposals should be developed for the identified problem areas.

**Keywords:** Sustainability · Bioengineering · Eco-technology · Circular economy · Biotechnology · Entrepreneurship

## 1 Introduction

Ecology and technology are seen as opposite concepts. Ecology refers to the natural environment and technology refers to the artificial environment [1]. Eco-tech concept refers to the maximum protection of ecology by technological means.

Today, one of the hottest topics on the world's agenda is sustainability and, in this context, the relationship between the environment and the economy [2]. Contrary to this parallelism in their definitions, these two fields of science developed completely separately from each other until economic and industrial development seriously brought environmental problems to the agenda since the 1970s. "Sustainable Development" came to the fore after emerging environmental problems were seen as undesirable consequences of economic growth [3].


V. Koval et al. (Eds.): ISCMEE 2022, Lect. Notes in Managem., Industrial Engineering, pp. 182–208, 2023.
https://doi.org/10.1007/978-3-031-23463-7_12

When the concept of sustainability, which forms the basis of sustainable development, is examined, sustainability can be defined as facilities supporting an ongoing process and ensuring its progress by preserving the conditions [4]. Sustainability is not a subject that should be considered as a concept, it is actually a lifestyle. This means that we can meet our own needs without compromising the needs of future generations.

Today, it is clear that natural resources must be possessed in order to implement the principle of sustainability. Because, due to the activities of people individually and in society and the selfishness of businesses, resources have been left alone with the problem of exhaustion. Although sustainability as a word evokes the conservation of resources and environmental measures, it is actually a comprehensive concept with economic, social and environmental dimensions [5]. In order to ensure sustainability, these three concepts must be managed in a balanced way:

**Environmental Sustainability**. It focuses on the fact that natural resources in nature will one day run out and that these resources should be used wisely. The use of natural resources as if they will not end in order to meet the economic need, and the intervention in the living rights of living things for development are the subjects covered by ecological sustainability [6].

**Social Sustainability**. Education, health, welfare, safe living, quality of life etc. It focuses on the sustainability of social conditions in fields. It enables the achievement of individual and group goals as well as enriching the relations between people [7].

**Economic Sustainability**. The sustainability of production and consumption dynamics is a very important concept for economic continuity and development [8]. According to the principles of sustainable development, society should not allow welfare to decline over time. While companies used to carry out sustainability on paper in the form of donations and aid campaigns in the past, today they will respond to the demands of the new generation, who have become much more sensitive and knowledgeable, as the effects of humanity on the world become more evident (climate change, global warming, plastic pollution, deforestation, endangered species, etc.) [9]. When the data of 10 different sustainability trends that will determine the business world of the 2022 Sustainability Trends Report are examined, it is stated that the actions of the industry against the risks stemming from climate change and sustainable supply chains increase the importance, and they will progress by considering the pandemic conditions. It is also expected that the environmental, social and corporate governance (Environmental, Social, and Corporate Governance - ESG) understanding will be increasingly integrated into investor strategies [10].

Considering all these factors, it is seen that sustainability is a common understanding that can be shaped not only by one or a few fields, but also by the joint efforts of the sector. In other words, the importance of sustainability has increased and sustainability has become a phenomenon that strives to increase the quality of health, life and education to a satisfactory level for all societies, while providing economic growth that will bring prosperity to societies without harming the environment. Sustainability is a long-term process. From food to agriculture, from textile and fashion to technology, from mining to transportation and many other processes and sectors, the necessity of continuity has arisen [11].

In our study, first of all, starting from the concept of sustainability in the world agenda, it is tried to determine the dimensions of the problems, including the problems on the way to achieving sustainable development. In addition, bioengineering applications on the principle of eco-technological approach are examined in many fields from food to agriculture, from biotechnology to health, in the main characteristic of the vision of "sustainable development". As a result, it is evaluated that the distinguishing feature of a multidisciplinary sustainable development is the vision of a common solution and a holistic approach, and it is argued that the restructuring demanded by this vision can be linked to the paradigm breaks of fields such as food-agriculture-biotechnology in the multidisciplinary scientific approach.

In this review, social, environmental and economic factors, which are the basic building blocks of sustainable development, have been evaluated from an engineering point of view; sustainable food, biotechnology and environmental approaches are discussed. However, especially in recent years, sustainable and innovative practices related to studies in related fields have been evaluated. Bio-technology applications in food industries, agricultural biotechnologies, biorefinery applications, environmental biotechnologies and health applications are discussed as sustainable bioengineering studies with an eco-biotechnological approach.

## 2 Green Economy

The main-stream economic theory, which was formed with the industrial revolution and continues its influence today, takes as data that human welfare is based on material welfare [12]. As the destructive effects of economic growth-oriented policies on the environment and society began to be felt more deeply, especially since the 1970s, classical growth and development policies began to be questioned in the World [13]. The green economy model, which is accepted as sustainable development and its complementary concept today, was put forward to get out of the economic, social and ecological crisis in the world economy and to maintain a safer life in the future (see Fig. 1). There are two basic concepts that have begun to produce. They are green economy and sustainable development. In fact, these concepts are not competing ones that can be substituted for each other, but the green economy is the complement of a sustainable life and growth-development processes both politically and economically, socially and ecologically [14].

The United Nations Environment Program (UNEP) defines the green economy as a growth strategy that not only eliminates environmental risks and ecological problems, but also increases human welfare and strives to achieve social equality. To put it more clearly, green economy can be thought of as an approach that includes reducing greenhouse gas emissions, increasing resource efficiency and social development [15]. The most important goal of the transition to a green economy, which establishes a very important link between economic growth and environmental sustainability, is to ensure increases in investment increases and economic growth, as well as in environmental quality and social inclusion. While doing this, the principles of different approaches that adopt a similar understanding in the field are also adopted. Circular economy is one of the examples that can be given in this sense [16]. The circular economy model is a mechanism that supports

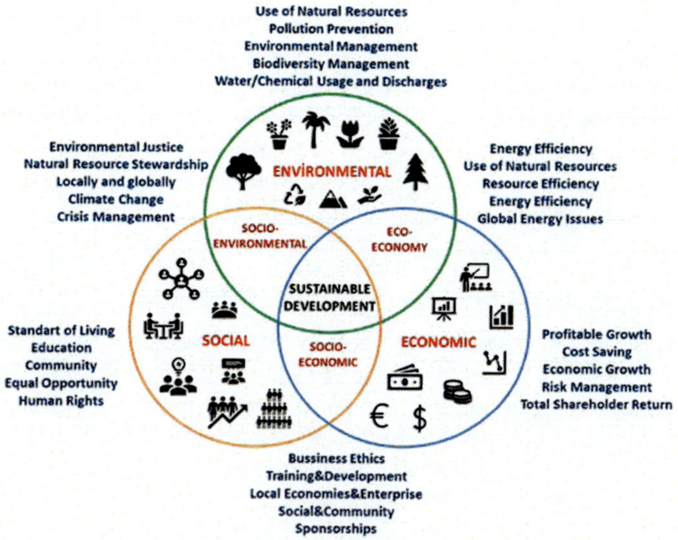


**Fig. 1.** Key elements of sustainable development.

the green economy and operates contrary to the linear economy principle. Therefore, especially the European Union countries have accelerated the transition processes from the linear economy model to the circular economy model [17]. Circular economy is a model that ensures the use of resources as long as possible, energy saving and reduction of waste by keeping the resources in cycle, depending on increasing the cooperation between the sectors (for example, one firm's use of the other firm's waste as raw material) in order to reduce the waste originating from the end of consumption or production and to provide resource efficiency. In the circular economy, wastes are considered as a resource for new processes, so resources are not depleted depending on use. Biological materials (bio-based materials such as food, natural fibers and wood) can be safely returned to the biosphere.

At this point, the issue of whether the raw material, which is the basis of production, will be handled with the support of natural resources comes to the fore. The field of bioeconomy appears as an approach that serves the mentioned criteria and also feeds the concepts of green economy and circular economy. This concept brings "What does biotechnology provide for the circular economy?" question to mind.

Bioeconomy supports issues such as circular supply and resource recovery on a business model basis. It is possible to collect the activities carried out in terms of serving the green economy in five basic areas [18]. These are as follows:

- Increasing the Use of Renewable Inputs
- Introducing New and Better Materials
- Providing Better Life Cycle Designs
- Compost (Fertilization)
- Better Reuse and Upcycling

In summary, the bio-based economy and the biotechnology that composes it are important for sustainability. Eco-technological initiatives are in a remarkable position among things to be done for field-related sustainability. Today, eco-enterprises are seen as an important factor of change in terms of sustainable development and the fact that many businesses start their activities as an environmentally friendly initiative at the establishment stage. The issue of eco-entrepreneurship should be put in front of entrepreneurs as an alternative with social awareness studies.

## 3 Food-Biotechnology and Environment Relationship

### 3.1 Sustainable Food

Food security is at the core of the food system. Food safety tries to regulate the control of the social, economic and environmental parameters that form the basis of the concept of sustainability on the food system. The increase in the human population and the increase in processed foods are progressing simultaneously. As a result, the food industry is growing steadily. The carbon footprint of the food industry, which is fueled by fossil fuels, has reached enormous proportions. Likewise, the agricultural industry causes irreparable damages in terms of greenhouse gas emissions. The right and sustainable food system must also take care of the ecosystem. The goals of a sustainable food system include supporting local producers, improving distribution conditions, and keeping nutritious food accessible to all. According to the reports presented by the United Nations Food and Agriculture Organization, only 2.7 billion tons of 4 billion tons of food produced in the world are consumed in a year. The rest, 1.3 tons, is thrown away. It is stated that 1/4 of the clean water resources are spent for the production of these discarded foods. These statistics show that there is a significant problem in the distribution of food. Because in a world where there is so much food waste, 821 million people cannot find food. In order to eliminate this imbalance, global food production must shift to the local. In order for people to find enough and nutritious food, it is important to produce food at the same standards all over the world. In summary, it should be noted that the basis of an efficient food system is sustainable food, economic production expenditures, fair distribution and healthy product alternatives.

### 3.2 Sustainable Biotechnology

Sustainable development requires the preservation and optimum use of existing sources in order to maintain an adequate resource base for future generations. The concept of sustainability takes a holistic perspective that blends ecological, social and economic components. According to this approach, while ensuring the continuity of production and diversity, it is necessary to meet the economic and social needs of our age without harming the opportunities of future generations [19, 20]. Sustainable biotechnology is a technology that harnesses the power of living organisms for sustainable improvements in human society. The current industry, based on science and technology developed in the 21st century, has revealed many different global environmental problems, including global warming and waste-plastic distribution. Therefore, "sustainable biotechnology" is proposed as a 21st century technology that harnesses the power of living organisms to recycle materials on Earth and reduce the use of underground resources [21].

Biotechnology can be defined as the science and technology involved in product and service development processes using biological agents. The potential role of biotechnology in industrial processes lays the groundwork for the examination of all processes within the scope of industrial sustainability. In this way, it is possible to make production processes greener by reducing greenhouse gas (GHG) emissions and environmental chemical pollution with industrial biotechnology applications [22]. Sustainable biotechnology practices include environmentally friendly production of industrial chemicals, drugs, materials and bio-energy. In this context, by using various substrates together with various enzymes and microorganisms with many bioengineering applications; New products with high added value can be obtained in sectors such as food, feed, chemistry, paper, cellulose, textile and energy [23, 24]. The raw materials used in the processes, instead of being obtained from fossil fuels, can form lignocelluloses containing starch and their residues and creating agricultural waste. The need for sustainable development, rapid technological developments and ongoing globalization/competition are the main drivers of industrial biotechnology. Compared to traditional processes, industrial biotechnology offers advantages such as minimizing costs and polluting gas emissions as well as reducing waste. In addition, thanks to the use of renewable resources in biotechnology applications, sustainable biotechnology has become a key technology in the development of the bioeconomy [25].

Genetic resources of plants, animals and microorganisms can be used as raw materials in biotechnology-based research, technology development, innovation development and the production of new products. The processes of collecting, protecting, evaluating and using genetic resources are greatly affected by biotechnological developments. With all molecular tools used in biotechnology applications, the processes of identification, isolation, cloning of target genes and production of organisms with desired characteristics by transferring them from one species to another have been accelerated. The ultimate goals of biotechnology in sustainable agriculture include detecting single nucleotide polymorphism, determining the functions of certain genes, assigning functions to unknown genes, and developing advanced transgenics with certain productive desirable traits. Genetic engineering techniques are widely used to modify living organisms or components in the development of functional products for different biological applications [26].

### 3.3 Sustainable Environment

Sustainability is vital for people and the environment as we move towards the future. In ensuring environmental sustainability; Conservation of biological diversity is essential in the protection and management of non-renewable and environmental resources, waste management, prevention and treatment of pollution. With biotechnology applications, the use of fossil fuels and, accordingly, greenhouse gas emissions can be reduced. In addition, more efficient energy production can be achieved in industries, biodegradable products can be developed, waste management and environmental pollution improvement studies can be carried out successfully [27].

Many environmental issues such as wastewater treatment and management, solid waste management, air pollution, industrial and hazardous waste treatment and management have become very important due to the intense increase in population and the corresponding increase in industrialization and urbanization. For this reason, it has become important for developing countries to produce sustainable solutions to all these environmental problems without compromising their economic development. Biotechnological developments in recent years have brought along sustainable bioengineering practices both for the removal of toxic pollutants and for the recovery of valuable resources from polluted environments such as wastewater streams. Environmental biotechnologies; It has a say in solving global problems with the use of conditioned biocatalysts to improve polluted areas, reduce environmental organic and inorganic pollutants and meet the world's fuel demand [28].

With the increase in global demand for food products reaching 98% until 2050, food demands are increasing at a significant rate, especially in countries where arable land is insufficient [29]. The role of agriculture in economic development is critical, as agricultural practices are common in developing countries. With the increase in population, consumption rates also increase, which intensifies the pressure on land areas. For this reason, it is necessary to carry out more intensive planting operations in agricultural areas. As a result, dependence on chemical fertilizers and pesticides used to meet food production requirements and increase yields in agricultural areas has also increased. The unconscious use of inorganic or synthetic fertilizers leads to loss of soil fertility due to the decrease in other resources such as water and energy. However, it also causes many negative effects such as washing, acidification and alkalization of the soil, pollution of water resources, destruction of microorganisms and irreversible destruction of the entire ecosystem [30]. In order to overcome all these problems, sustainable agricultural practices, which are considered in ensuring soil biodiversity and food security, encourage organic farming methods. With sustainable agricultural practices, it is possible to minimize the harmful effects on the ecosystem due to the use of chemical fertilizers and pesticides. Thus, it is possible to reach the agricultural sustainability target with the principle of "safe cultivation" for the production of safe and quality food without endangering the environment.

In a sustainable environmental context, the greatest contribution of biotechnology in agriculture is applications aimed at increasing crop yields. Thus, with biotechnological applications, primary energy flows can be manipulated and at the same time fossil fuel energy inputs to agricultural systems can be reduced. It can also contribute to reducing environmental problems such as deforestation and soil erosion [31]. Both food and fuel-energy sources are essential components of sustainability. Recovery and recycling of resources and disposal of hazardous waste are other environmentally beneficial aspects of biotechnology. Environmental biotechnologies for sustainable development are an important tool for improved manipulation of biotechnological and biogeochemical cycles, as they can expand the resource pool [32].

# 4 Field Related Innovative Sustainable Applications

Many countries have taken different steps according to their own qualifications to achieve the Sustainable Development Goals created by the United Nations, which aim to ensure that people live in peace and prosperity. Based on this, many researches are conducted, reports are written and striking data are revealed. Research linking sustainability in health with food and agriculture reveals alarming results. In this section, the eco-technological approach and the concept of sustainability are explained with applications in different fields.

## 4.1 Bioengineering in Food Industry

The main duties of the food industry can be listed as ensuring food safety and security, protecting the environment and natural resources, and maintaining the food supply in a sustainable way. According to the United Nations Food and Agriculture Organization (FAO) survey report, it has been reported that the population of the world will reach around 10 billion people by 2050 and the global food demand will increase by at least 60% [33]. Food products and devices developed to ensure food safety and sustainability have begun to gain an important place by integrating with the technology of the future. In order to meet the food demand of the rapidly increasing population, to protect the quality of the existing products and to adapt to the advancing technology in order to adapt to the advancing technology, the food sector works on applications to prevent waste with richer, alternative, sustainable and nanotechnological foods in cooperation with bioengineering applications.

The food sector is becoming a major sector benefiting from advances in biotechnology that can provide most of the elements needed for human nutrition in the near future. Processes that use microorganisms or metabolites of microorganisms are common in food production. Because, the common purpose in all of them is to apply the hygiene rules while producing the food. In this sense, biotechnology has become one of the research areas open to development in food engineering applications. Biotechnology supports bacteria, yeast and molds in the processing steps of the raw material until it becomes the final product, to help produce new products and improve the properties of existing products [34–38].

It is known that in some food production systems, biotechnological materials are used and the final product obtained may be a food product as well as a raw material used in pharmacy [39]. Another area where bioengineering is used in the food industry is the fight of agricultural raw materials with pests and increasing the yield of agricultural products. For this reason, biotechnology supports food engineering activities and the principle of economy in production in terms of improving the nutritional value of food and increasing the quality of raw material sources [40].

In the food industry, it is not possible to see applications that only add value to foods in the eco-technological approach. Biotechnology is also one of the most utilized areas in the evaluation of food waste. Waste management is one of the important components of the sustainability principle in practice and requires the management of wastes arising from production and services in accordance with the environment and human health. Until today, the general principle of waste management all over the world has been disposal or discharge into the environment with minimum cost and processing, and industrial wastes have been replaced rather than removed. While doing this, in order to render the waste harmless, it has been used to release it into large water bodies, incineration, chemical treatment processes, biological treatment processes and reuse. Recently, waste management has been approached with the principle of recycling and reuse, and in this perspective, waste materials mean new raw materials for various products. In determining the right method in waste management, the question of whether the amount of waste can be reduced or completely eliminated by modifying the production process, technology or product type; plays a guiding role in the emergence of the most economical solutions. For this purpose, eco-technologies (industrial ecology), which have become increasingly important in recent years, have emerged with integrated technologies that can be realized by using agro-industrial wastes. The basic principle on which this view is based is that industrial establishments and enterprises take biological ecosystems as an example while maintaining their activities. In a wide network of interactions in biological ecosystems, every substance produced is used by any organism in nature to support its own metabolism. Based on this principle, it is aimed to establish integrated industrial parks in which facilities that can establish a symbiotic relationship with each other in a certain region are together. Here, the waste of one enterprise becomes the raw material of another, and as technology develops, an evolution towards zero waste takes place. This view is gradually spreading and being adopted in our globalizing world. Different countries are trying to realize self-sufficient, ecological industrial parks within their own socio-economic conditions. Developing technological alternatives for waste minimizes the damage to the environment and ensures that the wastes are re-entered into the production cycle. Considering the evaluation of food industry wastes in integrated systems, the composition of the wastes (such as carbohydrates, protein, oil), the presence of toxic components (such as heavy metals, herbicides and insecticides), the availability, recyclability, price of the waste, the use of the produced products, the competitive situation, the investment cost, the socioeconomic Economic relations must be considered. Food industry wastes, which are highly susceptible to microbial development, can be included in energy production by using recycling and recovery methods. In this case, considering the humidity value, biomass can be used directly as fuel, as well as anaerobic conversions and biogas or ethanol can be produced as other fuel alternatives. The reuse possibilities of food industry wastes, the conversion of microbially metabolizable components into high value-added products with bioprocess applications, and the fact that these products are chemicals and food additives can be considered as a priority. Therefore, in this century, which is the century of biotechnology, modern strategies to approach this concept will certainly ensure that new food products with improved quality are obtained, but also new food pigments, food flavors, food preservatives, etc. will be produced. Along with the increasing world population, problems such as depletion of natural resources, health problems caused by environmental pollution, deterioration of ecological balance with the decrease of green cover, and global warming have been placed on our agenda. The food industry also has an important share in the amount of waste that is growing in our nature [41]. In the evaluation of these wastes, important developments have been made in biotechnological processes, and enzymes have found an important area of use with the industrialization of biological processes. In this context, if all the actions taken are in accordance with the sustainable development approach, the damage to the ecosystems will be reduced and natural resources will be protected.

## 4.2 Agricultural Biotechnologies

**Microbial Biotechnology**. By 2050, the inquiry for agricultural products is supposed to increase by at least 70%. Extensive agricultural production is required to encounter the food necessity of the growing world population. However, with the understanding of the importance of food safety over the years, awareness of the importance of sustainable agricultural applications in potential agricultural demands has increased [42]. Accordingly, agricultural practices are applied worldwide, aiming to preserve yield while protecting the biosphere. In this context, various methodologies are being developed to fulfill sustainable environmental and economic developments. Sustainable development methodology constitutes the intersection of the concepts of economy environment (agroecology), environmental society (environmental awareness) and community economy (standard of living) [27]. Sustainability aims to get effective processes for recycling food stuffs, controlling pesticides and pathogens, alleviating the unfavorable effects of abiotic stress variables that form the basis of human life, and the sustainability of global ecosystems. In the light of all this information, the usage of the root-associated microbiome is very critical to satisfy both economic and ecological sustainable problems [42]. Agricultural microbiology is in charge of exchanging know-how from general microbiology and microbial ecology to agricultural biotechnology. An analysis of the typical dissemination of microorganisms between plants, animals and soil-derived niches is absolutely necessary to regenerate the order of the microbiota in natural and agro-ecosystems [26, 30, 43].

Microbial biotechnology can be identified as the use of microorganisms and microbiological systems to develop or modify proceeding to improve the quality of human life [44]. It is also a critical field that supports developments in general basic research in agricultural sciences, human nutrition, functional foods, value-added products, food safety, plant and animal protection [26]. In recent years, tremendous crucial developments have been in progress in the area of microbial biotechnology, making it an important player in the worldwide market [45, 46]. Microorganisms; provides a wide variety of biotechnological compounds such as biopharmaceuticals, vitamins, antibiotics, single-cell proteins, organic acids, enzymes, pigments, biofertilizers, biopesticides, bio cement and biofuels for sustainable development. [45]. Microbial biotechnological applications are definitely the most sustainable and effective methods to encounter the growing challenges of a medicine, balanced nutrition, clean air, water, energy, and a healthy environment.

Microbial biotechnology tends to sustainable agriculture by decreasing the addiction to agrochemicals, especially pesticides, by regulating genes that provide resistance to biotic and abiotic stresses. Delicately chosen genes from related or unrelated genetic sources are combined into desired genotypes [26].

Agricultural biotechnology applications contribute to sustainable development in the following ways:

1. Increasing resistance to biotic stresses such as insect pests and diseases;
2. Increasing resistance to abiotic stresses such as drought, cold and flood;

3. Remediation of polluted soils;
4. Increasing crop productivity and quality;
5. Improved nutrient uptake and utilization efficiency with enhanced nitrogen fixation;
6. Advanced fermentation technologies;
7. Improvement of nutritional value;
8 Development of biotechnologies to produce energy from biomass [26, 30, 47].

The most promising strategy that can be implemented to get through all these stated targets is to replace all harmful pesticides and fertilizers used in agriculture with environmentally friendly symbiotic microorganisms. In this way, both the nutrition of the crops and the protection of the plant against biotic (pathogens and pesticides) and abiotic (pollution and climate change) challenges will be provided [26]. Bio-inoculants are strains of living microorganisms that colonize the rhizosphere and phyllosphere of plant tissues. Bacteria, fungi and algae are used as a bio-inoculant microbiological agent to efficiently induce plant growth when applied to soil, seeds or plant surfaces [48]. Microbial vaccine strains improve agricultural productivity, crop and livestock quality, herbicide tolerance, abiotic stress tolerance, and virus resistance [49, 50]. Bacteria or fungi that can perform profitable systems such as production of plant hormones, phosphate solubility, nitrogen fixation, sulfur oxidation, and decomposition of organic compounds are widely used as biofertilizers [51]. In addition to being environmentally friendly, biofertilizers also have the advantages of increasing agricultural production, increasing crop yield, reducing chemical additives and improving agricultural sustainability [52].

**Plant Biotechnology**. Many leguminous and non-legume plants have beneficial relationships with soil microorganisms. This microflora is often referred to as the rhizosphere microbiota and is beneficial to ensure plant health and soil sustainability. Rhizobia are nitrogen-fixing bacteria after settling in the root nodules of legumes such as soybean, cowpea, lentil, broad bean, pea, bean, and chickpea. Rhizobia cannot fix nitrogen independently and need a plant host. The nitrogen fixation process is executed symbiotically by rhizobial species such as *Rhizobium, Allorhizobium, Azorhizobium Mesorhizobium* and *Bradyrhizobium,* [53–55]. In addition to these, several free-living rhizobia such as *Azospirillum, Pseudomonas, Klebsiella, Azotobacter, Azomonas* and *Bacillus* carry into effect nitrogen fixation. Most leguminous plants such as *Alnus, Casuarina, Datisca* and *Parasponia,* are also capable of nitrogen fixation. Microorganisms such as *Agrobacterium, Arthrobacter, Burkholderia, Caulobacter, Chromobacterium, Erwinia, Flavobacterium, Micrococcus, Micromonospora, Serratia, Streptomyces* and *Thermobifida* can also promote plant growth apart from nitrogen fixation [46, 54–58]. Azolla, a heterosporous pteridophyte fern, is a genus of aquatic ferns from the Salviniaceae family. It is recognized a sustainable natural nitrogen source because it contains the $N_2$-fixing cyanobacteria *Nostoc azollae* [59].

The combination of many fungal species with higher plants in order to promote mycorrhiza provides increasing agricultural productivity and organic abundance of soil and aquatic ecosystems [57, 60]. The biological nitrogen fixation process helps to diminish the addiction to chemical fertilizers, thus controlling environmental pollution, creating a non-polluting and economical bioprocess. Rhizobacteria that promote plant growth with an environmentally friendly and ecologically sustainable management approach in agricultural applications function as bio-inoculators, also known as bio-inoculants, bio-fertilizers, rhizomediators, biopesticides, bioherbicides and biofungicides [61–64]. Although bio-inoculators increase production in agricultural plantations, it also provides protection against different abiotic stresses such as salt and heavy metals, weeds, fungal pathogens and insects [23, 54]. This kind of bio-inoculants have a crucial part in agricultural processes, that reduce the damage to environmental health with minimal investment costs.

In plant biotechnology, another field that deals with food safety, climate change adaptation and environmental sustainability in order to increase agricultural productivity is genetic engineering applications. This technique is beneficial to enhance crop protection, improve nutritional value, better flavor, fresher produce [65]. In genetic engineering techniques, genetic modifications that evaluate certain characteristics of an organism are carried out by employing the genetic material of a particular plant. In this technique, one or more well-characterized genes can be introduced into a plant, or genes from any species can be introduced into a plant. According to genetic manipulation strategies to prevent food losses and provide adequate nutrition to the world, the characteristics of genetically modified crops are pesticide resistance, herbicide tolerance and virus resistance [41].

The method used in gene editing, insertion, deletion or replacement of certain genes is the regularly clustered short palindromic repeats (CRISPR) technique [66]. CRISPR-Cas9 is the latest advancement and is much more efficient and economical than previously used methods. This scientific technique is very effective in altering multiple genes in a single plant, that substantially rising the potential of the biotechnology [67]. Genetic engineering techniques in agricultural biotechnology have been used profitably on large crops such as barley, maize, rice, soybeans, wheat and tomatoes [68].

### 4.3 Biorefinery Applications

**Conversion of Pollutants to Biofuels**. Renewable energy has become a important member in the sustainable strategy since the huge demand for energy necessarily intensify the depletion of non-renewable resources and ecological-social challenges. In this context, refinery approach is critical in renewable energy sustainable strategy, as increasing energy demands exacerbate the depletion of non-renewable resources [69].

As fossil fuel reserves are constantly being depleted, global energy requirement is growing at an alarming rate. From a climate change perspective, the usage of non-renewable energy resources causes the release of greenhouse gases that lead to rise in the amount of atmospheric $CO_2$. Studies have been conducted on biofuel production from microalgae as a potential lignocellulosic feedstock in the recent years [70].

Biofuels are alternative and renewable energy sources derived from living organisms or raw materials. Bioenergy production processes such as bioethanol, biodiesel and biomethane, in which biomass is used as a substrate, are of great interest worldwide. With the development of biogas, it has been replaced by natural gas and new possibilities for the use of biogas have also opened up. Bioethanol and biodiesel diminish the discharge of greenhouse gases and other unburned hydrocarbons into the environment [71]. When bioethanol is mixed with gasoline, emissions of greenhouse gases and other hazardous gases are decreased. Biodiesel fuel has a higher oxygen fraction content than diesel.

Therefore, biodiesel emits lower levels of particulate matter, unburned hydrocarbons and carbon monoxide, resulting in better combustion [72].

Biofuels obtained from biomass wastes such as agricultural and lignocellulosic residues are called second generation biofuels. Bioethanol production from the pre-treatment of dilute sulfuric acid of cassava straw [73], biofuel, bioplastic and bio sorbent production from potato peel [74], biofuel production from corn leaf by pyrolysis [75], bioethanol production by hydrolysis of horsetail, rye waste [76], biogas production by hydrolysis of banana peels [77], biomethane production by anaerobic degradation of sugar beet [78] from biomass are examples of studies on the production of biofuels.

**Resource Recovery from Biomass**. The biorefinery concept is useful and effective for all first, second and third generation biomass. First-generation biomass consist of sugar-cane, sugar beet, maize, soybean and rapeseed crops. Second-generation biomass consist of agricultural residues such as rice straw, wheat straw, wheat bran, sawdust, corn husk and grasses, agro-industrial wastes such as orange peel, coffee grounds, soybean oil and apple pulp and lignocellulosic biomass [79]. Third-generation biomass consist of marine biomass [80]. The utilization of these biological resources has the potential to supply to sustainable development. The factors that determine the efficiency of second-generation biomass are raw material supply chains, pretreatment techniques and conversion technologies. The raw material supply chain of the biorefinery includes substantial use of raw material resources including food products, agricultural wastes, lignocellulosic wastes and marine biomass at low cost [81].

Biorefinery strategies form the basis of the bio-based circular bioeconomy, which ensures the completion of the production and supply cycle. Biopolymer production using waste biorefineries is a solution-oriented approach, especially in waste management. Microbial bioplastic production bioprocesses using waste resources can also greatly reduce the excessive use of non-renewable resources [82]. The use of biorefinery strategies both promotes the environment-energy relationship and protects the ecosystem by reducing their carbon footprints [83, 84]. Microbial bioplastics such as polyhydroxybutyrate and polyhydroxyalkanoates are environmentally friendly bio-based products that can be produced by various bioprocesses based on the use of renewable resources [85]. In addition, microbial bioplastics can be produced directly with bioprocesses involving fermentation processes.

The evaluation of lignocellulosic biomass in the biorefinery approach includes process steps such as raw material preparation, pretreatment, confectionery and fermentation. With the biorefinery process of lignocellulosic wastes, thermochemical, chemical, biological and mechanical transformations can be realized. Lignin-based products (fuel, carbon fibers, polyurethane, aromatics), glycerol-based products (biodiesel, syngas, mono and polyesters, propanediol), cellulose-based products (paper, fiber, textile products) and hemicellulose-based products (C5 & C6 sugars, enzymes, biofuels, PHAs, Xylitol ...) can be formed [70]. For example; onion waste from the food processing industry can be used for the production of pigments, organic acids, polyphenols, polysaccharides and biofuels. Onion processing residues can be converted into high value-added products by different biorefinery processes such as enzymatic hydrolysis, fermentation and hydrothermal carbonization [86].

However, another biorefinery approach used is the processing of microalgal biomass. Microalgae are photosynthetic microorganisms that can grow rapidly in a variety of environments under a variety of nutrients, pH and temperature conditions. It can fix CO2 at a higher rate than terrestrial plants and use waste water streams and waste flue gases as low-cost nutrient sources [87]. At the same time, microalgal biomass constitutes a potential raw material source for the biorefinery approach, as it can be used to obtain a variety of high-value bioactive compounds suitable for the nutraceutical, pharmaceutical, energy and food industries [88]. In addition, proteins, pigments, lipids, carbohydrates, etc., which can be converted into various bioproducts in accordance with the biorefinery approach. They have various metabolic abilities to accumulate significant amounts of bioactive components such as [89]. Microalgae constitute a potential sustainable resource for the production of many valuable bio-based products. Microalgal biomass has 8–57% protein, 2–37% lipid and 4–65% carbohydrate content. In this context, microalgal carbohydrates can be used as high-value polysaccharides and low-cost carbon sources in fermentation processes. Long-chain fatty acids contained in microalgal lipids are used as health food supplements. In addition, the pigments, carotenoids and bioactive exopolymers contained in microalgae can be used in functional food products. All these microalgal bioactive components have important functions in supporting the immune system [90]. Microalgal proteins can also be used as food and feed supplements. Among microalgae with high protein content, especially *Scenedesmus spp., Dunaliella salina, Chlamydomonas rheinhardii* and *Chlorella pyrenoidosa* contain 44–57% protein. Microalgae with high carbohydrate content are *Haematococcus pluvialis, Scenedesmus dimorphus* and *Chlorella vulgaris,* contain 56–65% carbohydrates. Microags with high lipid content are *Scenedesmus sp., Botryococcus spp., Chlorella spp., Chlorococcum sp.* And *Nannochloropsis sp.,* contain 26–37% lipid [87]. *Botryococcus braunii,* which is rich in hydrocarbons, has a wide range of uses in the chemical and biopolymer industries. It also has great potential in the production of hydrocarbons and their conversion to ethylene and propylene. Supercritical $CO_2$ extraction method, which is used in the process of obtaining oil from microalgal biomass efficiently, can be used as a green and sustainable alternative [91]. However, it is important to integrate a sustainable microalgae production platform with subsequent extraction processes. While keeping algae alive with hydrostatic pressure applications during microalgae cultivation, processes for obtaining hydrocarbons are sustainable practices [92].

Based on the sustainable zero waste principle, microalgal biomass is turned into a variety of high added value products such as food and feed supplements, pharmaceuticals, biofuels, fertilizers and other waste-free products. In addition to these, microalgal biomass is also used for the throw of $CO_2$, the removal of pollutants from wastewater and the production of lipids, which are biodiesel raw materials [93]. Microalgal biorefinery applications are a sustainable strategic approach based on zero waste principle in reducing global warming caused by greenhouse gas emissions, bioremediation of wastewater, sustainable bioenergy and biochemical production [94].

### 4.4 Environmental Care Applications

**Water and Wastewater Treatment**. One of the main problems all over the world is water pollution. Considering the current world scenario, the vast majority of industrial wastes are discharged into lakes and rivers without pre-treatment. This situation causes very serious environmental pollution and visual problems [95]. Waste waters from some agricultural activities, raw materials and runoff also cause pollution of rivers and groundwater. Treatment processes have become increasingly complex due to the inadequacies in the maintenance of wastewater treatment plants and the implementation of strict regulations. Biologically designed wastewater treatment systems are used with the biodegradation of microorganisms such as bacteria, microalgae and fungi in the treatment of industrial wastewater streams, where the process parameters in the operating stages of wastewater treatment are carefully controlled [28, 96].

In the content of wastewater, there are generally high organic substances, nutrients, metals and chemicals with high added value. Due to the reduction of natural resources, the reuse of resources and the creation of new products with high added value from wastes, the evaluation of waste water has become a very important issue. The main target is on the circular economy of the "closed-loop" process through wastewater reuse and energy recovery, with the identification of biotechnologies used to recover available resources in wastewater treatment processes [97]. Among the wastewater treatment technologies, especially environmentally friendly and sustainable methods are microalgal treatment and treatment with microbial fuel cells. In addition, treated wastewater has applications in agriculture, aquaponics, fisheries and algae cultivation [78, 98, 99].

**Solid Waste Management**. Solid waste management is very important in ensuring long-term sustainability due to high population growth and irregular urbanization. Failure to use solid wastes causes uncontrolled degradation of environmental areas [100]. The biorefinery approach is critical for sustainable waste management in order to obtain value-added products by using solid wastes as raw materials [101]. The raw materials used as substrate in biorefinery systems are wastes obtained from various process steps of agriculture, forestry, industrial, domestic and organic waste sectors [71]. These include oil-based wastes such as inedible parts of the crop, organic municipal waste, domestic waste, industrial organic waste, used cooking oil waste, pulp, animal manures, and many other carbon-containing wastes [102].

The biorefinery approach in solid waste management processes deals with both providing non-renewable alternative energy sources from solid wastes and transforming them into high value-added bio products at the end of the process [103]. As an example of industrial applications; Solid wastes generated during the oil production process in industries where palm oil is processed have a high lignocellulose content. For this reason, various high value-added bio-based products such as many chemicals, biofuels and biomaterials can be produced with these food and agricultural wastes [100]. Another industrial example is biochar, which can be used as an alternative to coal and is a gas absorber; It can be obtained by slow pyrolysis and hydrothermal carbonization process [104]. With a similar process, fiberboard [105] and bio composites [106], which are effective on petrochemical-derived materials such as polymers and plastics, can also be produced. In addition, bio composites can be obtained by mechanical processing of hemp and flax wastes from agricultural wastes [107] and by methylation, carbonation, epoxidation processes of wastes from paper industries [108].

**Environmental Remediation**. Industrial and agricultural activities, poorly regulated waste disposal practices, oil spills and many other unpredictable human activities are the leading practices that cause environmental pollution. Many chemicals and toxins used in agricultural practices also deteriorate the quality of the soil, making the soil unsuitable for cultivation. Sludge, which is naturally found in most of the wastewater treatment systems in industrial enterprises, can be treated with stabilization, condensation, dewatering, drying and incineration applications. Especially in developing countries, low-cost and low-energy sludge treatment plants are preferred to be used in industries. Bioremediation applications, in which environmental pollutants are removed with the help of microorganisms, is an eco-biotechnological strategic approach to remove toxic pollutants from agricultural areas [109]. Bioremediation technology is a waste management application that uses microorganisms, plants or enzymes to separate organic and inorganic pollutants such as phenols, chlorophenols, petroleum hydrocarbons, polychlorinated biphenyls, organic solvents, azo dyes, pesticides and volatile substances from soils and wastewater [110]. Living organisms with enzymatic activity used in the bioremediation process can decompose the specified toxic pollutants by using toxic pollutants as a source in carbon and energy production in their metabolic activities [111, 112].

Another raw material used in bioremediation processes is microalgae biomass. In systems where microalgal cultivation is carried out, it is very difficult to support large-scale microalgal growth due to the lack of nutrient sources required for microalgal growth. In order to improve the sustainability and bio-economy of microalgae cultivation under these conditions, bioremediation practices are carried out to recover bioavailable nutrients from waste streams such as industrial wastewater, solid waste and flue gas. In this way, both biofuel production from microalgal biomass and environmental bioremediation processes can be performed [98, 113].

## 4.5 Health Care Applications

The development of the production and processing of biodegradable biomaterials, with the increase in microplastic pollution in the environment and the consequent increase in concerns about the sustainability of existing polymers, makes an important contribution to biotechnological approaches for sustainable development.

Cellulose, chitin and chitosan derived from natural sources are generally used as renewable, low-cost biopolymers. Since natural polymers are similar to the extracellular matrix and polymer structures in the human body, their use has started to become widespread, especially in translation and regenerative medicine. However, the importance of using biomaterials from natural sources is increasing due to the increasing need to find sustainable alternatives to existing synthetic polymers [114, 115].

In addition to the advantages of synthetic polymers such as low cost and compositional stability, their use in processes is generally limited due to their low biocompatibility and toxic degradation products. Accordingly, when we look at the production processes of synthetic polymers and the disposal of the by-products, the place of these polymers in the sustainable life cycle is a concern. Micro-plastic pollutants, which form the basis of environmental pollution, show the undesirable persistence of synthetic polymers in the environment [116]. However, the increase in the production of personal protective equipment such as fossil fuel-containing masks used in the SARS-CoV2 viral epidemic has once again revealed the microplastic problem [117].

One of the new biotechnological approaches in regenerative tissue engineering and biomedical applications is to produce new composites with superior properties by combining different biopolymers. Due to its low mechanical strength in the industrial application of chitosan, one of the natural polymers, its strength can be increased with the combination of cellulose. However, multifunctional, biodegradable biocomposites can be used by blending different biopolymers in various applications such as films, foams, fibers, filters and nanoparticles [118].

Cellulose is a natural polymer that is biodegradable, abundant in nature, and can be produced with sustainable and green technologies. All these characteristics of cellulose make it a useful, practical polymer in the production processes of bio-based hydrogel/or soft material. Cellulose-based hydrogels are especially used in biomedical fields such as wound healing, medical wound dressing, tissue engineering, industrial agricultural applications, industrial water treatment processes, and removal of environmental pollutants, heavy metals [119]. Nanocellulose, another green biomaterial, has biodegradability, sustainability, biocompatibility and specific physicochemical properties and has applications for controlled drug release [120].

In the field of regenerative medicine and tissue engineering, successful implantation of biomaterials can be performed, especially in structures such as heart valves, dental implants, vascular grafts, lenses, intraocular ligaments, tendons. However, pacemakers, biosensors, artificial skin, artificial hearts and medical devices can be successfully produced using biomaterials [121]. In addition, biomaterials can be used in order to replace the damaged and distorted parts of organs and tissues and to increase their functions. Some other biodegradable polymeric materials used as biomaterials due to their physical, chemical and biological properties are as follows; ceramic materials, some metals and their alloys, biopolymers and their composites, ceramic materials, composite materials, natural fiber reinforced bio composite materials and some other filling materials [122, 123].

Sustainable composites and nanocomposites can potentially be used in many areas, including the processing of waste from food processing and packaging units, agricultural and lignocellulosic waste, water treatment, environmental remediation applications, energy conversion and storage. The design of sustainable biomaterials has become popular as green and economical biotechnologies in the prevention and remediation processes of environmental pollution. When we look at the different usage areas of biomaterials; metals and their alloys are used in dental implants, orthopedic screws and stents. Ceramics are used in bone prostheses, heart valves, dental implants and joint prostheses. Composites are used in biosensors and implantable microelectrodes. Polymers are used in drug delivery systems, skin, cartilage, eye lens implants, contact lenses and surgical threads [121].

## 5 Results and Discussion

Humans and all organisms form part of ecosystems. To be able to continue our lives in the world, we live in and to live in prosperity; It depends on some of the services that ecosystems provide to us and the continuity of components such as organisms, soil, water and nutrients that ecosystems host. This structure, which is defined as ecosystem services, is very important for the sustainability of both the environment and all living things.

The climate crises and the problems caused by the use of energy systems based on fossil fuels have led to a better understanding of the importance of economic systems based on renewable resources and energy systems over the years. Achieving global nutritional conditions depends on farming more efficiently under controllable conditions and our ability to cope with extraordinary natural disasters. Increasing food crises in our world, decreasing agricultural productivity due to climate uncertainties, and energy systems dependent on fossil fuels in direct proportion to food prices constitute an important problem. On the other hand, our underground and surface resources, which we obtain from nature, continue to be depleted rapidly. Failure to provide ecosystem services regularly threatens not only the economic order, but also the life of the entire living population.

It is estimated that the gap between world demand and availability will continue to widen rapidly in the post-2030s period. Keeping biological and technical materials in their life cycle for an indefinite period; creation of new bio-based materials with new methods and technologies; stands out as the approach that needs to be advanced in this critical process. At this point, bioeconomy brings an important opportunity on the path of environmentally friendly growth; The paths of bioeconomy and circular economy intersect at the point of sustainability.

Circular economy, everything can be brought back to life (restorative), based on renewable energy (renewable), toxic chemicals are followed, reduced and eventually eliminated (toxic elimination), waste is destroyed with careful design philosophy, product and services are presented with a "user" point of view instead of "consumer" (user orientation). Its goal is to create new technologies and processes with long life cycles that eliminate material dependence and waste.

Biotechnology is a rapidly developing field and has various applications in sustainable agriculture. The genetic resources of plants, animals and microorganisms constitute the raw material for all research, technology development and creation of new products based on biotechnology. Today, food and agricultural practices are carried out on a large scale; different approaches are considered to meet sustainable environmental and economic developments, whose ultimate goal is to protect the biosphere.

From a bioengineering perspective, the benefits of biotechnological applications to the circular economy are as follows:

- Increasing the Use of Renewable Inputs (Bioprocesses using renewable raw materials)
- Introducing New and Better Materials (Biomaterials)
- Providing Better Life Cycle Designs (Biomimetic designs)
- Compost (Compostable materials)
- Better Reuse and Upcycling (Value added product from waste)

To summarize; the bio-based economy and its biotechnological applications are very important for sustainability. It is known that all over the world resources are gradually depleted. In this context, for the continuity of scientific research, it is a great necessity for different stakeholders to work together in a multidisciplinary manner, to think inspired by nature and to deal with biomimetic designs.

## 6 Conclusion

In this review, the importance of the concept of sustainability in the world; the relations of circular economy and bioeconomy approaches with entrepreneurship; Understanding the concepts of sustainable biotechnology, food and environment and innovative sustainable biotechnological applications with a bioengineering perspective are discussed. Sustainable development has become a remarkable approach that is considered more and more today with the increase in environmental pollution. The basis of the understanding of sustainability is the balance between the environment, food and bioeconomy. With the rapid population growth in the world, new agricultural production techniques have emerged in line with the increasing demand for agricultural products.

The focal point of sustainable development is new agricultural production techniques and biotechnologies. On the other hand, the rapid depletion of natural resources in the world and the damage to the ecological balance indicate a negative trend. In this case, the countries agreed that the use of renewable raw materials should be increased with the awareness that raw materials are depleted. One of the biggest challenges of the 21st century is to develop a sustainable bioeconomy system that uses environmentally friendly biological processes and renewable biological resources. With the concept of bioeconomy gaining importance all over the world at this point; also brought biotechnological applications to the agenda. Thanks to bioengineering practices, the sustainable management of biological resources and the competitiveness of the food-agricultural industries have been increased.

**Acknowledgements.** The authors are thankful to 2$^{nd}$ International Conference on Sustainable, Circular Management and Environmental Engineering (ISCMEE 2022) Organization Committee for access to the facilities.

## References

1. Neher, D.: Ecological sustainability in agricultural systems: definition and measurement. In: Integrating Sustainable Agriculture, Ecology, and Environmental Policy. Routledge (1992)
2. Agbedahin, A.V.: Sustainable development, education for sustainable development, and the 2030 agenda for sustainable development: emergence, efficacy, eminence, and future. Sustain. Dev. **27**, 669–680 (2019). https://doi.org/10.1002/sd.1931
3. Perrings, C.: Environment and development economics 20 years on. Environ. Dev. Econ. **19**, 333–366 (2014)
4. Klarin, T.: The concept of sustainable development: from its beginning to the contemporary issues. Zagreb Int. Rev. Econ. Bus. **21**, 67–94 (2018). https://doi.org/10.2478/zireb-2018-0005

5. Nikolaou, I.E., Tsalis, T.A., Evangelinos, K.I.: A framework to measure corporate sustainability performance: a strong sustainability-based view of firm. Sustain. Prod. Consum. **18**, 1–18 (2019). https://doi.org/10.1016/j.spc.2018.10.004
6. Oláh, J., Aburumman, N., Popp, J., Khan, M.A., Haddad, H., Kitukutha, N.: Impact of industry 4.0 on environmental sustainability. Sustainability **12**, 4674 (2020). https://doi.org/10.3390/su12114674
7. Sajjad, A., Shahbaz, W.: Mindfulness and social sustainability: an integrative review. Soc. Indic. Res. **150**(1), 73–94 (2020). https://doi.org/10.1007/s11205-020-02297-9
8. Curtis, S.K., Lehner, M.: Defining the sharing economy for sustainability. Sustainability. **11**, 567 (2019). https://doi.org/10.3390/su11030567
9. Barnes, S.J.: Out of sight, out of mind: Plastic waste exports, psychological distance and consumer plastic purchasing. Glob. Environ. Change. (2019)
10. Liew, W.T., Adhitya, A., Srinivasan, R.: Sustainability trends in the process industries: a text mining-based analysis. Comput. Ind. **65**, 393–400 (2014). https://doi.org/10.1016/j.compind.2014.01.004
11. Kates, R.W., Parris, T.M.: Long-term trends and a sustainability transition. Proc. Natl. Acad. Sci. USA **100**, 8062–8067 (2003). https://doi.org/10.1073/pnas.1231331100
12. Shivanna, K.R.: Climate change and its impact on biodiversity and human welfare. Proc. Indian Natl. Sci. Acad. **88**, 160–171 (2022). https://doi.org/10.1007/s43538-022-00073-6
13. Hirvilammi, T., Koch, M.: Sustainable welfare beyond growth. Sustainability **12**, 1824 (2020). https://doi.org/10.3390/su12051824
14. Pan, S.-Y., Gao, M., Kim, H., Shah, K.J., Pei, S.-L., Chiang, P.-C.: Advances and challenges in sustainable tourism toward a green economy. Sci. Total Environ. **635**, 452–469 (2018). https://doi.org/10.1016/j.scitotenv.2018.04.134
15. Merino-Saum, A., Clement, J., Wyss, R., Baldi, M.G.: Unpacking the green economy concept: a quantitative analysis of 140 definitions. J. Clean. Prod. **242**, 118339 (2020). https://doi.org/10.1016/j.jclepro.2019.118339
16. Corvellec, H., Stowell, A.F., Johansson, N.: Critiques of the circular economy. J. Ind. Ecol. **26**, 421–432 (2022). https://doi.org/10.1111/jiec.13187
17. Pardo & Schweitzer: A long-term strategy for a European circular economy—setting the course for success, 33 (2018)
18. Santagata, R., Ripa, M., Genovese, A., Ulgiati, S.: Food waste recovery pathways: challenges and opportunities for an emerging bio-based circular economy. A systematic review and an assessment. J. Clean. Prod. **286**, 125490 (2021). https://doi.org/10.1016/j.jclepro.2020.125490
19. Azam, M., Ftiti, Z., Hunjra, A.I., Louhichi, W., Verhoeven, P.: Do market-supporting institutions promote sustainable development? Evidence from developing economies. Econ. Model., 106023 (2022). https://doi.org/10.1016/j.econmod.2022.106023
20. Taghvaee, V.M., Nodehi, M., Saber, R.M., Mohebi, M.: Sustainable development goals and transportation modes: analyzing sustainability pillars of environment, health, and economy. World Dev. Sustain. **1**, 100018 (2022). https://doi.org/10.1016/j.wds.2022.100018
21. Varjani, S.J., Taherzadeh, M., Khanal, S., Pandey, A.: New horizons in biotechnology: advances in sustainable industrial and environmental bioprocesses and bioproducts. Ind. Crops Prod. **158**, 113000 (2020). https://doi.org/10.1016/j.indcrop.2020.113000
22. Sharma, S., Kumar Sharma, N.: Advanced materials contribution towards sustainable development and its construction for green buildings. Mater. Today Proc. (2022). https://doi.org/10.1016/j.matpr.2022.07.394
23. Singh, J., Yadav, A.N. (eds.): Natural Bioactive Products in Sustainable Agriculture. Springer, Singapore (2020). https://doi.org/10.1007/978-981-15-3024-1

24. Singh, R.P., Singh, P.K., Gupta, R., Singh, R.L.: Chapter 2 - biotechnological tools to enhance sustainable production. In: Singh, R.L., Mondal, S. (Eds.), Biotechnology for Sustainable Agriculture. pp. 19–66. Woodhead Publishing (2018). https://doi.org/10.1016/B978-0-12-812160-3.00002-7
25. Lokko, Y., Heijde, M., Schebesta, K., Scholtès, P., Van Montagu, M., Giacca, M.: Biotechnology and the bioeconomy—towards inclusive and sustainable industrial development. New Biotechnol. **40**, 5 (2018). https://doi.org/10.1016/j.nbt.2017.06.005
26. Umesha, S., Singh, P. K., Singh, R. P.: #Chapter 6 - microbial biotechnology and sustainable agriculture. In: Singh, R.L., Mondal, S. (Eds.), Biotechnology for Sustainable Agriculture, pp. 185–205. Woodhead Publishing (2018). https://doi.org/10.1016/B978-0-12-812160-3.00006-4
27. Bajpai, P.: #Chapter 2 - Industrial sustainability and biotechnology. In: Bajpai, P. (ed.) Biotechnology in the Chemical Industry, pp. 15–21. Elsevier (2020). https://doi.org/10.1016/B978-0-12-818402-8.00002-1
28. Boopathy, R., Rene, E.R., López, M.E., Annachhatre, A.P., Lens, P.N.L.: #Special issue on environmental biotechnologies for sustainable development. Int. Biodeterior. Biodegrad. **119**, 1–3 (2017). https://doi.org/10.1016/j.ibiod.2017.03.007
29. Bargaz, A., Lyamlouli, K., Chtouki, M., Zeroual, Y., Dhiba, D.: Soil Microbial resources for improving fertilizers efficiency in an integrated plant nutrient management system. Front. Microbiol. **9**, 1606 (2018). https://doi.org/10.3389/fmicb.2018.01606
30. Pratheesh, P.T., Lal, S., Tuvikene, R., Manickam, S., Sudheer, S.: #Chapter 2 - the contribution of microbial biotechnology to sustainable development in agriculture and allied sectors. In: Rastegari, A.A., Yadav, A.N., and Yadav, N. (eds.) New and Future Developments in Microbial Biotechnology and Bioengineering, pp. 17–28. Elsevier (2020). https://doi.org/10.1016/B978-0-12-820526-6.00002-6
31. Obaideen, K., Shehata, N., Sayed, E.T., Abdelkareem, M.A., Mahmoud, M.S., Olabi, A.G.: The role of wastewater treatment in achieving sustainable development goals (SDGs) and sustainability guideline. Energy Nexus. **7**, 100112 (2022). https://doi.org/10.1016/j.nexus.2022.100112
32. Mouat, A.R.: Sustainability in food-waste reduction biotechnology: a critical review. Curr. Opin. Biotechnol. **77**, 102781 (2022). https://doi.org/10.1016/j.copbio.2022.102781
33. FAO, FAO Statistical Yearbook: World food and agriculture, Food and Agriculture Organization of the United Nations, Rome (2013). http://www.fao.org/docrep/018/i3107e/i3107e00.htm
34. Woodley, J.M.: Towards the sustainable production of bulk-chemicals using biotechnology. New Biotechnol. **59**, 59–64 (2020). https://doi.org/10.1016/j.nbt.2020.07.002
35. Korcz, E., Varga, L.: Exopolysaccharides from lactic acid bacteria: Techno-functional application in the food industry. Trends Food Sci. Technol., 110 (2021). https://doi.org/10.1016/j.tifs.2021.02.014
36. Ciuła, J., Gaska, K., Iljuczonek, Ł., Generowicz, A., Koval, V.: Energy efficiency economics of conversion of biogas from the fermentation of sewage sludge to biomethane as a fuel for automotive vehicles. Architecture, Civil Engineering, Environment. **12**, 131–140 (2019). https://doi.org/10.21307/acee-2019-029
37. Bocso, N.-S., Butnariu, M.: The biological role of primary and secondary plants metabolites. Crit. Rev. Food Sci. Nutr. **5**, 1–7 (2022). https://doi.org/10.31579/2637-8914/094
38. Gomes, R.J., Borges, M. de F., Rosa, M. de F., Castro-Gómez, R.J.H., Spinosa, W.A.: Acetic acid bacteria in the food industry: systematics, characteristics and applications. Food Technol. Biotechnol. **56**, 139–151 (2018). https://doi.org/10.17113/ftb.56.02.18.5593
39. Ali, A., et al.: Non-thermal processing technologies for the recovery of bioactive compounds from marine by-products. LWT **147**, 111549 (2021). https://doi.org/10.1016/j.lwt.2021.111549

40. Harfouche, A.L., Petousi, V., Meilan, R., Sweet, J., Twardowski, T., Altman, A.: Promoting ethically responsible use of agricultural biotechnology. Trends Plant Sci. **26**, 546–559 (2021). https://doi.org/10.1016/j.tplants.2020.12.015
41. Crippa, M., Solazzo, E., Guizzardi, D., Monforti-Ferrario, F., Tubiello, F.N., Leip, A.: Food systems are responsible for a third of global anthropogenic GHG emissions. Nat. Food. **2**, 198–209 (2021). https://doi.org/10.1038/s43016-021-00225-9
42. Barea, J.M.: Future challenges and perspectives for applying microbial biotechnology in sustainable agriculture based on a better understanding of plant-microbiome interactions. J. Soil Sci. Plant Nutr. **15**, 261–282 (2015)
43. Pole, A., Srivastava, A., Zakeel, M.C.M., Sharma, V.K., Suyal, D.C., Singh, A.K., Soni, R.: #12 - Role of microbial biotechnology for strain improvement for agricultural sustainability. In: Soni, R., Suyal, D.C., Yadav, A.N., and Goel, R. (Eds.), Trends of Applied Microbiology for Sustainable Economy, pp. 285–317. Academic Press (2022). https://doi.org/10.1016/B978-0-323-91595-3.00001-X
44. Okafor, N., Okeke, B.C.: Modern Industrial Microbiology and Biotechnology. CRC Press (2017)
45. Rana, K.L., Kour, D., Kaur, T., Devi, R., Yadav, N., Subrahmanyam, G., Kumar, A., Yadav, A.N.: Chapter 8 - Biotechnological applications of seed microbiomes for sustainable agriculture and environment. In: Rastegari, A.A., Yadav, A.N., Yadav, N. (Eds.), New and Future Developments in Microbial Biotechnology and Bioengineering, pp. 127–143. Elsevier (2020). https://doi.org/10.1016/B978-0-12-820526-6.00008-7
46. Yadav, A.N., Kour, D., Kaur, T., Devi, R., Guleria, G., Rana, K.L., Yadav, N., Rastegari, A.A.: Chapter 18 - Microbial biotechnology for sustainable agriculture: current research and future challenges. In: Rastegari, A.A., Yadav, A.N., Yadav, N. (Eds.), New and Future Developments in Microbial Biotechnology and Bioengineering. pp. 331–344. Elsevier (2020). https://doi.org/10.1016/B978-0-12-820526-6.00020-8
47. Lyson, T.A.: Advanced agricultural biotechnologies and sustainable agriculture. Trends Biotechnol. **20**, 193–196 (2002). https://doi.org/10.1016/S0167-7799(02)01934-0
48. Soni, R., Kumar, V., Suyal, D.C., Jain, L., Goel, R.: Metagenomics of Plant Rhizosphere Microbiome. In: Singh, R.P., Kothari, R., Koringa, P.G., Singh, S.P. (eds.) Understanding Host-Microbiome Interactions - An Omics Approach, pp. 193–205. Springer, Singapore (2017). https://doi.org/10.1007/978-981-10-5050-3_12
49. Goel, R., Soni, R., Suyal, D.C. (eds.): Microbiological Advancements for Higher Altitude Agro-Ecosystems & Sustainability. RB, Springer, Singapore (2020). https://doi.org/10.1007/978-981-15-1902-4
50. Kushwah, S., Singh, A.K., Chowdhury, T., Soni, R.: Comparative Effect of Selected Bacterial Consortia on Yield and Yield Attributes of Chickpea (CICER ARIETINUM L.), 3 (2021)
51. Verma, M., Mishra, J., Arora, N.K.: Plant Growth-Promoting Rhizobacteria: Diversity and Applications. In: Sobti, R.C., Arora, N.K., Kothari, R. (eds.) Environmental Biotechnology: For Sustainable Future, pp. 129–173. Springer, Singapore (2019). https://doi.org/10.1007/978-981-10-7284-0_6
52. Mahanty, T., et al.: Biofertilizers: a potential approach for sustainable agriculture development. Environ. Sci. Pollut. Res. **24**(4), 3315–3335 (2016). https://doi.org/10.1007/s11356-016-8104-0
53. Mus, F., Crook, M.B., Garcia, K., Garcia Costas, A., Geddes, B.A., Kouri, E.D., Paramasivan, P., Ryu, M.-H., Oldroyd, G.E.D., Poole, P.S., Udvardi, M.K., Voigt, C.A., Ané, J.-M., Peters, J.W.: Symbiotic nitrogen fixation and the challenges to its extension to Nonlegumes. Appl. Environ. Microbiol. **82**, 3698–3710 (2016). https://doi.org/10.1128/AEM.01055-16

54. Yadav, A.N.: Plant Microbiomes for Sustainable Agriculture: Current Research and Future Challenges. In: Yadav, A.N., Singh, J., Rastegari, A.A., Yadav, N. (eds.) Plant Microbiomes for Sustainable Agriculture. SDB, vol. 25, pp. 475–482. Springer, Cham (2020). https://doi.org/10.1007/978-3-030-38453-1_16
55. Yadav, A.N.: Microbiomes of Wheat (Triticum aestivum L.) Endowed with Multifunctional Plant Growth Promoting Attributes. Mol. Microbiol., 15 (2019)
56. Ahemad &Kibret: Mechanisms and applications of plant growth promoting rhizobacteria: Current perspective - ScienceDirect, https://www.sciencedirect.com/science/article/pii/S1018364713000293. Accessed 2 September 2022
57. Yadav, A.N., Mishra, S., Singh, S., Gupta, A. (eds.): Recent Advancement in White Biotechnology Through Fungi. FB, Springer, Cham (2019). https://doi.org/10.1007/978-3-030-10480-1
58. Yadav, A.N., Sachan, S.G., Verma, P., Tyagi, S.P., Kaushik, R., Saxena, A.K.: Culturable diversity and functional annotation of psychrotrophic bacteria from cold desert of Leh Ladakh (India). World J. Microbiol. Biotechnol. **31**(1), 95–108 (2014). https://doi.org/10.1007/s11274-014-1768-z
59. Bhuvaneshwari, K., Singh, P.K.: Response of nitrogen-fixing water fern Azolla biofertilization to rice crop. 3 Biotech **5**(4), 523–529 (2014). https://doi.org/10.1007/s13205-014-0251-8
60. Igiehon, N.O., Babalola, O.O.: Rhizosphere Microbiome Modulators: Contributions of Nitrogen Fixing Bacteria towards Sustainable Agriculture. Int. J. Environ. Res. Public. Health. **15**, 574 (2018). https://doi.org/10.3390/ijerph15040574
61. Rana, K.L., et al.: Endophytic Fungi: Biodiversity, Ecological Significance, and Potential Industrial Applications. In: Yadav, A.N., Mishra, S., Singh, S., Gupta, A. (eds.) Recent Advancement in White Biotechnology Through Fungi. FB, pp. 1–62. Springer, Cham (2019). https://doi.org/10.1007/978-3-030-10480-1_1
62. Rana, K.L., et al.: Biodiversity of Endophytic Fungi from Diverse Niches and Their Biotechnological Applications. In: Singh, B.P. (ed.) Advances in Endophytic Fungal Research. FB, pp. 105–144. Springer, Cham (2019). https://doi.org/10.1007/978-3-030-03589-1_6
63. Rana vd.: Chapter 16-Agriculturally important microbial biofilms: Biodiversity, ecological significances, and biotechnological applications - ScienceDirect, https://www.sciencedirect.com/science/article/pii/B9780444642790000165?via%3Dihub. Accessed 1 September 2022
64. Vishwakarma, K., Kumar, N., Shandilya, C., Mohapatra, S., Bhayana, S., Varma, A.: Revisiting plant–microbe interactions and microbial consortia application for enhancing sustainable agriculture: a review. Front. Microbiol., 11 (2020)
65. Zilberman, D., Holland, T.G., Trilnick, I.: Agricultural GMOs—what we know and where scientists disagree. Sustainability. **10**, 1514 (2018). https://doi.org/10.3390/su10051514
66. Samanta, M.K., Dey, A., Gayen, S.: CRISPR/Cas9: an advanced tool for editing plant genomes. Transgenic Res. **25**(5), 561–573 (2016). https://doi.org/10.1007/s11248-016-9953-5
67. Globus, R., Qimron, U.: A technological and regulatory outlook on CRISPR crop editing. J. Cell. Biochem. **119**, 1291–1298 (2018). https://doi.org/10.1002/jcb.26303
68. Araki, M., Ishii, T.: Towards social acceptance of plant breeding by genome editing. Trends Plant Sci. **20**, 145–149 (2015). https://doi.org/10.1016/j.tplants.2015.01.010
69. Duan, Y., et al.: Sustainable biorefinery approaches towards circular economy for conversion of biowaste to value added materials and future perspectives. Fuel **325**, 124846 (2022). https://doi.org/10.1016/j.fuel.2022.124846
70. Culaba, A.B., Mayol, A.P., San Juan, J.L.G., Vinoya, C.L., Concepcion, R.S., Bandala, A.A., Vicerra, R.R.P., Ubando, A.T., Chen, W.-H., Chang, J.-S.: Smart sustainable biorefineries for

lignocellulosic biomass. Bioresour. Technol. **344**, 126215 (2022). https://doi.org/10.1016/j.biortech.2021.126215
71. Sharma vd.: Trends in mitigation of industrial waste: Global health hazards, environmental implications and waste derived economy for environmental sustainability | Elsevier Enhanced Reader, https://reader.elsevier.com/reader/sd/pii/S0048969721074349?token=CA1B54FE7361E7C6A97CAA6B112548E66172D24DE899DBA3AD635640D3E35C446C9C6790467E183414A154D2688269E9&originRegion=eu-west-1&originCreation=20220902093203. Accessed 2 September 2022. https://doi.org/10.1016/j.scitotenv.2021.152357
72. Shekofteh, M., Mesri Gundoshmian, T., Jahanbakhshi, A., Heidari-Maleni, A.: Performance and emission characteristics of a diesel engine fueled with functionalized multi-wall carbon nanotubes (MWCNTs-OH) and diesel–biodiesel–bioethanol blends. Energy Rep. **6**, 1438–1447 (2020). https://doi.org/10.1016/j.egyr.2020.05.025
73. Sivamani, S., Baskar, R., Chandrasekaran, A.P.: Response surface optimization of acid pretreatment of cassava stem for bioethanol production. Environ. Prog. Sustain. Energy. **39**, e13335 (2020). https://doi.org/10.1002/ep.13335
74. Ebrahimian, F., Denayer, J.F.M., Karimi, K.: Potato peel waste biorefinery for the sustainable production of biofuels, bioplastics, and biosorbents. Bioresour. Technol. **360**, 127609 (2022). https://doi.org/10.1016/j.biortech.2022.127609
75. Uzoejinwa, B.B., He, X., Wang, S., El-Fatah Abomohra, A., Hu, Y., Wang, Q.: Co-pyrolysis of biomass and waste plastics as a thermochemical conversion technology for high-grade biofuel production: Recent progress and future directions elsewhere worldwide. Energy Convers. Manag. **163**, 468–492 (2018). https://doi.org/10.1016/j.enconman.2018.02.004
76. Mihajlovski, K., Buntić, A., Milić, M., Rajilić-Stojanović, M., Dimitrijević-Branković, S.: From Agricultural waste to biofuel: enzymatic potential of a bacterial isolate streptomyces fulvissimus CKS7 for bioethanol production. Waste Biomass Valorization **12**(1), 165–174 (2020). https://doi.org/10.1007/s12649-020-00960-3
77. Martínez-Ruano, J.A., Caballero-Galván, A.S., Restrepo-Serna, D.L., Cardona, C.A.: Techno-economic and environmental assessment of biogas production from banana peel (Musa paradisiaca) in a biorefinery concept. Environ. Sci. Pollut. Res. **25**(36), 35971–35980 (2018). https://doi.org/10.1007/s11356-018-1848-y
78. Shilpi, S., Lamb, D., Bolan, N., Seshadri, B., Choppala, G., Naidu, R.: Waste to watt: Anaerobic digestion of wastewater irrigated biomass for energy and fertiliser production. J. Environ. Manage. **239**, 73–83 (2019). https://doi.org/10.1016/j.jenvman.2019.02.122
79. Hassan, S.S., Williams, G.A., Jaiswal, A.K.: Moving towards the second generation of lignocellulosic biorefineries in the EU: drivers, challenges, and opportunities. Renew. Sustain. Energy Rev. **101**, 590–599 (2019). https://doi.org/10.1016/j.rser.2018.11.041
80. Konda, N.V.S.N.M., Singh, S., Simmons, B.A., Klein-Marcuschamer, D.: An investigation on the economic feasibility of Macroalgae as a potential feedstock for Biorefineries. BioEnergy Res. **8**(3), 1046–1056 (2015). https://doi.org/10.1007/s12155-015-9594-1
81. Patel, N., Feofilovs, M., Blumberga, D.: Evaluation of bioresource value models: sustainable development in the agriculture biorefinery sector. J. Agric. Food Res. **10**, 100367 (2022). https://doi.org/10.1016/j.jafr.2022.100367
82. Dahiya, S., Kumar, A.N., Shanthi Sravan, J., Chatterjee, S., Sarkar, O., Mohan, S.V.: Food waste biorefinery: sustainable strategy for circular bioeconomy. Bioresour. Technol. **248**, 2–12 (2018). https://doi.org/10.1016/j.biortech.2017.07.176
83. Leong, E.C.: Editorial. Int. J. Geotech. Eng. **15**, 1 (2021). https://doi.org/10.1080/19386362.2021.1876432
84. Talan, A., Pokhrel, S., Tyagi, R.D., Drogui, P.: Biorefinery strategies for microbial bioplastics production: Sustainable pathway towards circular Bioeconomy. Bioresour. Technol. Rep. **17**, 100875 (2022). https://doi.org/10.1016/j.biteb.2021.100875

85. Ferreira, J.A., Mahboubi, A., Lennartsson, P.R., Taherzadeh, M.J.: Waste biorefineries using filamentous ascomycetes fungi: present status and future prospects. Bioresour. Technol. **215**, 334–345 (2016). https://doi.org/10.1016/j.biortech.2016.03.018
86. Sagar, N.A., et al.: Onion waste based-biorefinery for sustainable generation of value-added products. Bioresour. Technol. **362**, 127870 (2022). https://doi.org/10.1016/j.biortech.2022.127870
87. Yadav, G., Sen, R.: Sustainability of Microalgal Biorefinery: Scope, Challenges, and Opportunities. In: De, S., Bandyopadhyay, S., Assadi, M., Mukherjee, D.A. (eds.) Sustainable Energy Technology and Policies. GET, pp. 335–351. Springer, Singapore (2018). https://doi.org/10.1007/978-981-10-7188-1_15
88. Nishshanka, G.K.S.H., Anthonio, R.A.D.P., Nimarshana, P.H.V., Ariyadasa, T.U., Chang, J.-S.: Marine microalgae as sustainable feedstock for multi-product biorefineries. Biochem. Eng. J., 108593 (2022). https://doi.org/10.1016/j.bej.2022.108593
89. Koyande, A.K., Chew, K.W., Rambabu, K., Tao, Y., Chu, D.-T., Show, P.-L.: Microalgae: a potential alternative to health supplementation for humans. Food Sci. Hum. Wellness. **8**, 16–24 (2019). https://doi.org/10.1016/j.fshw.2019.03.001
90. Silva, S.C., Ferreira, I.C.F.R., Dias, M.M., Barreiro, M.F.: Microalgae-derived pigments: a 10-year bibliometric review and industry and market trend analysis. Molecules **25**, 3406 (2020). https://doi.org/10.3390/molecules25153406
91. Yildiz-Ozturk, E., Yesil-Celiktas, O.: Supercritical $CO_2$ extraction of hydrocarbons from Botryococcus braunii as a promising bioresource. J. Supercrit. Fluids. **130**, 261–266 (2017). https://doi.org/10.1016/j.supflu.2017.06.019
92. Yildiz-Ozturk, E., Ilhan-Ayisigi, E., Togtema, A., Gouveia, J., Yesil-Celiktas, O.: Effects of hydrostatic pressure and supercritical carbon dioxide on the viability of Botryococcus braunii algae cells. Bioresour. Technol. **256**, 328–332 (2018). https://doi.org/10.1016/j.biortech.2018.02.041
93. Cheirsilp, B., Maneechote, W.: Insight on zero waste approach for sustainable microalgae biorefinery: Sequential fractionation, conversion and applications for high-to-low value-added products. Bioresour. Technol. Rep. **18**, 101003 (2022). https://doi.org/10.1016/j.biteb.2022.101003
94. Lu, X., et al.: Sustainable development of microalgal biotechnology in coastal zone for aquaculture and food. Sci. Total Environ. **780**, 146369 (2021). https://doi.org/10.1016/j.scitotenv.2021.146369
95. Attri, S.D., Singh, S., Dhar, A., Powar, S.: Multi-attribute sustainability assessment of wastewater treatment technologies using combined fuzzy multi-criteria decision-making techniques. J. Clean. Prod. **357**, 131849 (2022). https://doi.org/10.1016/j.jclepro.2022.131849
96. Usman, M., Adeel, S., Amjad, Z., Bokhari, T.H., Akram, N., Anam, S.: Chapter 15 - Biotechnology: the sustainable tool for effective treatment of wastewater. In: Dar, G.H., Bhat, R.A., Qadri, H., Hakeem, K.R. (Eds.), Microbial Consortium and Biotransformation for Pollution Decontamination, pp. 347–380. Elsevier (2022). https://doi.org/10.1016/B978-0-323-91893-0.00007-9
97. Kundu, D., et al.: Valorization of wastewater: a paradigm shift towards circular bioeconomy and sustainability. Sci. Total Environ. **848**, 157709 (2022). https://doi.org/10.1016/j.scitotenv.2022.157709
98. Geremia, E., Ripa, M., Catone, C.M., Ulgiati, S.: A review about microalgae wastewater treatment for bioremediation and biomass production—a new challenge for Europe. Environments **8**, 136 (2021). https://doi.org/10.3390/environments8120136
99. Lu, Q., Xiao, Y., Chen, F.: Chapter three - application of microalgae biotechnology for the sustainable development of aquaculture. In: Li, Y., Zhou, W. (Eds.), Advances in Bioenergy, pp. 117–163. Elsevier (2021). https://doi.org/10.1016/bs.aibe.2021.06.001

100. Shah, A.V., Singh, A., Sabyasachi Mohanty, S., Kumar Srivastava, V., Varjani, S.: Organic solid waste: Biorefinery approach as a sustainable strategy in circular bioeconomy. Bioresour. Technol. **349**, 126835 (2022). https://doi.org/10.1016/j.biortech.2022.126835
101. Liu, T., Ren, X., Soundari, P.G., Chen, H., Awasthi, S.K., Varjani, S., Pandey, A., Zhang, Z., Awasthi, M.K.: 7 - waste Biorefinery development toward circular Bioeconomy with a focus on life-cycle assessment. In: Pandey, A., Tyagi, R.D., Varjani, S. (Eds.), Biomass, Biofuels, Biochemicals, pp. 199–230. Elsevier (2021). https://doi.org/10.1016/B978-0-12-821878-5.00014-3
102. Varjani, S., Shah, A.V., Vyas, S., Srivastava, V.K.: Processes and prospects on valorizing solid waste for the production of valuable products employing bio-routes: a systematic review. Chemosphere **282**, 130954 (2021). https://doi.org/10.1016/j.chemosphere.2021.130954
103. Filote, C., Santos, S.C.R., Popa, V.I., Botelho, C.M.S., Volf, I.: Biorefinery of marine macroalgae into high-tech bioproducts: a review. Environ. Chem. Lett. **19**(2), 969–1000 (2020). https://doi.org/10.1007/s10311-020-01124-4
104. Zhou, Y., et al.: Production and beneficial impact of biochar for environmental application: a comprehensive review. Bioresour. Technol. **337**, 125451 (2021). https://doi.org/10.1016/j.biortech.2021.125451
105. Simon, V., et al.: VOC and carbonyl compound emissions of a fiberboard resulting from a coriander biorefinery: comparison with two commercial wood-based building materials. Environ. Sci. Pollut. Res. **27**(14), 16121–16133 (2020). https://doi.org/10.1007/s11356-020-08101-y
106. Karmee: A spent coffee grounds based biorefinery for the production of biofuels, biopolymers, antioxidants and biocomposites | Elsevier Enhanced Reader, https://reader.elsevier.com/reader/sd/pii/S0956053X17307882?token=D70526BA887CAA38B943590C72E3A734AE9AD4B49E9631BB3B81A31E27A08B4C37B2CBA43D54EE18048C173A1788140A&originRegion=eu-west-1&originCreation=20220902122012. Accessed 2 September 2022. https://doi.org/10.1016/j.wasman.2017.10.042
107. Pallesen, B.E., Andersen, T.L.: Reinforced Biocomposites from flax and hemp. J. Ind. Hemp. **7**, 61–81 (2002). https://doi.org/10.1300/J237v07n01_06
108. Yamini, G., Shakeri, A., Zohuriaan-Mehr, M.J., Kabiri, K.: Cyclocarbonated lignosulfonate as a bio-resourced reactive reinforcing agent for epoxy biocomposite: from natural waste to value-added bio-additive. J. CO2 Util. **24**, 50–58 (2018). https://doi.org/10.1016/j.jcou.2017.12.007
109. Yaashikaa, P.R., Kumar, P.S.: Bioremediation of hazardous pollutants from agricultural soils: a sustainable approach for waste management towards urban sustainability. Environ. Pollut. **312**, 120031 (2022). https://doi.org/10.1016/j.envpol.2022.120031
110. Duodu, M.G., Singh, B., Christina, E.: Chapter 10 - Waste management through bioremediation technology: an eco-friendly and sustainable solution. In: Samuel, J., Kumar, A., and Singh, J. (eds.) Relationship Between Microbes and the Environment for Sustainable Ecosystem Services, Vol. 2, pp. 205–234. Elsevier (2022). https://doi.org/10.1016/B978-0-323-89937-6.00007-3
111. Enerijiofi, K.E.: Chapter 19 - Bioremediation of environmental contaminants: a sustainable alternative to environmental management. In: Saxena, G., Kumar, V., and Shah, M.P. (Eds.), Bioremediation for Environmental Sustainability, pp. 461–480. Elsevier (2021). https://doi.org/10.1016/B978-0-12-820524-2.00019-5
112. Patel, K., Yadav, S., Verma, S.K., Pandey, M., Aseri, G.K., Sohal, J.S., Sharma, D., Khare, N.: Chapter 23 - Improvement of ecological sustainability of in situ bioremediation through coordinated approaches. In: Singh, J., Sharma, D. (Eds.) Microbial Resource Technologies for Sustainable Development, pp. 445–456. Elsevier (2022). https://doi.org/10.1016/B978-0-323-90590-9.00005-5

113. Premaratne, M., et al.: Resource recovery from waste streams for production of microalgae biomass: a sustainable approach towards high-value biorefineries. Bioresour. Technol. Rep. **18**, 101070 (2022). https://doi.org/10.1016/j.biteb.2022.101070
114. Kostag, M., Jedvert, K., El Seoud, O.A.: Engineering of sustainable biomaterial composites from cellulose and silk fibroin: fundamentals and applications. Int. J. Biol. Macromol. **167**, 687–718 (2021). https://doi.org/10.1016/j.ijbiomac.2020.11.151
115. Reis vd.: Natural-Based Polymers for Biomedical Applications - 1st Edition, https://www.elsevier.com/books/natural-based-polymers-for-biomedical-applications/reis/978-1-84569-264-3. Accessed 3 September 2022
116. Shamshina, J.L., Zavgorodnya, O., Choudhary, H., Frye, B., Newbury, N., Rogers, R.D.: Search of stronger/cheaper chitin nanofibers through electrospinning of chitin–cellulose composites using an ionic liquid platform. ACS Sustain. Chem. Eng. **6**, 14713–14722 (2018). https://doi.org/10.1021/acssuschemeng.8b03269
117. Vanapalli, K.R., et al.: Challenges and strategies for effective plastic waste management during and post COVID-19 pandemic. Sci. Total Environ. **750**, 141514 (2021). https://doi.org/10.1016/j.scitotenv.2020.141514
118. Díez-Pascual, A.M.: Synthesis and applications of biopolymer composites. Int. J. Mol. Sci. **20**, 2321 (2019). https://doi.org/10.3390/ijms20092321
119. Kundu, R., Mahada, P., Chhirang, B., Das, B.: Cellulose hydrogels: green and sustainable soft biomaterials. Curr. Res. Green Sustain. Chem. **5**, 100252 (2022). https://doi.org/10.1016/j.crgsc.2021.100252
120. Das, S., Ghosh, B., Sarkar, K.: Nanocellulose as sustainable biomaterials for drug delivery. Sens. Int. **3**, 100135 (2022). https://doi.org/10.1016/j.sintl.2021.100135
121. Biswal, T., BadJena, S.K., Pradhan, D.: Sustainable biomaterials and their applications: a short review. Mater. Today Proc. **30**, 274–282 (2020). https://doi.org/10.1016/j.matpr.2020.01.437
122. Idrees, M.: A review on application of biomaterial in medical sciences. J. Sci. Health Sci. **4**, 390–393 (2014)
123. Behera, A.: Classification of biomaterials used in medicine. Int. J. Adv. Appl. Sci. IJAAS **1**, 125–129 (2012)

# Enablers of Sustainable Knowledge Management in Higher Education Institutions: A Case from Turkey

Melisa Ozbiltekin-Pala[1](✉), Muruvvet Deniz Sezer[2], Yesim Deniz Ozkan-Ozen[1], and Yigit Kazancoglu[1]

[1] Logistics Management Department, Yasar University, 35100 Izmir, Turkey
melisa.ozbiltekin@yasar.edu.tr
[2] Business Administration Department, Yasar University, 35100 Izmir, Turkey

**Abstract.** The amount of information that emerges with the rapidly developing technology is increasing daily, revealing the need for knowledge management. Knowledge management updates the ever-increasing information capacity in organizations, makes the information available, defines the processes necessary to reach the required information, and enables the knowledge needed to be shared. Especially in higher education institutions, sustainable knowledge management is essential for developing a long-term sustainable culture. Therefore, this study aims to determine the enablers of sustainable knowledge management in higher education institutions. As a result, fostering sustainable learning is determined as the best criterion for the case of higher education institutors.

**Keywords:** Knowledge Management · Sustainability · Best Worst Method · Sustainable development

## Abbreviations

| | |
|---|---|
| BWM | Best worst method |
| HEIs | Higher education institutions |
| KM | Knowledge management |
| SDG | Sustainable development goal |

## 1 Introduction

Due to technological developments, the increasing value of information and awareness of the sustainable development goals have led to an increased usage of knowledge management (KM) in sustainability [1]. KM has a critical role in achieving sustainable competitive advantage with its unique characteristics such as intangibility and complex concept. Organizations realize that they can only gain strategic benefits by utilizing and effectively utilizing new knowledge. Thus, many organizations and institutions aim to


V. Koval et al. (Eds.): ISCMEE 2022, Lect. Notes in Managem., Industrial Engineering, pp. 209–215, 2023.
https://doi.org/10.1007/978-3-031-23463-7_13

propose new strategies and policies by developing practical tools for knowledge creation [2]. HEIs contribute to the KM process through knowledge creation, transfer, and sharing [3].

KM can be considered fundamental for SDGs [3]. Thus, sustainable KM stands out as a new development paradigm aiming to simultaneously achieve economic, environmental, and social SDGs [4]. Considering the complexity of the concept of sustainability and KM, integrating these two concepts needs to be managed effectively [5]. As a motivation, many HEIs aim to elaborate their valuable resources for developing sustainable competitive advantage [6]. Thus, this study aims to investigate enablers of sustainable KM in HEIs from the point of Turkey.

This study proposes the following research questions:

- **Research Question 1:** What are the enablers of sustainable KM in HEIs from the point of Turkey?
- **Research Question 2:** What are the most important criteria for achieving sustainable KM in HEIs?

To answer these mentioned research questions, after determining enablers of sustainable KM in HEIs, the BWM approach is used to evaluate criteria in HEIs to develop sustainable KM. This study is structured as follows. Section 2 includes a literature review related to sustainable KM in HEIs. The case study, implementation, and results of this study are discussed in Sect. 3. Finally, Sect. 4 covers discussions and conclusions.

## 2 Sustainable KM in HEIs

HEIs have a significant impact on achieving sustainability; therefore, they need to have the capacity to integrate KM applications into their implementations [7, 8]. Although KM is a popular topic in the literature, incorporating it with sustainability is still a promising field [9]. On the other hand, literature on sustainable KM in HEIs is minimal [10]. From this point of view, proposing enablers for sustainable KM in HEIs are expected to be beneficial in presenting a starting point for more comprehensive applications.

One of the early studies [11] related to KM for sustainable development in HEIs mainly focused on campus sustainability. In this study, the potential benefits of KM in sustainable HEIs are discussed in detail, and it is stated that the usefulness of KM in other sectors has similar impacts on HEIs. Furthermore, [12] conducted an exploratory analysis for sustainable KM in HEIs and proposed a framework for improving the system's performance.

In a more recent study, [12] presented a sustainable framework for KM in HEIs with different clusters. In the proposed framework, technological development, relational learning, organizational learning, and individual learning dimensions are integrated with capital dimensions (i.e., human, organizational, relational, and innovation) by presenting critical indicators for each intersection. Study [7] primarily focused on Sustainable Development Goal (SDG) target 4.7, which is "Education for Sustainable Development and Global Citizenship," to achieve KM in HEIs. This study presents barriers and enablers of integrating KM into HEIs by considering SDH and suggests practical implications.

As it can be understood from the brief literature review related to sustainable KM in HEIs, current studies lack in providing enablers for further implementations. From this point of view, seven enablers are presented in this study to make evaluations and are shown in Table 1. The previous studies support these enablers.

**Table 1.** Sustainable KM enablers for HEIs.

| Enablers | Sources |
|---|---|
| Incorporate SDG into the curriculum (E1) | [7, 13] |
| Fostering sustainable learning (E2) | [11, 12] |
| Investing intellectual capital (E3) | [12] |
| Achieving green campus (E4) | [11] |
| Developing a sustainable management culture (E5) | [7, 14] |
| Incorporate sustainable knowledge and expertise into practice (E6) | [11, 12] |
| Digital applications for transferring sustainable knowledge (E7) | [7, 15] |

A case study is conducted in the following section to evaluate these enablers for a sample study. This study's results are expected to guide future implementations in this field.

## 3 Case Study

This study is implemented in one of Turkey's leading universities with more than 600 employees and more than 10000 students. The university hosts many studies within the scope of sustainability and circular economy. Moreover, throughout the university, courses on green and sustainability are given in both engineering and business faculties.

Five researchers from the university participated in the implementation of the study. These researchers consist of experts in sustainability and circular economy. The characteristics of these researchers are given in Table 2.

**Table 2.** The characteristics of researchers.

| Researchers | Profession | Faculty | Working years |
|---|---|---|---|
| Researcher 1 | Sustainability | Engineering | 10 |
| Researcher 2 | Circular Economy | Engineering | 8 |
| Researcher 3 | Circularity and Sustainability | Business Administration | 5 |
| Researcher 4 | Sustainable Consumption | Art and Design | 4 |
| Researcher 5 | Sustainability | Business Administration | 7 |

## 3.1 Implementation and Results

### 3.1.1 Bwm

Decision makers need to use multi-criteria decision-making (MCDM) methods to evaluate many criteria and sub-criteria [16]. It is a method that compares the best criterion with other criteria and all other criteria with the worst criterion [17]. With BWM, decision-makers do not need to make pairwise comparisons between all criteria [18]. They need to define the most and most minor desirable criteria and then make pairwise comparisons between the best/worst criteria and other criteria [16]. BWM, which also helps to control the consistency rate, was developed by Rezaei (2015) [19].

BWM is implemented in this study. Before implementation of this study, the steps of BWM are given as follows;

1st Step: A criteria set is created.

2nd Step: Decision makers determine the best and worst criteria according to their own opinions from the set of criteria.

3rd Step: A pairwise comparison matrix is created (best to others). The pairwise comparison vector is calculated as in Eq. 1.

$$C_b = (c_{b1}, c_{b2}, \ldots, c_{bn}) \tag{1}$$

where $c_{bj}$ is the preference of the best criteria b over the criteria j.

4th Step: A pairwise comparison matrix is created (worst to others). The pairwise comparison vector is calculated as in Eq. 2.

$$C_w = (c_{w1}, c_{w2}, \ldots, c_{wn}) \tag{2}$$

where $c_{bj}$ is the preference of the worst criteria w over the criteria j.

5th Step: Weights of these criteria is calculated.

$$(w_1^*, w_2^*, w_3^*, w_n^*)$$

To calculate optimal weights, Eq. 3 is used.

$$\text{Mimax}\{|w_b - c_{bj}w_b|, |w_j - c_{jw}w_w|\} \tag{3}$$

$\sum_j W_j = 1$
$w_j \geq 0$, for all j

Then, the model is solved in a linear programming as follows;
minmax $\varepsilon_j$
$|w_B - a_{Bj}w_j| \leq \varepsilon$
$|w_j - a_{jw}w_w| \leq \varepsilon$
$\sum_j W_J = 1$
$w_j \geq 0$, for all j

### 3.1.2 Results of the Study

Among the sustainability KM enablers in the HEIs, six criteria are determined by the decision- makers in the context of best and worst. As a result of the determination of the best and worst criteria, firstly, it is ensured to determine the importance levels with the evaluation matrix rated from best to the others (1–9 scale), and then the importance levels are determined with the evaluation matrix rated from others to the worst (1–9 scale). As a result of the implementation made by following the BWM steps, the weights of the enablers were obtained as shown in the table below (Table 3).

**Table 3.** Weights of enablers of sustainable knowledge management.

| Enablers of sustainable knowledge management | Weights |
|---|---|
| Incorporate SDG into the curriculum (E1) | 0.24 |
| Fostering sustainable learning (E2) | 0.29 |
| Investing intellectual capital (E3) | 0.09 |
| Achieving green campus (E4) | 0.06 |
| Developing a sustainable management culture (E5) | 0.09 |
| Incorporate sustainable knowledge and expertise into practice (E6) | 0.05 |
| Digital applications for transferring sustainable knowledge (E7) | 0.18 |

As a result of the implementation, the average consistency rate is determined as 0.12, which is consistent for the implementation. As shown in Table 3, the top three criteria are determined as "*fostering sustainable learning*", "*incorporate SDG into the curriculum*" and "*Digital applications for transferring sustainable knowledge*", respectively. The worst criteria are "*incorporate the concept into practice*" for the case organization.

## 4 Discussion and Conclusion

Sustainable KM is critical for developing sustainable advantages in the HEIs. Thus, it is crucial to identify the necessary factors to achieve sustainability and KM together [20–22]. This study aims to analyze the enablers of sustainable KM in HEIs.

According to the results, the most significant three criteria for sustainable KM in HEIs are determined as "fostering sustainable learning", "incorporating SDG into the curriculum" and "Digital applications for transferring sustainable knowledge", respectively. Studies which are [12, 20, 21] highlighted that fostering sustainable learning is a critical enabler for supporting the development of crucial skills such as creative thinking and KM in HEIs.

In alignment with this study, [7] stated that incorporating SDG into the curriculum and digital applications for transferring sustainable knowledge can be a crucial enabler for KM-higher education for sustainable development by enhancing SDGs' performance. Similarly, [13, 22] indicated the importance of incorporating SDG into the curriculum to achieve strategic advantage. Moreover, according to [15], digital applications for transferring sustainable knowledge are inevitably necessary to understand learning as a concept and achieve SDGs [23, 24].

As a limitation of this study, enablers are developed based on the case university. Thus, the results cannot be generalized. Besides, new enablers of sustainable KM in HEIs may emerge. In future work, different MCDM methods can be applied to enhance the study's validity. It is possible to generalize the results by applying a large-group questionnaire. Comparisons can be carried out with other universities.

## References


1. Di Vaio, A., Palladino, R., Pezzi, A., Kalisz, D.: The role of digital innovation in knowledge management systems: a systematic literature review. J. Bus. Res. **123**, 220–231 (2021)
2. Kaba, A., Ramaiah, C.K.: Demographic differences in using knowledge creation tools among faculty members. J. Knowl. Manag. **21**(4) (2017)
3. Martins, V.W.B., Rampasso, I.S., Anholon, R., Quelhas, O.L.G., Leal Filho, W.: Knowledge management in the context of sustainability: literature review and opportunities for future research. J. Clean. Prod. **229**, 489–500 (2019)
4. van Reijsen, J., Helms, R., Batenburg, R., Foorthuis, R.: The impact of knowledge management and social capital on dynamic capability in organizations. Knowl. Manag. Res. Pract. **13**(4), 401–417 (2015)
5. Bucci, M., El-Diraby, T.E.: The functions of knowledge management processes in urban impact assessment: the case of Ontario. Impact Assess. Project Appr. **36**(3), 265–280 (2018)
6. Mahdi, O.R., Nassar, I.A., Almsafir, M.K.: Knowledge management processes and sustainable competitive advantage: an empirical examination in private universities. J. Bus. Res. **94**, 320–334 (2019)
7. Adhikari, D.R., Shrestha, P.: Knowledge management initiatives for achieving sustainable development goal 4.7: higher education institutions' stakeholder perspectives. J. Knowl. Manag., (ahead-of-print) (2022)
8. Zhao, S., Liu, X., Andersson, U., Shenkar, O.: Knowledge management of emerging economy multinationals. J. World Bus. **57**(1), 101255 (2022)
9. Yu, S., Abbas, J., Álvarez-Otero, S., Cherian, J.: Green knowledge management: scale development and validation. J. Innov. Knowl. **7**(4), 100244 (2022)
10. de Bem Machado, A., Secinaro, S., Calandra, D., Lanzalonga, F.: Knowledge management and digital transformation for Industry 4.0: a structured literature review. Knowl. Manag. Res. Pract. **20**(2), 320–338 (2022)
11. McNeil, R.: Application of knowledge management for sustainable development in institutions of higher education. Dalhousie J. Interdisciplinary Manag. **7**, 1–13 (2011)
12. Hussein, T., Anumolu, G.: An Exploratory analysis for sustainable knowledge management system in higher educational institutions. Adv. Econ. Bus. Manage. (AEBM) **3**(4), 380–386 (2016)
13. Franco, I., Saito, O., Vaughter, P., Whereat, J., Kanie, N., Takemoto, K.: Higher education for sustainable development: actioning the global goals in policy, curriculum and practice. Sustain. Sci. **14**(6), 1621–1642 (2019)
14. Abad-Segura, E., González-Zamar, M.D., Infante-Moro, J.C., Ruipérez García, G.: Sustainable management of digital transformation in higher education: global research trends. Sustainability **12**(5), 2107 (2020)

15. Schulz, J., Köhler, T., Weith, T.: Knowledge Management for Sustainability: The Spatial Dimension of Higher Education as an Opportunity for Land Management. In: Weith, T., Barkmann, T., Gaasch, N., Rogga, S., Strauß, C., Zscheischler, J. (eds.) Sustainable Land Management in a European Context. HI, vol. 8, pp. 227–248. Springer, Cham (2021). https://doi.org/10.1007/978-3-030-50841-8_12
16. El Baz, J., Tiwari, S., Akenroye, T., Cherrafi, A., Derrouiche, R: A framework of sustainability drivers and externalities for Industry 4.0 technologies using the best-worst method. J. Cleaner Prod., 130909 (2022)
17. Rezaei, J., Arab, A., Mehregan, M.: Analyzing anchoring bias in attribute weight elicitation of SMART, Swing, and best-worst method. Int. Trans. Oper. Res., 1–31 (2022)
18. Beemsterboer, D., Hendrix, E.M., Claassen, G.D.: On solving the best-worst method in multi-criteria decision-making*. IFAC-PapersOnLine **51**, 1660–1665 (2018)
19. Rezaei, J.: A concentration ratio for non-linear best worst method. Int. J. Inf. Technol. Decis. Mak. **19**(3), 891–907 (2020)
20. Quarchioni, S., Paternostro, S., Trovarelli, F.: Knowledge management in higher education: a literature review and further research avenues. Knowl. Manag. Res. Pract. **20**(2), 304–319 (2020)
21. De La Torre, E.M., Perez-Encinas, A., Gomez-Mediavilla, G.: Fostering sustainability through mobility knowledge, skills, and attitudes. Sustainability **14**(3), 1349 (2022)
22. Ferrer-Estévez, M., Chalmeta, R.: Integrating sustainable development goals in educational institutions. Int. J. Manage. Educ. **19**(2), 100494 (2021)
23. Gómez-Marín, N., Cara-Jiménez, J., Bernardo-Sánchez, A., Álvarez-de-Prado, L., Ortega-Fernández, F.: Sustainable knowledge management in academia and research organizations in the innovation context. Int. J. Manage. Educ. **20**(1), 100601 (2022)
24. Arsawan, I.W.E., Koval, V., Rajiani, I., Rustiarini, N.W., Supartha, W.G., Suryantini, N.P.S.: Leveraging knowledge sharing and innovation culture into SMEs sustainable competitive advantage. Int. J. Product. Perform. Manag. **71**, 405–428 (2022)

Check for updates

# Seaport Business Actions to Ensure Clean and Affordable Energy

Aylin Çalışkan[1], Cansu Akçiçek[2](✉), and Yücel Öztürkoğlu[1]

[1] Faculty of Business, Yaşar University, İzmir 35100, Turkey
[2] Graduate School, Yaşar University, İzmir 35100, Turkey
cansu.akcicek@yasar.edu.tr

**Abstract.** Many different sectors are obliged to implement the 17 goals established by the United Nations General Assembly, both for their own life cycles and for the future of our world. Each goal has its own goals and plans. Although different applications are made for these targets on a sectoral basis, a common language must be developed for each target. In this study, the 7th goal of the Sustainable Development Goals (SDGs), the goal of providing access to economic, sustainable and clean energy for everyone, was emphasized. In the study, the maritime transport sector, which has a large share in the logistics sector in terms of both economic and environmental damage, has been selected. In this context, the sustainability, environment, corporate social responsibility and annual reports of the 33 biggest European ports with a gross weight handling volume were examined in line with the SDG 7 target and a content analysis was made on these reports. According to the results of the analysis, the contribution and approach of Europe's largest ports to Clean and Affordable Energy, the 7th sustainable development goal of the United Nations, emerged.

**Keywords:** Sustainable Port · Clean Marine · Renewable Energy

## 1 Introduction

The Sustainable Development Goals (SDGs) were established by the United Nations in 2015, which address environmental, social and economic problems that the world has been struggling with for years and which could lead to critical consequences if no action is taken [1]. In the previous years, similarly, goals that could be a solution to social problems were determined under the name of Millennium Development Goals (MDGs). The MDGs, which consist of eight goals covering social issues such as poverty, hunger, disease and gender inequality, had an important place in the fight against poverty, especially between 2000 and 2015. However, problems such as climate change and environmental pollution that the world is facing clearly show that environmental targets are needed as well as social targets [2]. At the United Nations Conference on Sustainable Development (Rio + 20) in June 2012, member states adopted that new targets should be established on existing MDGs, taking into account environmental situations as well. Thus, at the United Nations (UN) Sustainable Development Summit held in New York,


V. Koval et al. (Eds.): ISCMEE 2022, Lect. Notes in Managem., Industrial Engineering, pp. 216–228, 2023.
https://doi.org/10.1007/978-3-031-23463-7_14

the agenda consists of 17 SDGs was published in 2015 [3, 4]. In addition to 17 goals, 169 targets have been determined in total and it is planned to reach these targets by 2030.

The SDGs include goals of peace, prosperity and partnership, as well as objectives that concern people and the environment. One of the environmental goals is to ensure access to affordable, sustainable, clean energy for all which is the 7$^{th}$ goal of SDGs. There are five main goals, and the first one is to provide that everyone has access to affordable, reliable and modern energy services. As an indicator of this, the ratio of the population that has access to electricity and the population that is dependent on clean fuels and technology is taken into account [5]. According to SDGs report, progress in electrification has slowed. In 2010, there are 1.2 billion of people without electricity, this number has decreased to 733 million people by 2020. However, it is estimated that 679 million people still do not have access to electricity in 2030, the target date, due to the slowdown in progress [6]. The second target is about increasing the use of renewable energy significantly and it is measured by the share of renewable energy in total energy consumption. Although there has been awareness and many developments regarding renewable energy consumption in recent years, the share of renewable energy such as solar energy, wind energy, geothermal energy, in total energy consumption is 17.7% as of 2019. Third target of SDG-7 is doubling the global development rate in terms of energy efficiency. Finally, until 2030, in addition to increasing investment opportunities to facilitate access to research and clean energy technology, expanding the infrastructures of especially underdeveloped and developing countries are among the other targets [5]. However, it is seen that the international financial investment flow that will help the renewable energy consumption of developing countries decreased from $24.7 in 2017 to $10.9 in 2019. Therefore, it seems that more work needs to be done to reach the targets by 2030 under the SDGs.

In maritime transport, which is one of the most important building blocks of trade and supply chain, it is important to carry out port activities in an efficient and sustainable manner. Especially considering the land and ship activities of the ports, this situation has led to the necessity of actions to prevent environmental pollution. Optimization of port services, use of energy-efficient equipment, and reduction of heating consumption can be given as examples of activities to increase energy efficiency in port operations. In addition, using more electric vehicles in operations to reduce fuel consumption, encouraging gasification, saving energy in lighting through the use of new technologies are some of the energy-based improvements in port operations as well. In this context, port operations should be considered within the scope of SDG-7 in terms of access to affordable, sustainable, clean energy and consumption.

This study is conducted based on a qualitative content analysis. The objective is to create extensive port-related sustainability practices that may be implemented in order to help the fulfillment of the clean and affordable energy goal of sustainable development. The achievement of the SDGs depends on the active participation of many industries and economic participants, which makes this study crucial.

## 2 Literature Review

Since the SDGs were published, there have been many studies on this subject in the literature. The quality and quantity of the studies in the literature have also diversified, as there are many different sectors that each goal can be related to. In the light of these studies, Meschede [7] conducted a bibliometric study grouping scientific studies on the SDGs and obtained important findings. According to the results, studies in the literature so far have focused more on SDG 3 (health and well-being). Another literature review [8] identifies the targets of the Sustainable Development Goals, the obstacles to their implementation, and also present some areas for future attention.

This study is about affordable and clean energy, which is the 7th goal of the Sustainable Development Goals, SDG-7 themed studies also have been reviewed in the literature. It has been discovered that there are many studies that analyze the relationships between SDGs. A study of McCollum et al. [9], the interactions between SDG-7 and the other SDGs have been studied. For instance, consumption of renewable resources and increasing energy efficiency are strongly related to the mutual benefits of keeping global warming below 2 °C together with the 13th SDG climate action. With the increase in the use of renewable energy within the scope of SDG-7, reducing fuel consumption that is harmful to the environment and contributing to global warming can make an important contribution to climate action. Likewise, increasing awareness on climate change can contribute to the increase in investments required for energy efficiency. Some knowledge gaps are also mentioned in their studies for future studies. Similarly, the synergy effect between SDGs was investigated in terms of renewable electricity and a strong interaction was found between SDG-7 and SDG-8 using panel data [10]. As a study conducted on the basis of SDG-7, the difficulties faced by Bangladesh in energy efficiency and the progress made are explained and the data is presented [11]. Moreover, a literature study was also conducted in which the contribution of the education sector to affordable and clean energy was investigated [12]. Thus, in addition to the general overview studies in the literature regarding SDG-7, empirical studies were also carried out, and these studies are summarized in Table 1.

Since this study is related to seaports and energy efficiency, studies carried out within the scope of seaport and sustainable development goals in the literature are also examined. Balić et al. [16] analyzed the main scientific contributions in the field of sustainable passenger ports. Focusing on the development of a sustainable port, the study presents a 10-year literature review by grouping the results according to the research area and the method used. As a result, they found that environmental impact is the most studied area of research and quantitative analysis is the most used methodology. As a different perspective, Hiranandani [17] has done a qualitative multi-case study for sustainable development goals in seaports. It is aimed to compare the sustainability policies and practices of seaports in different locations and to investigate the difficulties encountered. While different environmental issues were mentioned in terms of sustainability practices, renewable energy and energy efficiency were also mentioned in the theme of land or resource use. In another study, considering port activities that negatively affect the environment, the impact of port authorities and policies for sustainable transport

**Table 1.** SDG-7 related studies in literature.

| Author(s) | Main theme | Methods used |
|---|---|---|
| McCollum et al. [9] | The interactions between SDGs | Overview study |
| Katekar et al. [10] | Major challenges to attain SDG-7 | Content analysis |
| Swain and Karimu [11] | Examining synergy effect between SDGs with renewable electricity | Panel data and two-step estimation |
| Salvia and Brandli [12] | Role of universities to contribute to SDG-7 | Systematic literature review |
| Santika et al. [13] | Interconnections between energy and SDGs | Mixed approach (qualitative and quantitative) |
| Chovancová and Vavrek [14] | Progress evaluation of EU countries towards achieving SDG-7 | CV-TOPSIS technique |
| de Miguel Ramos and Laurenti [15] | Examining synergies and trade-offs between SDGs in Spain | Regression model |

growth was investigated [18]. **In addition, some studies have addressed the difficulties** encountered in reaching the SDGs in ports and the reasons for the obstacles encountered [19, 20]. However, it is recognized that there is a gap in the literature for studies focusing on both seaports and SDG-7. Therefore, in this study to consolidate the actions and practices of port industry linked to SDG-7, the biggest European ports' sustainability disclosures were analyzed through content analysis and the results were presented in the following sections.

## 3 Empirical Study

In this study, a content analysis method is used to gather and examine the documentation's content [21]. The aim of the study is to reveal current and possible port actions that contribute to the clean and affordable energy goal. For this reason, qualitative content analysis, which deals with the content and meanings of words in the documents, was preferred [22].

## 4 Data Set

Secondary data was used to conduct content analysis as the data set of the study. In this context, using the data set used in study of Çalışkan [23], the 28 leading European ports with TEU handling volume [24] and the 33 leading European major ports with gross weight handling volume [25] in 2020 were combined and the data is given in Table 2. The sustainability statements of 33 ports were obtained from their websites, annual reports, corporate social responsibility reports and sustainability reports.

**Table 2.** Data set of the study [24, 25].

| Top Port Authorities (volume in TEU and gross weight) | Country | The 28 leading European ports 2020 handling volume in TEU (.000) (rank) | The 33 leading European ports 2020 handling volume in gross weight (.000) (rank) | Source of sustainability statements* | Pages |
|---|---|---|---|---|---|
| Aliaga | Turkey | | 64,922 (10) | CSR 2020 | 24 |
| Ambarli | Turkey | 3,105 (10) | 33,285 (27) | SR 2020–2021 | 108 |
| Amsterdam | Netherlands | | 103,913 (4) | SR 2017 | 33 |
| Antwerpen | Belgium | 11,676 (2) | 214,030 (2) | SR 2021 | 67 |
| Barcelona | Spain | 3,313 (9) | 54,709 (15) | SR 2020 | 181 |
| Botas | Turkey | | 66,945 (8) | Website | 8 |
| Bremerhaven | Germany | 4,850 (7) | 47,586 (19) | SR 2021 + ER 2020 | 243 |
| Dunkerque | France | | 42,558 (22) | ER 2020 | 16 |
| Felixstowe | United Kingdom | 3,838 (8) | | ER 2020–2021 | 16 |
| Gdansk | Poland | 1,800 (16) | 45,522 (20) | CSR 2021 | 30 |
| Genova | Italy | 2,176 (13) | 49,695 (18) | Website | 5 |
| Gothenburg | Sweden | 763 (24) | 38,891 (25) | SR 2021 | 62 |
| Hamburg | Germany | 9,282 (3) | 117,152 (3) | SR 2017–2018 | 123 |
| Kocaeli/Izmit | Turkey | 1,715 (18) | 71,359 (7) | Website | 7 |
| Iskenderun | Turkey | | 61,895 (11) | Website | 2 |
| La Spezia | Italy | 1,659 (19) | | SR 2019 | 62 |

*(continued)*

**Table 2.** (*continued*)

| Top Port Authorities (volume in TEU and gross weight) | Country | The 28 leading European ports 2020 handling volume in TEU (.000) (rank) | The 33 leading European ports 2020 handling volume in gross weight (.000) (rank) | Source of sustainability statements* | Pages |
|---|---|---|---|---|---|
| Las Palmas | Spain | 1,007 (22) | | Website | 14 |
| Liverpool | United Kingdom | 885 (23) | | SR 2020 | 12 |
| London | United Kingdom | 1,731 (17) | 54,035 (17) | ER 2021 | 21 |
| Medway | United Kingdom | 113 (28) | | SR 2020 | 12 |
| Mersin | Turkey | 1,854 (15) | | Website | 2 |
| Milford Haven | United Kingdom | | 34,951 (26) | ER 2019 | 50 |
| Piraeus | Greece | 5,646 (4) | 56,825 (14) | SR 2020 | 48 |
| Riga | Latvia | | 30,625 (29) | Website | 5 |
| Rotterdam | Netherlands | 13,493 (1) | 439,631 (1) | AR 2021 | 22 |
| Sines | Portugal | 1,423 (21) | 38,907 (24) | SR 2020 | 111 |
| Southampton | United Kingdom | 1,880 (14) | 33,151 (28) | SR 2021 | 82 |
| Tallinn | Estonia | | 19,635 (32) | AR 2021 | 134 |
| Taranto | Italy | | 17,609 (33) | AR 2019 | 2 |
| Trieste | Italy | | 60,333 (12) | Website | 1 |
| Valencia | Spain | 5,421 (5) | 65,308 (9) | SR 2019 | 77 |
| Wilhelmshaven | Germany | | 28,867 (30 | SR 2021 | 68 |
| Zeebrugge | Belgium | 477 (27) | | Website | 1 |

* SR: Sustainability report; ER: Environmental report; CSR: Corporate social responsibility report; AR: Annual report [23]

## 5 Content Analysis

Three stages were applied in process of content analysis. In the first stage, an initial node set has been created according to SDG 7. In order to reveal major themes, goals, actions and concepts based on the SDG 7, "UNSDG Business Reporting" guide developed by GRI and United Nations Global Compact were used. 17 keywords and phrases relevant to 7th goal was coded via NVivo 11 Plus software. In the second stage, sustainability statements of European ports were read in detail and affordable and clean energy related all items were coded on the NVivo 11 as the new sub-nodes under the SDG 7. The third and last stage included the axial coding process. In this process, codes were read, categorized, and summarized according to the appropriate SDG category. Axial coding is crucial for grouping highly similar codes and removing meaningless ones. In the axial coding stage, the 69 codes identified in the preceding two phases were reduced to 45.

## 6 Results

The content analysis results reveal 5 main port related actions to contribute SDG 7. These are using renewable energy, providing cleaner marine fuels in a safe and efficient manner, saving energy, encouraging third parties for using clean energy, and conducting a climate change adaptation plan.

Under the main heading of renewable energy, there exist the types of environmentally friendly and renewable energy sources such as wind and solar power. Apart from the energy sources, it includes the activities of ports in terms of producing, managing, recovering, reusing, developing, and sharing renewable energy sources. The second main heading of SDG 7, saving energy, involves three sub actions: Reducing fuel consumption, saving energy in operations, and lighting. Minimizing fuel usage is a vital contributor in achieving quality air and ports have some actions to contribute on this. For example, ports promote gasification by offering service through necessary Liquefied Natural Gas and Compressed Natural Gas station infrastructures. More, for the employers and partners, they may promote natural gas as a mobility fuel. They may convert port equipment to run on renewable energies. Numerous methods exist for reducing energy use in port operations. For instance, investing, employing and improving energy efficiency in port equipment, ACs, heating, buildings, etc. Furthermore, optimizing port services in terms of time, energy usage, and cost. Energy protecting in lighting is achieved by modernizing the public lighting network, using demand-oriented lighting, and with new tech for example LEP (Laser Excited Phosphor) that produces strong, efficient light with an emphasis on beam distance. The final action is to support customers and partners who are pursuing green energy efforts by providing Environmental Ship Index discounts, environmental concessions, promoting LNG as a fuel substitute, supplying services like LNG bunkering and shore power that help cut greenhouse gas emissions, and awarding clean shipping. This objective also provides for the supply of safe and efficient cleaner (marine) fuels as well as a strategy for coping with climate change.

Results reflected in the Fig. 1 are also more clearly demonstrated with the Table 3 visual below.

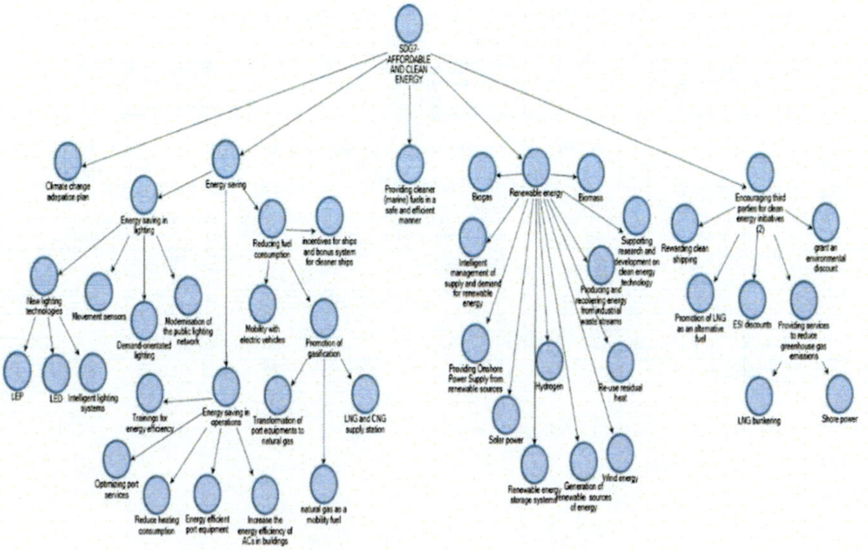


**Fig. 1.** Map diagram of port related SDG 7 actions.

## 7 Conclusions

The Sustainable Development Goals are a universal call to action that includes the goals that the member states of the United Nations aim to achieve by the end of 2030. With less than a decade remaining, how close industry-based companies are to these goals is a matter of debate. In this study, the general situation of target 7 in the sector was evaluated based on the maritime transport sector. In line with the SDG 7 target, the sustainability reports of the 33 European ports with the largest handling volume were examined and a content analysis was made on these reports. Processes of initial, open, and axial coding showed actions that ports may take to adopt clean and affordable energy. The content produced can be utilized by ports to report on their sustainability efforts. In future studies, the relationship of SDG 7 with other SDGs can be examined. Because the results of this study show that several SDGs are connected to each other. For example, energy reduction initiatives, using and generating renewable energy concern also SDG 12 (Responsible consumption and production), SDG 13 (Climate Action), SDG 9 (Industry, innovation, and infrastructure), and SDG 11.

**Table 3.** Port-related SDG 7 actions.

| Renewable energy | Climate change adaptation plan | Energy saving | | | Providing cleaner marine fuels | Encouraging third parties for clean energy initiatives | |
|---|---|---|---|---|---|---|---|
| Wind energy | | Reducing fuel consumption | Promotion of gasifi-cation | Transfor-mation of port equipment to natural gas | | Reward-ing clean shipping | |
| Solar power | | | | Natural gas as a mobil-ity fuel | | Providing services to reduce green-house gas emissions | Shore power |
| Biomass | | | | LNG and CNG sup-ply station | | | LNG bun-kering |
| Biogas | | | Mobility with electric vehi-cles | | | Promotion of LNG as an alterna-tive fuel | |
| Hydrogen | | | Bonus system for cleaner ships and motives for ships | | | Grant an environ-mental discount | |
| Re-use residual heat | | Energy saving in operations | Trainings for energy efficiency | | | ESI discounts | |
| Supporting research and development on clean energy technology | | | Reduce heating consump-tion | | | | |

*(continued)*

**Table 3.** (*continued*)

| Renewable energy | Climate change adaptation plan | Energy saving | | | Providing cleaner marine fuels | Encouraging third parties for clean energy initiatives | |
|---|---|---|---|---|---|---|---|
| Renewable energy storage systems | | | Optimizing port services | | | | |
| Providing Onshore Power Supply from renewable sources | | | Increase the energy efficiency of ACs in buildings | | | | |
| Producing and recovering energy from industrial waste streams | | | Energy efficient port equipment | | | | |
| Intelligent management of supply and demand for renewable energy | | Energy saving in lighting | New lighting technologies | LEP | | | |
| Generation of renewable sources of energy | | | | LED | | | |

(*continued*)

**Table 3.** *(continued)*

| Renewable energy | Climate change adaptation plan | Energy saving | | | Providing cleaner marine fuels | Encouraging third parties for clean energy initiatives | |
|---|---|---|---|---|---|---|---|
| | | | | Intelligent lighting systems | | | |
| | | | Movement sensors | | | | |
| | | | Moderni-zation of the public lighting network | | | | |
| | | | Demand-orientated lighting | | | | |

## References


1. Sustainable Development Goals, https://www.undp.org/sustainable-development-goals. Accessed 26 August 2022
2. Sachs, J.D.: From millennium development goals to sustainable development goals. The Lancet **379**(9832), 2206–2211 (2012)
3. Sustainable Development Goals History, https://sdgs.un.org/goals#history. Accessed 26 August 2022
4. Caliskan, A., Ozturkoglu, O., Ozturkoglu, Y.: Ranking of Responsible Automotive Manufacturers According to Sustainability Reports Using PROMETHEE and VIKOR Methods. Advanced Sustainable Systems, 2100301 (2022)
5. Targets and Indicators, SDG-7, https://sdgs.un.org/goals/goal7#targets_and_indicators (2022). Accessed 28 August 2022
6. Overview, SDG-7, https://sdgs.un.org/goals/goal7 (2022). Accessed 28 August 2022
7. **Meschede, C.: The sustainable development goals in scientific literature: a bibliometric** overview at the meta-level. Sustainability **12**(11), 4461 (2020)
8. Caiado, R.G.G., Leal Filho, W., Quelhas, O.L.G., de Mattos Nascimento, D.L., Ávila, L.V.: A literature-based review on potentials and constraints in the implementation of the sustainable development goals. J. Clean. Prod. **198**, 1276–1288 (2018)
9. McCollum, D., Gomez Echeverri, L., Riahi, K., Parkinson, S.: SDG7: Ensure access to affordable, reliable, sustainable and modern energy for all. Int. Council Sci. (2017)
10. Katekar, V.P., Deshmukh, S.S., Elsheikh, A.H.: Assessment and way forward for Bangladesh on SDG-7: affordable and clean energy. Int. Energy J. **20**(3A) (2020)
11. Swain, R.B., Karimu, A.: Renewable electricity and sustainable development goals in the EU. World Dev. **125**, 104693 (2020)
12. Salvia, A.L., Brandli, L.L.: Energy sustainability at universities and its contribution to SDG 7: a systematic literature review. In: Leal Filho, W., Salvia, A.L., Pretorius, R.W., Brandli, L.L., Manolas, E., Alves, F., Azeiteiro, U., Rogers, J., Shiel, C., Do Paco, A. (Eds.), Universities as Living Labs for Sustainable Development, pp. 29–45. Springer, Singapore (2020)
13. **Santika, W.G., Anisuzzaman, M., Bahri, P.A., Shafiullah, G.M., Rupf, G.V., Urmee, T.: From** goals to joules: a quantitative approach of interlinkages between energy and the sustainable development goals. Energy Res. Soc. Sci. **50**, 201–214 (2019)
14. Chovancová, J., Vavrek, R.: On the road to affordable and clean energy: assessing the progress of European Countries toward meeting SDG 7. Pol. J. Environ. Stud **31**, 1587–1600 (2022)
15. de Miguel Ramos, C., Laurenti, R.: Towards the Agenda 2030: a quantitative analysis of synergies and trade-offs between the SDGs of Spain during 2000–2019. Sustainability (2020)
16. Balić, K., Žgaljić, D., Ukić Boljat, H., Slišković, M.: The port system in addressing sustainability issues—a systematic review of research. J. Marine Sci. Eng. **10**(8), 1048 (2022)
17. Hiranandani, V.: Sustainable development in seaports: a multi-case study. WMU J. Marit. Aff. **13**(1), 127–172 (2013). https://doi.org/10.1007/s13437-013-0040-y
18. Kotowska, I.: Policies applied by seaport authorities to create sustainable development in port cities. Transp. Res, Procedia **16**, 236–243 (2016)
19. Katuwawala, H.C., Bandara, Y.M.: System-based barriers for seaports in contributing to sustainable development goals. Maritime Bus. Rev. (2022)
20. Beleya, P., Veerappan, G., Ding, W.J., Tan, J.: Challenges in attaining sustainable development goals in port Klang: port management perspective. Int. J. Supply Chain Manag. **9**(1), 349–355 (2020)
21. Lawrence, N.W.: Basics of social research: Qualitative and quantitative approaches, 2nd edn. Pearson Education Inc., Boston (2007)

22. Elo, S., Kyngäs, H.: The qualitative content analysis process. J. Adv. Nurs. **62**(1), 107–115 (2008)
23. Caliskan, A.: Seaports participation in enhancing the sustainable development goals. J. Cleaner Prod., 134715 (2022)
24. Eurostat 2021a. https://ec.europa.eu/eurostat/databrowser/view/mar_mg_am_pvh/default/table?lang=en (2021). Accessed 13 May 2021
25. Eurostat 2021b. https://ec.europa.eu/eurostat/databrowser/view/mar_mg_am_pwhc/default/table?lang=en (2021). Accessed 13 May 2021

# A Theorem on the Recycling Paradox

D. Yu. Cherevatskyi[1], R. G. Smirnov[2](✉), O. V. Lyakh[1], and M. O. Soldak[1]

[1] Institute of Industrial Economics of National Academy of Sciences of Ukraine, 2 Maria Kapnist Street, Kyiv 03057, Ukraine

[2] Department of Mathematics and Statistics, Dalhousie University, 6316 Coburg Road, Halifax, NS 15000 B3H 4R2, Canada

roman.smirnov@dal.ca

**Abstract.** Significant differences in the scale of metallurgical slag processing in different countries has led us to observe and formulate a paradox for a company that is engaged in recycling in the framework of a market economy. The theorem about the existence of the paradox of waste processing is formulated and proven. A mathematical model based on the original logarithmic one-factor production function of the enterprise is used. An analysis of the interaction of factors made it possible to determine the optimal conditions for the processing of production waste. The elasticity of the production function, the price of waste and the price of the product ratio predetermine the profitability for the company of processing either its own waste or those involved in the processing of waste from other enterprises on an agency basis. Based on the research results, options for building efficient circular industrial ecosystems are proposed. The institutional alternative is a market recycling company or an enterprise as part of a vertically integrated structure. An example is given from the practice of a large metallurgical plant.

**Keywords:** Industrial waste · Recycling · Production function · Sustainability · Maximization of profits

## 1 Introduction

The circular economy, along with blockchain and the Internet of things, has been established as one of the main components of the Fourth Industrial Revolution. In essence, it is based on replacing the existing management scheme that can be briefly described as "take, make, waste" with the more sustainable formula of 3R: "reduce, reuse, recycle", which means reduce (consumption), use repeatedly, recycle (waste) [1]. In the industrial sectors of the old economy this means the dominance of the third component. The circular economy in the context of waste management is not only an important strategy for reducing pressure on the environment, it has got also a recognized potential for creating jobs and attracting investments [2]. However, the circular economy still co-exists with a phenomenon of little use of recycling opportunities in industrial enterprises. For example, in China, slag, which is a typical byproduct of high-temperature metallurgical processes, is recycled by 29.5%, although in Japan by comparison the volume of


V. Koval et al. (Eds.): ISCMEE 2022, Lect. Notes in Managem., Industrial Engineering, pp. 229–237, 2023.
https://doi.org/10.1007/978-3-031-23463-7_15

recycling of steelmaking slag reaches 98.4%, in Europe—87.0%, in the United States—84.4% [3]. The Brazilian steel industry is currently actively investing in research and development of processing of steelmaking slag [4]. Moreover, it is safe to say that metallurgy is no exception, there is an opinion that enterprises built on the principles of "take, make, waste" are not configured at all and do not seek to process their own production waste [5]. The low level of the circular economy in the world—no more than 9% [6]—indicates that the motives of society regarding the recycling of industrial waste do not coincide with the motives of enterprises. The main goal of this work is to prove the theorem on the existence of an economic paradox in the practice of enterprises engaged in recycling of industrial waste.

## 2 The Theorem

Consider an enterprise (firm) that processes (recycles) waste of its own production and seeks to get maximize profits from it, namely, we have the following optimization problem (see [7, 8] for more details and references):

$$\pi = pf(z_t) - wz_t \mapsto \max_{z \geq 0}, \tag{1}$$

where π is the profit that the firm receives measured in USD, $p$—the price of the Consumer Product, USD/t, $f(z_t)$ is a single-input production function describing the transformation of a resource into a useful product and is given by

$$f(z_t) = K \cdot Ln(z) + S, \tag{2}$$

where K, S are the parameters of the logarithmic function. It is our contention that the logarithmic production function most adequately describes the existence of a zone with zero waste processing productivity and compatible with the economic law of diminishing marginal returns. Here $z_t$ is the independent variable of the production function that represents the volume of production waste generated during the period of time t that is suitable for recycling, $z_{1t}$ is the amount of waste recycled during the time period $t$ (see below), $w$ is the price of mixtum compositum, or composite product which in our case is the mixture of human labor, electricity, fuel, production waste as raw materials used in recycling, various materials, services, etc.), the cost of which coincides with the annual volume of expenses incurred by the enterprise [9]. Another name for a composite product is the polyresource equivalent [10]. Next, we have

$$z_t w = C_t, \tag{3}$$

where $C_t$ is the sum of the company's recycling costs for the period of time $t$. Furthermore, the company in question is a price taker, i.e., the prices for the product and resources available are set. We assume that the normalized production function of the enterprise is given by

$$\mathrm{s} = \mathrm{k} \cdot \mathrm{Ln(r)} + 1, \tag{4}$$

where $s$ is the annual output of a useful product, unity fractions, k > 0 is the coefficient of the production function, $r$—the composite product used, unity fractions. We have

$$s = \frac{q_t}{N}; \; r = \frac{Z_t}{Z_N}, \tag{5}$$

where $q_t$ is the product output for the period $t$, tonne; $N$—the maximum production capacity of the enterprise for processing production waste into a useful product for the period t, tonne; $z_N$—the volume of production waste that is in circulation during the period of time t while the operation of the production facility to produce a useful product is at full capacity, tonne. Next, we have

$$r_0 = \frac{z_t - z_{1t}}{z_N}, \tag{6}$$

where $r_0$ is the normalized value of industrial waste remaining unused during recycling for the period of time *t; $p_1$*—the cost of producing useful products under conditions $s = 1$ (when production is operating at full capacity—the price of a unit of output), USD. Next, $w_1$ is the cost of the composite product consumed under the conditions of operation of production at full capacity, the price of the unit of composite product (under conditions s = 1), USD. We wish to prove the existence of the following paradox: the less a company (firm) that operates within the framework of a market economy and strives to maximize profits processes waste from its own production, the more profitable it is to recycle the waste generated by other industrial enterprises, that is, the greater the value of $r_0$ is, the greater the value of $r_{opt}$ is and, if:

$$r_{opt} > 1, \tag{7}$$

it makes sense to recycle the waste generated by other companies.

## 3 Proof of the Theorem

Figure 1 contains the graphs of the normalized production function of the enterprise in question. The point ($r_0$; 0) corresponds to the inventory of unused industrial waste during recycling. Two companies whose production functions are shown in Fig. 1 receive the same amount of production waste for the period of time $t$. However, the company with $r_{01}$ recycles more waste than the company represented by the parameter $r_{02}$, since $r_{01} < r_{02}$. The second company (represented by the right graph) has a more elastic production function than the first company. Indeed, the same increment of the argument yields a greater increase in output than in the production function of the second company. In the theory of electric machines, such characteristics are called "soft", as opposed to "hard", that is, inelastic. In mathematical terms, the monotonously increasing production function defined by the initial point ($r_{02}$, 0) grows faster than the production function that originates at the point ($r_{01}$, 0). The production function parameter $k$ is a measure of the elasticity of the function: the larger $k$ is, the faster the production function grows. At the same time, we have

$$k = f(r_0). \tag{8}$$

The result of the production activity of the enterprise in question is determined by the profit function given by

$$\pi = p_1 s - w_1 r, \tag{9}$$

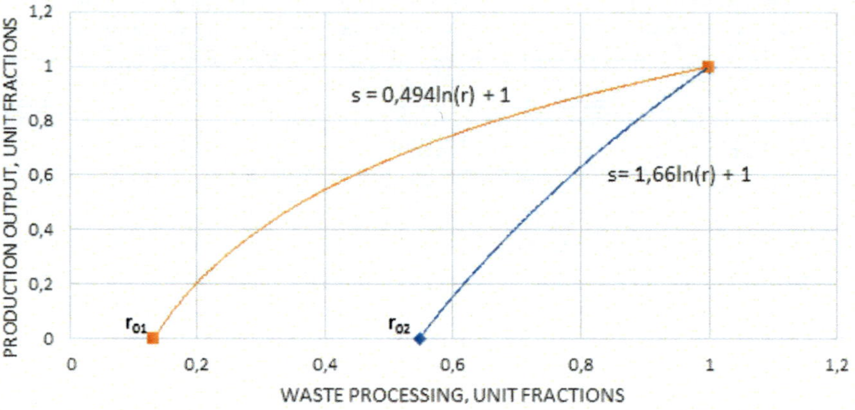


**Fig. 1.** The normalized production function of an industrial enterprise.

while its ultimate goal is to maximize profit, i.e.,

$$\frac{\pi}{p_1} = k \cdot \ln(r) + 1 - \frac{w_1}{p_1} r \mapsto \underset{r>0}{max}. \tag{10}$$

In order to determine the point at which the profit function attains its maximum, we employ the second derivative test. First, we find the critical point, i.e., the point at which the first derivative of the profit function (9) is zero. Solving the equation

$$\frac{d\left(\frac{\pi}{p_1}\right)}{dr} = \left[k \cdot \ln(r) + 1 - \frac{w_1}{p_1} r\right]' = 0, \tag{11}$$

we obtain

$$r_{opt} = k\frac{p_1}{w_1}, \tag{12}$$

where $r_{opt}$ is the value corresponding to the optimality conditions. To assure that (12) is the point at which the profit function (9) attains its maximum, we verify that the second derivative at this point is negative. Indeed, we find

$$\frac{d^2\left(\frac{\pi}{p_1}\right)}{dr^2} = \left(kr^{-1} - \frac{w_1}{p_1}\right)' = -kr^{-2} < 0. \tag{13}$$

Therefore, since k > 0, the second derivative is everywhere < 0, including the stationary point (11). Thus, we conclude that the maximum profit occurs when the relation $r(k)$ is determined in a certain way by $w_1$ and $p_1$.

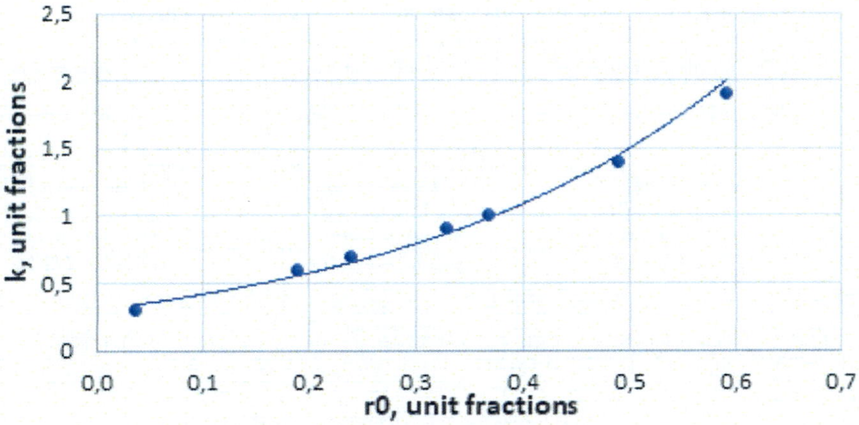


**Fig. 2.** The dependence of $k$ on $r_0$.

Figure 2 shows the exponential dependence of the value $k$ on the value $r_0$ obtained by calculation and graphical methods, in particular, with the aid of Excel, using the "solution search" module.

$$k = 0,303e^{3,198r_0}. \tag{14}$$

From which we get, using (12),

$$r_{opt} = 0,303e^{3,198r_0}\frac{p_1}{w_1}. \tag{15}$$

Figure 3 shows changes in the optimal volume of waste processing relative to changes in the price ratio in the case of a rigid (i.e., demonstrating slow growth) production function (for $r_0 = 0.076$; $k = 0.389$). If $w_1/p_1 = 0.9$, the optimal amount of waste recycling is 0.432. In the case when price ratio is 0.4, it makes sense to recycle large volumes of waste resources.

Figure 4 also shows the dependence of the optimal volume of waste processing on changes in the price ratio, but in the case of a soft (i.e., fast growing) production function ($r_0 = 0.5$; $k = 1.352$). It is obvious that even under unfavorable market conditions ($w_1/p_1 = 0.9$), the optimal volume of waste processing is 1.502, which is more than one, that is, the zone of optimality goes beyond the production function of the enterprise. To ensure that the zone of optimal recycling does not go beyond the production capabilities of the enterprise ($r_{\text{opt}} \leq 1$) for a certain ratio of prices for products and resources, one should adhere to the restriction

$$r_0 \leq 0,312\ln\left(\frac{w_1}{p_1}\right) + 0,373. \tag{16}$$

Figure 5 illustrates this relationship. The more favorable the market conditions for the enterprise (e.g., resource prices are significantly lower than the prices of manufactured goods) are, the more rational conditions for the recycling of industrial waste. At

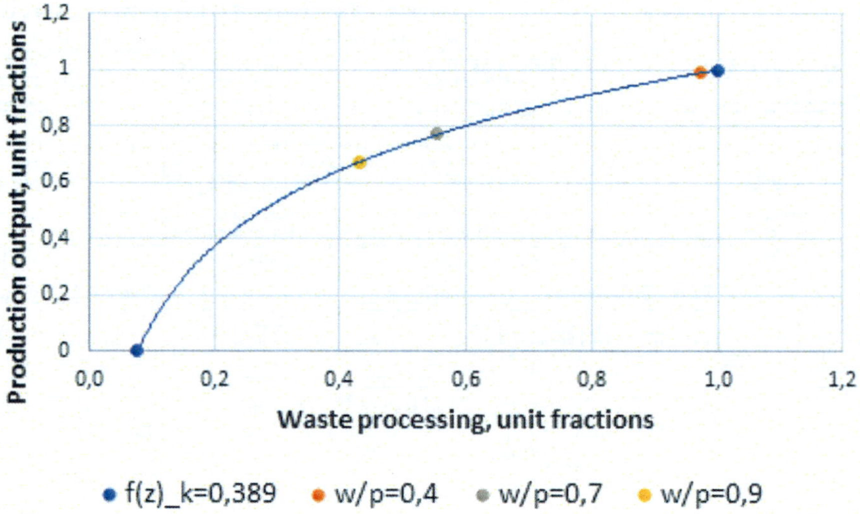


**Fig. 3.** Dependence of optimal production waste processing volumes on the price ratio (the case of slow growing production function).

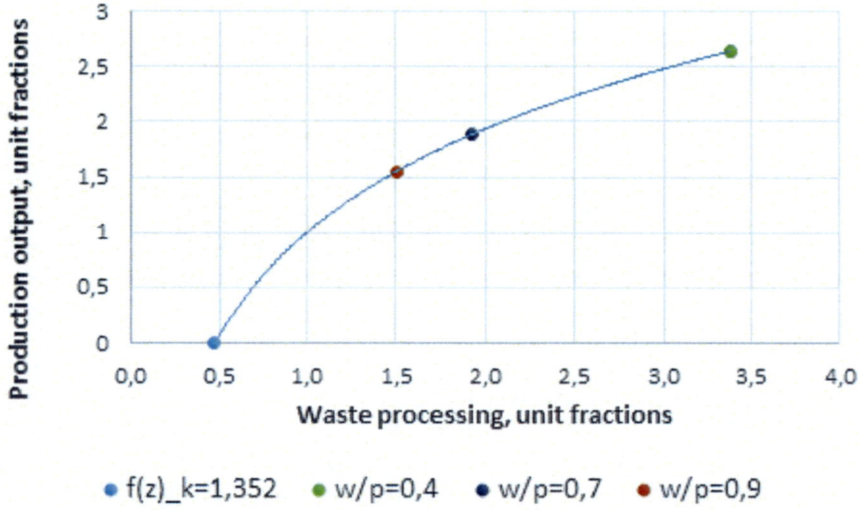


**Fig. 4.** Dependence of optimal production waste processing volumes on the price ratio (the case of fast-growing production function).

certain price ratios, the graph of the function $r_0 = f(w_1/p_1)$ even passes into the zone of negative values, which can be interpreted as the insufficiency of the enterprise's own waste reserves for processing and the expediency of purchasing them from third-party suppliers. The maximum value of $r_0$ should not be greater than 0.373 at all to maintain the optimality mode within the company's production capabilities. If $r_0 \ll 1$ (high

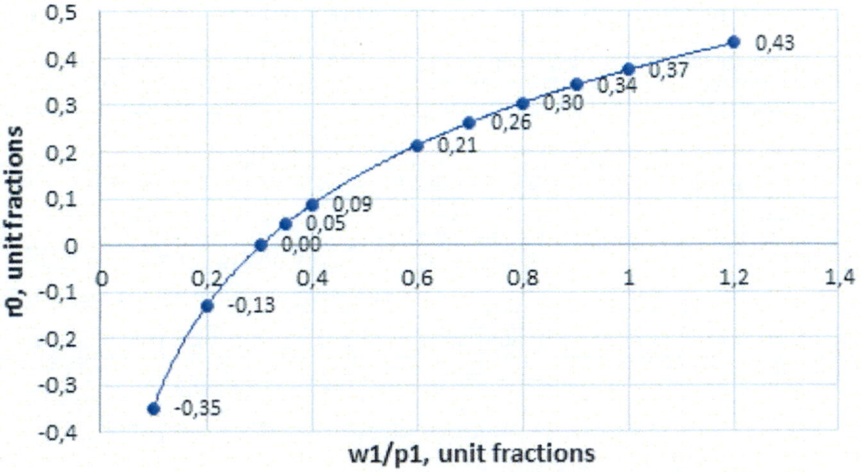


**Fig. 5.** Dependence of $r_{opt} = 1$ on the value of $r_0$ and the ratio $w_1/p_1$.

degree of waste recycling) and $w_1/p_1 \ll 1$ (favorable market conditions), the enterprise is able to operate at full capacity (s = 1), the more unfavorable the market conditions are, the less rational recycling is in the specified zone where $f(w_1/p_1) \geq r_0$. In the zone $r_0 > f(w_1/p_1)$, the $r_{opt}$ value becomes greater than one, which can be interpreted as the development of resources that do not belong to a waste recycling company. Thus, we have demonstrated that it makes sense for an enterprise that processes a relatively small amount of its own waste on a market basis to intensify the processing of waste generated by other companies. Quod erat demonstrandum.

## 4 Results

Therefore, we conclude that the paradox studied in this paper can be overcome by (a) a company engaged in recycling and working within the framework of a market economy if it serves one producer of industrial waste, but receives high added-value ($w_1/p_1 << 1$); (b) a recycling company which serves a group of waste producers; (c) a company that does not operate within a framework of a market economy, but is an integral part of a vertically integrated system (holding). An example that vividly illustrates our results is the practice of slag recycling facility at the Mariupol Ilyich Iron and Steel Works (MIISW) in Ukraine. According to the work [11] in 2004, which turned out to be the year of maximum production of pig iron and steel, as well as processing of slag. In 2004 the plant produced almost 2.8 million tonnes of blast furnace slag of which 2.1 million tonnes (74%) were processed and granulated and then sold as product to consumers. However, it must be mentioned that due to the current market conditions, recycling of slag became unprofitable which led to the curtailment of recycling and intensive growth of dumps. In 2021, for example, the export price of slag produced at the MIISW was \$3.5/tonne, which led to a price ratio (*w/p*) of more than 0.91 meaning that even achieving the results of 2004 with $r_0 = 0.26$ was unattainable [12].

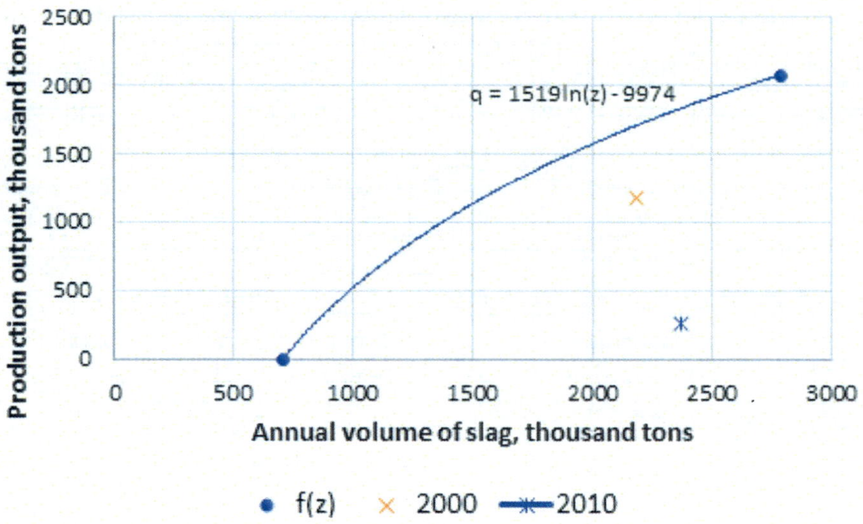


**Fig. 6.** Production function for processing blast furnace slags according to the data from 2000, 2004, and 2010.

Currently, the activity of slag processing is reduced to the extraction of scrap metal from slag masses using two US-made AMCOM complexes. In 2011, almost 2.7 million tonnes of blast furnace and metallurgical slag were processed at the MIISW alone, which made it possible to obtain about 290,000 tonnes of recycled metal. Such recycling saves iron ore raw materials, agglomerate, and cast iron [13]. This particular recycling production no longer operates based on the market principles, but on the basis of a vertically-integrated structure—the AMKOM complexes are jointly used in the slag dump zone ("Balka Grekovataya" district), that is, they are part of the Metinvest Holding, which is the owner of both the MIISW and the "Azovstal" iron and steel works. Figure 6 shows a graph of the production function for processing blast furnace slag in 2004 and the results of recycling activities in other years.

## 5 Conclusions

An original mathematical model based on a one-factor logarithmic production function of a recycling firm was used to prove the profit-maximizing firm's paradox theorem. The proved theorem gives grounds to assume that a waste processing enterprise can operate within the framework of a market economy either if it produces products that are expensive relative to resource prices, or if it does not process its own waste, but is an agent of other enterprises that generate waste to be recycled.

## References

1. Kabirifar, K., Kamyar, M., et al.: Construction and demolition waste management contributing factors coupled with reduce, reuse, and recycle strategies for effective waste management: a review. J. Clean. Prod. **263**, 121265 (2020)
2. Van Beukering, P., Kuik, O., Oosterhuis, F.: Chapter 31—the economics of recycling. Editor(s): Ernst Worrell, Markus A. Reuter, Handbook of Recycling, Elsevier, 479–489 (2014)
3. Guo, J., Bao, Y., Wang, M.: Steel slag in China: treatment, recycling, and management. Waste Manage. **78**, 318–330 (2018)
4. Goncalves, D.R., Fontes, W.C., Mendes, J.C., Silva, G.J., Peixoto, R.A.: Evaluation of the economic feasibility of a processing plant for steelmaking slag. Waste Manag. Res. **34**(2), 107–112 (2016)
5. Ryumina, E.V.: Why businesses are unwilling and unable to protect the environment: a quantitative analysis]. Ekonomicheskaya nauka sovremennoj Rossii. **3**, 6–74 (2009). [In Russian]
6. The Circular Gap Report 2019. https://www.legacy.circularity-gap.world/2019
7. Nerlove, M.: Estimation and identification of Cobb-Douglas production functions. Rand McNally, Chicago (1965)
8. Smirnov, R.G., Wang, K.: In search of a new economic model determined by logistic growth. E. J. Appl. Math. **31**, 339–368 (2020)
9. Cherevatskyi, D. Yu.: Compositum mixtum of a one-factor production function. Ekonomika promyslovosti **4**(92), 118–128 (2020) [In Ukrainian]
10. Cherevatskyi, D., Atabyekov, O.I.: Production function of a coal mine and economic efficiency of its operation. Econ. Visn. Donbas. **4**(30), 56–62 (2012)
11. Nazyuta, L., Smotrov, A.V., Gubanova, A.V., Kornev, G.V.: The structure of formation and recycling of technological waste at full-cycle metallurgical enterprises. Energotekhnol. Resursosberezhenie **4**, 44–54 (2011). [In Russian]
12. The port of Mariupol resumed transshipment of blast furnace slag. https://gmk.center/ua/news/mariupolskij-port-vidnoviv-perevalku-domennih-shlakiv/
13. The MIISW n 2011 processed 2.66 million tons of slag and received more than 288,000 tonnes of recyclable materials. http://www.ux.ua/ua/issuer-m/51/9885/20120216-mmk-im-ilicha-v-2011g-pererabotal.aspx

# Circular Economy: Ukraine's Reserves and the Consequences of the Global Recession

Tetyana Calinescu[1], Ganna Likhonosova[1](✉), and Olena Zelenko[2]

[1] National Aerospace University "Kharkiv Aviation Institute", 17, Chkalova St, Kharkiv 61070, Ukraine
a.likhonosova@gmail.com

[2] Volodymyr Dahl East Ukrainian National University, 74 Gagarin Ave, Dnipro 49010, Ukraine

**Abstract.** Development of alternative energy sources, creation of production facilities for processing household and industrial waste are far from a list of directions of strategic cooperation between Ukraine and the EU within the framework of building circular economy. The purpose of the study is to determine the reserves of Ukraine regarding the further development of the principles of the circular economy and the consequences of overcoming the recession in the post-war period, using globalization trends. The results of the study demonstrate that the main driving forces on the way to building a circular economy are: state authorities focused on the implementation of sustainable "green development" strategies; specialists capable of developing and implementing circular development projects; programs for financing waste processing projects. In the course of the research, an actual model for the implementation of the circular economy in Ukraine under martial law was determined and proposed for implementation, which takes into account objective prerequisites. The relevance of the circular economy concept is confirmed through its relationship with the concept of sustainable development. The main indicators for monitoring the implementation of circular economy principles are also defined, which are relevant in the given operating conditions, but will require revision in the future.

**Keywords:** Circular Economy · Recession · Crisis · Sustainable Development

## 1 Introduction

### 1.1 Principles of Circular Economy of Ukraine

The circular economy is currently conquering all spheres of the traditional linear economy in Ukraine, where products are manufactured, used and disposed of. It is believed that such a circular approach is based on the principles of 3-R [1] or 10-R [2, 3]. Although, from the point of view of modern Ukrainian economic development, it is always possible to add certain features to the characteristics of use, expand their concepts and increase their list, such as (Fig. 1).

The last principle (recycle) is the most developed in the economy of Ukraine. Moreover, all these principles are supplemented by:


V. Koval et al. (Eds.): ISCMEE 2022, Lect. Notes in Managem., Industrial Engineering, pp. 238–251, 2023.
https://doi.org/10.1007/978-3-031-23463-7_16

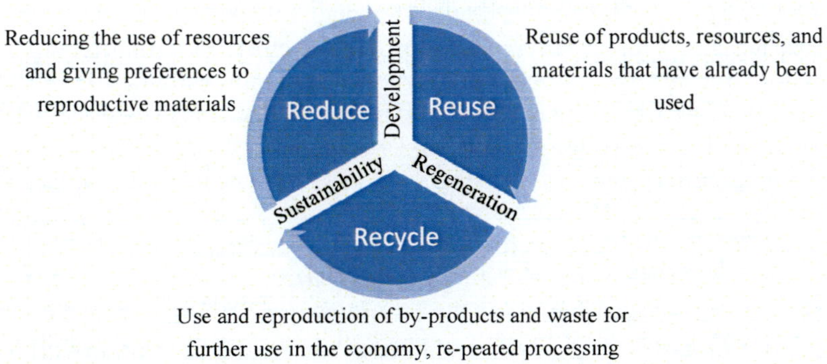


**Fig. 1.** Principles of cyclical economic.

Refuse – rejection of excessive use of raw materials.

Rethink – revision of directions of use of materials and products.

Repair – recondition and maintenance of a defective product with its further use for its main purpose. This approach is very useful for application at the moment of hostilities currently taking place in Ukraine.

If all the above-mentioned principles can be attributed to traditional technologies used to ensure environmental protection, then the ones listed below have tendencies towards **innovation, specifically to its components such as invention, initiation and diffusion,** namely:

Refurbish is the modernization of an old product.

Remanufacture – changing product parameters, using parts of an old product in a new product.

Repurpose – repurposing a used product into another area.

Recover – energy production from materials and products.

Recession is a non-critical stoppage of economic growth, the use of existing circular technologies (reuse of resources, materials, and other products) and may predict a transition to another growing phase of the cyclical development of the economy.

Undoubtedly, the isolation of these principles is quite conditional, they are quite closely related and can be used both in general (collectively, providing a certain synergistic effect), and each separately, supporting separate mechanisms of the country's economic development. However, the path to a circular economy has already begun in Ukraine. Within the framework of those reforms and adjustments to the requirements of the European Union (EU) regarding the provision of full membership to Ukraine, there is a departure from traditional foundations and a transition to new development strategies, which involve the search for certain reserves in the circular economy, which will help withstand the negative effects of the state of war in the country, the stoppage of production and the need for transformation to new models and features of its development.

However, this process is long and time-consuming not only for the economy of Ukraine, but also for all economies of the world, because according to reports presented at the World Economic Forum in Davos, only 9% of materials in the world are reused.

According to the National Institute of Strategic Studies, the development of alternative energy sources, the creation of production capacities for the processing of household and industrial waste are relevant areas of cooperation with the EU in Ukraine in this field; development of organic food products and others. However, the introduction and implementation of these measures in Ukraine have certain obstacles, and accordingly, the prospects for the development of the circular economy are quite illusory, because only the market for secondary raw materials and its processing is opaque [4] (more than 50% of operations carried out there are in shadows) due to the lack of tariffs for processing raw materials, strategies for managing and handling waste, uncontrolled burning of garbage (practically the same amount of waste is burned and processed – 3%) and the creation of landfills (up to 27 thousand new ones appear every year). And these problems will only accumulate, because the country is now in a state of war, and has no opportunity to fulfill its immediate tasks and responsibilities regarding the preservation of the environment. Therefore, there is a need to determine the reserves of Ukraine regarding the further development of the principles of the circular economy and the consequences of overcoming the recession in the post-war period, using the trends of globalization, European and world experience.

### 1.2 An Avalanche Recession: Ukrainian Meaning

If someone were to say that the times of mass lockdowns, behavioral restrictions will not seem so bad to us in 2022, many experts would probably be surprised. Large-scale hostilities began between the Russian Federation and Ukraine, and this, as is known and banal, requires large financial costs and economic efforts. Considering that Ukraine in this conflict is supported by the vast majority of European countries, as well as the United States, Great Britain, all of them also have to spend huge resources to maintain the defense capability of Ukraine. In addition, the sanctions imposed against Russia are reflected not only on the revenue side of the budget of Russia itself, but also on the economies of the countries that support them. European top officials are no longer shy about talking about it directly. In June, German Chancellor Olaf Scholz thanked German industrialists for supporting sanctions against Russia, despite the fact that these restrictions also damage the national economy [5]. At the same time, O. Scholz previously emphasized that sanctions against Russia would remain until a "fair" agreement with Ukraine [6]. The chancellor also noted that only Ukraine itself will be able to determine what general conditions can be considered "fair", this cannot and will not be dictated by anyone in Europe.

In addition 2022 may show people not so bad compared to the looming crisis [7]. According to IMF, the outlook, compared to our last update in April 2022, has deteriorated significantly, citing a wider spread of inflation, an increase in interest rates, a slowdown in economic growth even in China, and tougher sanctions related to military actions in Ukraine [7].

Against the background of these events, the urgency of implementing and spreading the principles of the circular economy is only growing.

The need for an even greater economy of energy carriers, the use of secondary raw materials, and the minimization of waste from the society of the leading countries of the world are markers that testify to the real basis for further steps in the direction of the formation of a circular economy.

## 2 Literature Review

It should be noted that the topic of using the circular economy in Ukraine remains open, debatable and relevant for more than one decade, this was reflected in author's works, starting from 1987 [8–11]. And despite the fact that there are already significant steps on the way to the transition to a circular economy (circular economy, reuse) [2, 12–16], which is based on both the European development experience and national traditions, the topic remains unsolved.

Specialists are actively engaged in the problem and models of implementation within the framework of the activities of the international economic forum [17], as well as in the context of scientific research at Harvard University [18]. Researchers emphasize the economic expediency of implementing the principles of the circular economy, emphasizing the significant saving of financial resources, as well as improving the quality of secondary materials, which usually significantly lose their original properties after recycling [17]. The matrix of the circular economy [18], which offers a practical toolkit for making managerial decisions depending on the chosen strategy and the quadrant to which the company belongs, also deserves attention.

However, there are still many issues and discourses that require detailed research and practical implementation both at the level of the state and at the level of an individual business (entrepreneur), namely: availability of the necessary knowledge and incentives in the production market; insufficient awareness of the population regarding the needs and requirements of the circular economy; imperfect state regulation; difficulties in financing this kind of business, the presence of systemic change strategies, etc. Therefore, the path to the future mature and balanced circular economy constantly remains relevant and in demand, based on the new conditions of Ukraine's development.

## 3 Research Methodology

### 3.1 Ukraine's New Model of the Circular Economy Implementation

The model of the circular economy that is being formed today in Ukraine is based on various concepts of the development of the world economy, based on the need to cover, combine and coordinate a wide range of economic, ecological, social, technical, political, and informational issues that will contribute to thrift and restorative aspects of the development of the entire Ukrainian society. The key point underlying the development of the circular economy is the concept of sustainability, which justifies such main aspects as [19, 20]:

1. Resource limitations in the conditions of global demand for resources and growing shortage of critically important types of natural and water resources.

2. Innovative technological development [19], without the implementation of which it is impossible to introduce new business models, the formation of innovative approaches to recycling, replacement and reuse of resources, the use of new IT technologies, etc.
3. Socio-economic development in the context of growing urbanization, which requires the development, implementation, support, collection and transformation of various goods and materials into other types of resources, which is economically beneficial and environmentally safe.
4. The ecological direction, neglect of which leads to environmental pollution, harms the health of the population, agriculture, soils, water resources, and causes global environmental threats. Therefore, the internationalization of these externalities will allow to minimize damages and restore natural ecosystems to a safe level.

The European experience of introducing a circular economy determines new trends in the formation of strategic priorities of national development, where the main drivers in the implementation of these concepts should be [7]:

(1) the state, which should change the structure of public procurement and focus on ecological "green" types of production;
(2) bodies of legislative and executive power, which, using the experience of European countries, should form and legislate similar legal and institutional bases for the functioning of the circular economy;
(3) economic entities anticipating changes in the market situation and global trends, modernize production and invest in processing;
(4) new business models (simple and complex) related to the replacement of components, raw materials and materials, as well as the formation of new production links, complete or partial change of the technological process;
(5) labor demand, which requires new production design professionals;
(6) programs for financing projects for waste processing or from other areas of waste disposal.

Each country goes its own way to implementing the principles of circular economy. Ukraine in modern conditions faces the problem of implementing this concept in the special conditions of martial law. Strategically important sectors of the economy require priority changes and radical transformations: the defense industry, the construction industry, energy infrastructure, and the small and medium-sized business sector (Fig. 2).

The presented model of implementation of the circular economy in the modern realities of the functioning of the economic system of Ukraine is adapted to the specific needs of the country, provides for the concentration of efforts on critically important problems, and provides for comprehensive public-private partnership both in strategically important areas of development and beyond them.

The last few crises occurring at the same time have prompted the IMF to revise growth forecasts twice in a short period. The first time this was done in January 2022 due to the omicron and again in April 2022 due to the invasion of Ukraine and the imposed sanctions.

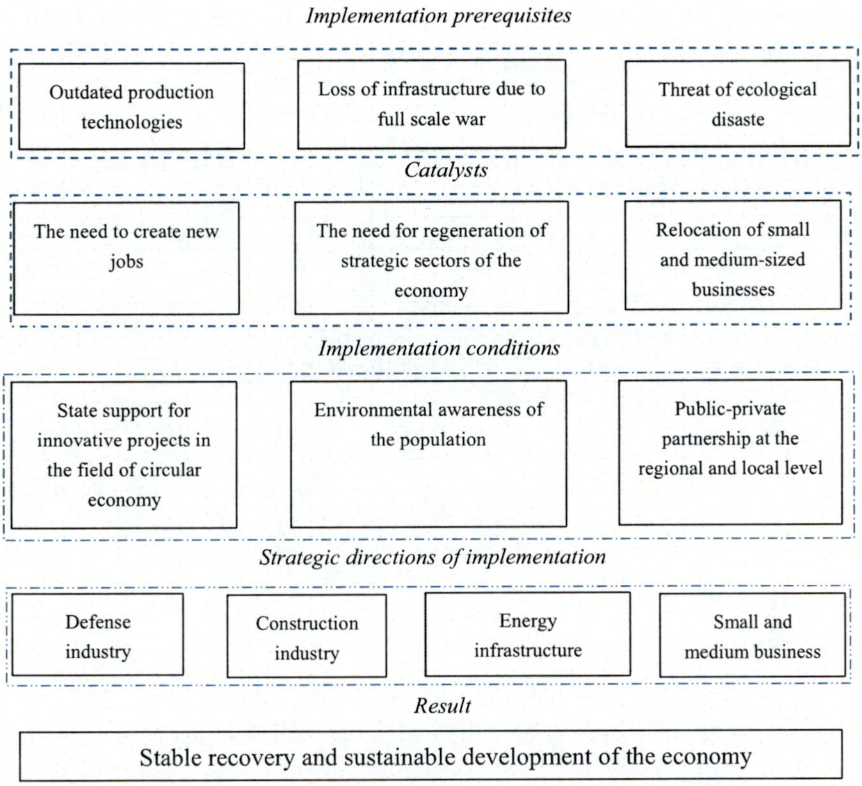


**Fig. 2.** Ukraine's model of the circular economy implementation.

### 3.2 World Growth Forecasts: Worrying Expectations

The growth forecast fell from 4.9% to 3.6% [21]. However, fears are connected not so much with changing the percentage of growth (in fact, 3.6% is also a significant increase), but more with the risk of fragmentation of the economy and the formation of overt and covert trade blocs.

However, already in the last statement of the head of the IMF, there was great concern about the financial situation. It's about a recession. According to K. Georgieva, it will be a difficult year in 2022, but 2023 may be much more difficult due to the consequences of 2022 [21]. The risks of a global recession have increased significantly [21]. The IMF plans again, for the third time this year, to lower its growth forecast for the global economy. Last time this figure was 3.6%.

In fact, a recession is a manifestation of a sustained significant slowdown in economic growth, which can last for months or even years. When a recession approaches, the economy becomes unmanageable, enters a vicious circle, the pace of movement in which is accelerating with each stage. It all starts with a decrease in government spending, then there is a decrease in production, after an increase in unemployment, then the population loses its income, as a result, there is a decrease in consumer demand, after which retail sales decrease and, as a result, unnecessary production is leveled (Fig. 3).

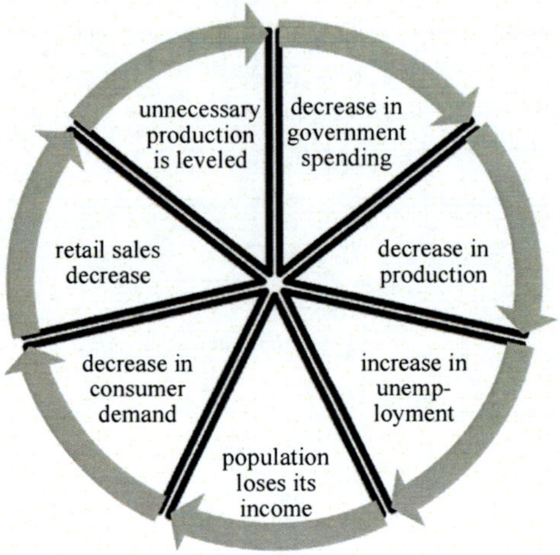


**Fig. 3.** Cyclical economic development during the recession phase.

As is known, a recession can turn into a depression. As one of the most famous examples of the manifestation of depression, the Great Depression in the United States in 1920–30 was characterized by a fall in GDP by more than 10% and an unemployment rate of 25%, while it lasted almost 10 years. Economists began to warn of an impending global recession in the spring of 2022. The macroeconomic forecast of The Institute of International Finance expert speaks of the collapse of the economies of Russia and Ukraine, as well as very low growth rates in the rest of the world. A rapid tightening of **U.S. financial conditions risks further destabilizing the global growth picture, already** shaken by Russia's invasion of Ukraine and the spread of the micro-strain of coronavirus in China. Taking into account these factors, the experts of the Institute of International Finance have sharply reduced the forecast for economic growth in 2022 [22]. We are in solidarity with experts from this Institute and analysts from the Financial Times, who also stated in the spring of 2022 that the global economy was facing a large-scale recession.

# 4 Results

## 4.1 The Relationship Between the Priorities of the Circular Economy and the Principles of Sustainable Development

Based on the application of the principles and goals of sustainable development, the following immediate priorities can be proposed by the UN programs that should be resolved (see Table 1).

**Table 1.** Principles of sustainable development and their connection with circular economy priorities.

| Principles of sustainability | Tasks of the circular economy | Development priorities |
|---|---|---|
| Long-term development and promotion of population needs | Maintaining the health of the population, which depends on reducing emissions, improving the sanitary conditions of the population and water indicators | Alternative and efficient use of water |
| Determination of restrictions on the use of natural resources and taking into account the level of technical support and the ability to restore the biosphere | Responsible consumption and sorting of processing raw materials, preservation of land and sea ecosystems | Use of renewable technologies and materials on a biological basis |
| Encouraging and realizing the needs of the population, taking into account the standard of living of the population | Absence of the poor, decent work and economic growth | Management, enrichment, preservation, conservation, reconstruction, renewal, preservation and continuation of what has already been done |
| Consideration of ecological possibilities according to the used energy | Preservation of ecosystems and formation of renewable energy | Production of energy from waste, evaluation of the level of energy efficiency |
| Changes in indicators of the demographic state in accordance with the use of the potential of global ecosystems | The use of recyclable and non-recyclable resources in order to increase the effective recycling of resources and reduce emissions | Effective use of materials (non-toxic, non-critical and reusable) subject to secondary processing |

*Source* developed using [13, 23]

The main goals aimed at fulfilling all the tasks of the circular economy are: (1) creation and maintenance of living conditions; (2) public health; (3) connections between different communities (4) involvement and participation of citizens in the development of their country (territory).

In this context, the main indicators of successful implementation of circular economy principles can be considered:

- increasing the average level of wages and the average income per capita of the population;
- decrease in the level of morbidity of the population;
- positive dynamics of environmental indicators of the surrounding environment;
- restoration of transport, housing, social infrastructure, etc.

It should be noted that the presented indicators are relevant for these conditions. Further development will require a review of indicators and the definition of new strategic guidelines for life activities.

### 4.2 Rising Food Prices: Global Dependencies

Among the main negative trends for the global economy, they also noted the decline in China's economic growth due to the spread of COVID-19, the recession of the US economy, the price crisis in Europe and the threat of a food crisis and even mass starvation in developing countries. Israeli financial analyst, adviser to the world's major hedge funds Avi Tiomkin [24] said in an interview with Glob that the global economy, led by the United States, is heading into a deep recession. The world economy is slipping into a very severe recession, and today there are changes in the inflationary process [24]. The analyst adds that the 2008 economic crisis was never fully resolved. It was stopped only temporarily by lowering interest rates around the world down to zero or even negative values, as well as aggressive fiscal policies in most countries.

Elon Musk, the founder of Tesla, also warned about a recession in the United States: at what point is a recession inevitable? As for it will be in the near future [25]. Inflation rates are also certain indicators of the state of the economy. So, in May, consumer prices in the US increased by 8.6% on an annualized basis. This is the highest level since December 1981. The last time inflation in the United States exceeded the value recorded in May was in December 1981, when the level was 8.9%. The most significant rise in price in May: energy products, prices for which increased by 34% compared to the same month in 2021. Food became more expensive by 10%, the strongest increase since March 1981. An almost similar situation is emerging in the EU: annual inflation in the eurozone accelerated to a record 8.6% in June after 8.1% in May. The main drivers of price growth in the eurozone were: energy 41%, food, alcohol, tobacco, which added 8.9% and 7.5% respectively to the price.

The Russian Federation is also expecting a recession in 2022–2023. On April 29, 2022, the Bank of the Russian Federation published an official forecast for economic growth in the face of tough sanctions, according to which in 2022 GDP will decline by 8–10%. The decline is mainly due to supply-side factors [26]. In 2023, the Russian economy will move gradually to growth in the context of structural adjustment, but the change in GDP in general for the year will be −3% to 0%.

## 5 Discussion

### 5.1 Economic Realities of Ukrainian Metallurgy and Food Market

Global trends will also be reflected in the indicators of Ukraine's economic stability. It would be strange to expect that the global crisis will bypass Ukraine, which is also conducting hostilities. The spectrum of economic problems continues to expand in the country. One of the main ones is a possible hyperinflation due to the issuance of the hryvnia. In 2022, the head of the National Bank of Ukraine (NBU), Kyrylo Shevchenko, pays special attention to this issue. Based on world experience, the Head of the National

Bank of Ukraine emphasizes that the ways to minimize the negative consequences of the war in Ukraine cannot be only in the issue of the monetary unit (hryvnia). Otherwise, such a policy of the National Bank of Ukraine will lead to hyperinflation in the country. According to the head of the National Bank of Ukraine, since February 24, 2022, the bank has printed about UAH 225 billion. High inflation can potentially quickly devalue the income and savings of the general population. Short-term benefits are received only by the owners of real assets, currencies, other valuables, exporters of products and services. Ukraine may face negative economic consequences from the uncontrolled emission of the hryvnia in scale similar to those that occurred in the 1990s [27].

At the same time, in fact, one of the main industrial sectors of Ukraine is in stagnation. Historically, there are two key export-oriented industries in Ukraine: metallurgy and agriculture. In 2021, Ukraine produced 23.5 million tons of steel goods, 80% of this production was exported. It was 70% of the entire country's exports. In 2022, this figure dropped to the level of 8 million tons, and exports also decreased accordingly. The whole world watched the tragedy of Mariupol. Of course, the company is out of order. It was 50% of the entire industry. Agriculture: exports of corn, wheat, production has decreased by at least 2 times, 68% of arable land is not being cultivated. Gas production decreased by 20%. In general, over the entire portfolio of production and exports of Ukraine, the reduction is more than 2 times [28].

At the end of June 2022, the MetInvestr group announced that on 01.07 it was suspending the extraction of raw materials from the Yuzhny and Ingulets mining and processing plants [29]. From mid-July, the Northern Mining and Processing Plant will also suspend work. The company decided to take such radical steps due to a significant reduction in demand, consumption of raw materials, logistical problems caused by hostilities, as well as falling iron and steel prices and the high cost of metal products. Experts note that even in 2021, the cost of iron ore raw materials increased by 30%. A sharp jump in energy prices added another 30% to the cost of raw materials. The logistical problems caused by the hostilities actually leveled the industry's opportunities, as they caused an increase in prices for the transportation of raw materials.

The situation with steel production is not much better than with the extraction and export of ore. Against the backdrop of economic problems and huge difficulties with the export of products, since July 2022, Arcelormittal Kryvyi Rih has transferred a significant part of its employees, including top management and foreign workers, to 2/3 of the load due to the inability to pay full wages [30].

The company also stated that in addition to the blockade of ports, they were prompted to take such a step by an increase in the cost of tariffs for cargo transportation by 70% by Ukrzaliznytsia [31]. Ukrainian farmers also found themselves in a bad situation. According to experts, the total export volume of the current crop, together with the remains of last year, will be about 65–70 million tons [13]. At the current export rate of 2 million tons per month, Ukraine will export the available volume of grain and oilseeds in the next 32–35 months. Farmers note that the prospect of overflowing warehouses with grain, the inability to sell it, the lack of funds to purchase resources for the next season can lead to catastrophic consequences. As a result, up to 10–15 million hectares of fields in the controlled territory of Ukraine may not be sown for the 2023 harvest. As a result, it is safe to say that the former world as we knew it is changing dramatically and rapidly. 2022 becomes the period of the beginning of the economic recession in many countries. Such objective conditions become the basis for the application of measures of strict state economy. Taking into account the principles of the circular economy, this will lead to new irreconcilable positions among the population, an increase in unemployment, the emergence of waves of populism and new social upheavals. On the one hand, it is interesting to observe the historical changes taking place in all areas of life support, on the other hand, it is very difficult to be a subject on whom the corresponding consequences fall like "stones of surprise". Thus, before we can even hope to start moving towards socio-economic development, we need to learn lessons, assess domestic issues in the country and, of course, give encouraging signals to the population.

The path of transformation will not be easy, but the circumstances in which Ukraine found itself in 2022 will become catalysts for change. Ukraine already has a number of documents that regulate the processes of positive changes: the National Waste Management Strategy until 2030; Strategy of the State Environmental Policy of Ukraine for the Period until 2030; The Concept of Implementation of State Policy in the Field of Climate Change for the Period up to 2030. Adherence to the deadlines for the implementation of the measures presented in the documents will certainly have to be reviewed, but the implementation schedule can be made up at the expense of the latest technologies, which are planned to be used for the restoration of the affected and de-occupied territories. It is very telling that the government, having already developed a plan for the reconstruction of the country, foresees the program of energy independence and the "green" course among the most important and the largest in terms of financing.

## 6 Conclusion

The circular economy is the driver of a new stage of industrial development. Conceptually, the circular economy provides for achieving maximum economic efficiency and reducing the negative impact on the environment in compliance with the socio-economic and ecological requirements of sustainable development.

The implementation of various concepts and approaches to stabilize the functioning of the circular economy is carried out depending on the development of the country's economy, taking into account environmental aspects and principles in the development of production processes regarding the organization of clean production with a low level of emissions, the implementation of waste prevention systems, as well as strengthening consumer responsibility.

All aspects of the implementation of the circular economy have a multi-vector effect depending on the sphere of activity and the production chain, which requires systematic actions at different levels of management: state, regional and local. For Ukraine, even in wartime conditions, it is very important to focus at least on the implementation of the simplest, but key environmental problems, such as: effective management and waste management and the use of circular innovative business models, which allows to ensure a quality and safe life for its population.

The policy of Ukraine in the sphere of the circular economy should be formed and implemented comprehensively, taking into account global trends and requirements, focusing on local problems, which will be very important for the reconstruction of the post-war economy. And in the future, it is worth paying attention to evaluating the effectiveness of the implementation of circular processes, adapting the experience of EU countries in the context of identifying "bottlenecks" and developing a mechanism for their elimination.

The proposed model of circular economy implementation in the modern realities of Ukraine takes into account the objective conditions, needs and priorities of the country's further recovery. Its implementation will make it possible to restore critical infrastructure, create new jobs, and reconfigure the economic system in accordance with new operating conditions in order to eliminate global environmental threats.

**Acknowledgment.** The conducted research is carried out within the framework of the scientific topic, which is carried out at the Department of Finance, Accounting and Taxation in National Aerospace University «Kharkiv Aviation Institute» on the topic of the scientific research (research and design) work "Accounting and financial mechanism of innovative interaction of economic entities in the context of the transformation of the knowledge economy and socio-behavioral restrictions" (State Registration No. 0121U108324).

## References

1. Nechitaylo, D.: From a clean sheet: how does the circular economy work and how beneficial is it, https://www.epravda.com.ua/rus/columns/2020/09/2/664626/2
2. Ruda, M.V., Myrka, Ya.V.: Circular business models in Ukraine. Management and entrepreneurship in Ukraine: stages of formation and development problems **1**, 107–121 (2020). https://doi.org/10.23939/smeu2020.01.107
3. Ruda, M., Yaremchuk, T., Bortnikova, M.: Circular economy in Ukraine: adaptation of European experience. Management and entrepreneurship in Ukraine: stages of formation and development problems **3**(1), 212–222 (2021). https://doi.org/10.23939/smeu2021.01.212
4. Circular economy as a direct industrial modernization: advanced international information, http://confmanagement.kpi.ua/proc/article/view/230935
5. Speech by Olaf Scholz at a conference of German industrialists, https://inbusiness.kz/ru/last/sholc-nakonec-to-priznal-chto-sankcii-protiv-rossii-vredyat-ekonomike-germanii. Accessed 21 June 2022
6. Interview of Olaf Scholz to ZDF TV channel. 2022. Information agency InterFax Ukraine, https://interfax.com.ua/news/political/839736.html. Accessed 17 June 2022
7. Shalal, A.: Thomson Reuters, https://www.reuters.com/markets/us/imf-chief-says-cannot-rule-out-possible-global-recession-2022-07-06/. Accessed 15 July 2022
8. Rozovsky, B.G., Zharkova, I.A., Pashchenko, T.V.: Organization of environ-mental protection in the region. Kyiv (1992)
9. Goncharov, V.M., Pashchenko, T.V., Kharkivskiy, B.T., others: Industrial ecology and her economic aspect. Kiev (1996)
10. Calinescu, T.V.: Socio-economic factors influx of the environment on the development of the population of the regions. Econ. Sci. **12**, 105–109 (2008)
11. Calinescu, T.V.: Accounting and analytical support for the environmental and economic security of an enterprise in a crisis. Actual problems, directions and mechanisms for the development of the productive forces of the North - 2020: Collection of articles of the seventh All-Russian scientific and practical conference (with international participation) (September 9 - 11, 2020), Syktyvkar). Part 1: at 2 parts. Syktyvkar: Komi Republican Printing House LLC 1, 257–263 (2020)

12. Gryshova, I.Yu., Nesterova, K.S.: The concept of a circular economy in the context of sustainable development. Ekonomika APK **4**, 88–94 (2021). https://doi.org/10.32317/2221-1055.202104088
13. Zalunin, M.: Circular economy as a prerequisite for insuring sustainable development. Black Sea Econ. Stud. **47**(1), 196–201 (2019) https://doi.org/10.32843/bses.47-69
14. Varfolomieiev, M., Churikanova, O.: Circular economy as an integral way of Ukraine's future in the aspect of globalization. Effective Econ. **5** (2020). https://doi.org/10.32702/2307-2105-2020.5.200. http://www.economy.nayka.com.ua/
15. Derii, Zh., Butenko, N., Zosymenko, T.: Implementation of the concept of circular economy: problems and prospects. Problems and prospects of economics and management **1**(25), 54–62 (2021). https://doi.org/10.25140/2411-5215-2021-1(25)-54-62
16. Zlotnik, M.: Circular economy as a part of the organizational corporate strategy. Management and entrepreneurship in Ukraine **3**(1), 183–192 (2021). https://doi.org/10.23939/smeu2021.01.183
17. Jensen, H.H.: 5 circular economy business models that offer a competitive advantage. World Economic Forum. January 27, 2022. URL: https://www.weforum.org/agenda/2022/01/5-circular-economy-business-models-competitive-advantage/
18. Dumas, A.C., Wassenhove, L.N. Van.: The Circular Business Model. Harvard Business Review. July-August 2021. URL: https://hbr.org/2021/07/the-circular-business-model
19. Nahara, M.: Circular economy: genesis, structure features. Economy and state **10**, 68–73 (2021). https://doi.org/10.32702/2306-6806.2021.10.68
20. Calinescu, T.: A role of innovative diffusion in development of the socio-economic systems. Time description of economic reforms **2**(46), 31–43 (2022). https://doi.org/10.32620/cher.2022.2.05
21. Shalal, A.: IMF chief 'cannot rule out' possible global recession. The Jakarta Post. URL: https://www.thejakartapost.com/paper/2022/07/07/imf-chief-cannot-rule-out-possible-global-recession.html. Accessed 7 July 2022/
22. Experts saw the risk of a global recession due to China and Ukraine. News, https://voina.news/2022/05/12/eksperty-uvideli-risk-globalnoy-retsessii-iz-za-kitaya-i-ukrainy/. Accessed 7 July 2022
23. Shinkaruk, A.: Circular Economy and Place. Why are we becoming popular? https://gwaramedia.com/reyuz-resajkling-rekonstrukcziya-pryrodnogo-materialu-yak-praczyuyut-czyrkulyarni-mista/
24. Tiomkin, A. The US is leading the world into recession. Globes. News, https://en.globes.co.il/en/article-avi-tiomkin-the-us-is-leading-the-world-into-recession-1001415209. Accessed 15 July 2022
25. Elon Musk warned about the risk of a recession in the US. Socio-political online news-paper USAVAT, https://musavat.com/ru/news/ilon-mask-predupredil-o-riske-recessii-v-ssha_899863.html?d=1. Accessed 23 June 2022
26. Inflation expectations and consumer sentiment. Information and analytical commentary of the Central Bank of the Russian Federation **6**(66), 1–7 (2022), http://www.cbr.ru/collection/collection/file/42129/infl_exp_22-06.pdf. Accessed 18 July 2022
27. Shevchenko, K. Issuance of money is not an option. Other methods are needed to sup-port the economy. NABU A banker's opinion, https://nabu.ua/ua/kirilo-shevchenko.html. Accessed 7 June 2022
28. Kiryanova, Yu. The war caused losses to the main businesses, http://iz.com.ua/zaporoje/virobnitstvo-vsih-biznesiv-smart-holdingu-cherez-viynu-vpalo-bilshe-nizh-udvichi-ale-mi-zhivi-gendirektorka-na-vef. Accessed 23 June 2022

29. Pidgola, Yu., Mogurov V. In Kryvyi Rih, the work of two GZCs has been suspended, the workers are idle. What the human rights defender and the employee of the enter-prise say. Social news, https://suspilne.media/256432-u-krivomu-rozi-prizupinili-robotu-dva-gzk-robitniki-u-prostoi-so-kazut-pravozahisnik-ta-pracivnik-pidpriemstva/. Accessed 2 July 2022
30. Kipelov, A. Plants stop: what happens in metallurgy in Ukraine and why it is danger-ous for the whole country. New edition, https://novoeizdanie.com/zavody-stoyat-chto-proishodit-metallurgiej-v-ukraine-i-pochemu-eto-opasno-dlya-vsej-strany/. Accessed 7 June 2022
31. **Agrarians call on government officials to prevent the crisis of the argo sector. Ukrainian** Club of Agrarian Business, https://ucab.ua/ua/pres_sluzhba/novosti/agrarii_zaklikayut_uryadovtsiv_uperediti_krizu_agrosektoru. Accessed 30 June 2022

# Author Index




V. Koval et al. (Eds.): ISCMEE 2022, Lect. Notes in Managem., Industrial Engineering, pp. 253–254, 2023.
https://doi.org/10.1007/978-3-031-23463-7